Tropical Medicine

Dedication

This book is dedicated to Dr Dion Bell, author and later editor of the first four editions of this book; a gifted teacher of tropical medicine, and an inspiration to generations of doctors working in the tropics.

LECTURE NOTES ON

Tropical Medicine

Edited by

GEOFF GILL
MA, MSc, MD, FRCP, DTM&H
Reader in Tropical Medicine
Liverpool School of Tropical Medicine

NICK BEECHING
MA, FRCP, FRACP, DCH, DTM&H
Senior Lecturer in Infectious Diseases
Liverpool School of Tropical Medicine

Fifth Edition

Blackwell
Science

Blackwell Publishing, Inc., 350 Main Street, Malden, Massachusetts 02148-5020, USA
Blackwell Publishing Ltd, 9600 Garsington Road, Oxford OX4 2DQ, UK
Blackwell Science Asia Pty Ltd, 550 Swanston Street, Carlton, Victoria 3053, Australia

First published 1981
Second edition 1985
Reprinted with corrections 1987
Third edition 1990
Reprinted 1991, 1992, 1993
Fourth edition 1995
Reprinted 1996 (twice), 1997, 1998, 1999, 2000, 2001
Fifth edition 2004
4   2007

Library of Congress Cataloging-in-Publication Data

Lecture notes on tropical medicine.—5th ed. / edited by Geoff Gill, Nick Beeching.
        p.;   cm.
Includes bibliographical references and index.
  ISBN: 978-0-632-06496-0
  1. Tropical medicine.
  [DNLM: 1. Tropical Medicine. WC 680 L471 2003] I. Gill, Geoffrey V.   II. Beeching, N.

  RC961.L42   2003
616.9'883—dc22

                                                                    2003015301

ISBN: 978-0-632-06496-0

A catalogue record for this title is available from the British Library

Set in 9/11.5 Gill Sans by SNP Best-set Typesetter Ltd, Hong Kong
Printed and bound in India by Replika Press Pvt. Ltd

Commissioning Editor: Vicki Noyes
Managing Editor: Geraldine Jeffers
Editorial Assistant: Nic Ulyatt
Production Editor: Karen Moore
Production Controller: Kate Charman

For further information on Blackwell Publishing, visit our website:
http://www.blackwellpublishing.com

Contents

Contributors, ix

Preface, xi

List of Abbreviations, xiii

New drug names, xv

Part 1: A General Approach to Syndromes/Symptom Complexes

1 Gastrointestinal Presentations, 3

2 Respiratory Presentations, 11

3 Neurological Presentations, 17

4 Febrile Presentations, 26

5 Dermatological Presentations, 32

6 The Patient with Anaemia, 36

7 A Syndromic Approach to Sexually Transmitted Infections, 40

8 Splenomegaly in the Tropics, 49

Part 2: Major Tropical Infections

9 Malaria, 55

10 Visceral Leishmaniasis, 73

11 Cutaneous Leishmaniasis, 80

12 Tuberculosis, 84

13 HIV Infection and Disease in the Tropics, 98

14 Filariasis and Onchocerciasis, 112

15 African Trypanosomiasis, 120

16 South American Trypanosomiasis (Chagas' Disease), 127

17   Schistosomiasis, 129

18   Leprosy, 141

Part 3: Other Tropical Diseases

Gastrointestinal

19   Amoebiasis, 153

20   Bacillary Dysentery, 160

21   Cholera, 163

22   Giardiasis and Other Protozoal Infections, 167

23   Intestinal Cestode Infections (Tapeworms) Including
     Cysticercosis, 171

24   Soil-Transmitted Helminths, 174

25   Viral Hepatitis, 180

26   Liver Flukes, 190

27   Hydatid Disease, 193

Respiratory

28   Pneumonia, 196

29   Lung Flukes, 206

30   Tropical Pulmonary Eosinophilia, 209

Neurological

31   Pyogenic Meningitis, 211

32   Cryptococcal Meningitis, 219

33   Encephalitis, 221

34   Acute Flaccid Paralysis, 226

35   Spastic Paralysis, 229

36   Rabies, 232

37   Tetanus, 237

Fever

38   Brucellosis, 240

39   Typhoid and Paratyphoid Fevers, 245

40 Arboviruses, 252

41 Viral Haemorrhagic Fevers, 254

42 Dengue and Yellow Fever, 262

43 Relapsing Fevers, 267

44 Rickettsial Infections, 270

45 Leptospirosis, 272

46 Melioidosis, 275

Miscellaneous

47 Tropical Ulcer, 277

48 Buruli Ulcer, 279

49 Myiasis, 282

50 Cutaneous Larva Migrans, 284

51 Scabies and Lice, 286

52 *Strongyloides stercoralis*, 288

53 Guinea Worm Infection (Dracunculiasis), 292

54 Histoplasmosis, 295

55 Other Fungal Infections, 297

56 Haemoglobinopathies and Red Cell Enzymopathies, 300

57 Haematinic Deficiencies, 304

58 Bites and Stings, 307

59 Non-Communicable Diseases, 312

60 Refugee Health, 328

Index, 337

A colour plate section faces p. 146

Contributors

IMELDA BATES, Senior Lecturer, Liverpool School of Tropical Medicine, Pembroke Place, Liverpool L3 5QA [Chapters 6, 8, 56, 57]

NICK BEECHING, Senior Lecturer, Liverpool School of Tropical Medicine, Pembroke Place, Liverpool L3 5QA [Chapters 1, 18, 25, 31, 38]

TOM BLANCHARD, Senior Lecturer, Liverpool School of Tropical Medicine, Pembroke Place, Liverpool L3 5QA [Chapter 4]

MARTIN DEDICOAT, Wellcome Trust Training Fellow, Liverpool School of Tropical Medicine, Pembroke Place, Liverpool L3 5QA [Chapter 13]

NEIL FRENCH, Wellcome Career Fellow, Liverpool School of Tropical Medicine, Pembroke Place, Liverpool L3 5QA [Chapters 2, 28]

CHARLIE GILKS, Professor, Department of HIV/AIDS, World Health Organization, Geneva, Switzerland [Chapter 13]

GEOFF GILL, Reader, Liverpool School of Tropical Medicine, Pembroke Place, Liverpool L3 5QA [Chapters 44, 47, 49–52, 54, 59]

RACHEL KNEEN, Specialist Registrar, Royal Liverpool Children's Hospital NHS Trust, Alder Hey, Liverpool L12 2AP [Chapters 3, 33, 34]

DAVID LALLOO, Clinical Director and Senior Lecturer, Liverpool School of Tropical Medicine, Pembroke Place, Liverpool L3 5QA [Chapters 9, 15, 16, 32, 37, 46, 58]

DIANA LOCKWOOD, Consultant Physician and Leprologist, Hospital for Tropical Diseases, Mortimer Market Centre, Capper Street, London WC1E 6AU [Chapter 18]

MALCOLM MOLYNEUX, Professor, Liverpool School of Tropical Medicine, Pembroke Place, Liverpool L3 5QA [Chapter 9]

FRED NYE, Consultant Physician, Royal Liverpool University Hospital, Liverpool L7 8XP [Chapters 36, 45]

TIM O'DEMPSEY, Senior Lecturer, Liverpool School of Tropical Medicine, Pembroke Place, Liverpool L3 5QA [Chapters 10, 11, 19, 20, 22–24, 26, 27, 43, 48, 60]

CHRIS PARRY, Senior Lecturer, Department of Medical Microbiology and Genitourinary Medicine, University of Liverpool, Duncan Building, Liverpool L69 3GA [Chapter 39]

PAUL SHEARS, Senior Lecturer, Liverpool School of Tropical Medicine, Pembroke Place, Liverpool L3 5QA [Chapter 21]

TOM SOLOMON, Lecturer, Department of Neurological Sciences, University of Liverpool, Liverpool L9 7LJ [Chapters 3, 33–35, 40–42]

BERTIE SQUIRE, Senior Lecturer, Liverpool School of Tropical Medicine, Pembroke Place, Liverpool L3 5QA [Chapters 12, 17]

MIRIAM TAEGTMEYER, Lecturer, Wellcome Trust Laboratories, PO Box 43640, Nairobi, Kenya [Chapter 7]

GEORGE WYATT, Emeritus Consultant, Liverpool School of Tropical Medicine, Pembroke Place, Liverpool L3 5QA [Chapters 5, 14, 29, 30, 53, 55]

Preface

The first edition of this book was published just over 22 years ago in 1981. It was conceived and entirely written by the late Dr Dion Bell of the Liverpool School of Tropical Medicine. Its forthright, highly practical and entertaining style made it rapidly a highly successful 'classic'. The next two editions involved a similar format with expansion and updating. By the time of the fourth edition, however, in 1994, the spectrum of tropical disease was rapidly expanding — HIV/AIDS in particular had become of major importance. Other authors (also from the Liverpool School) were therefore included to cover selected topics.

Since Dion Bell's retirement, we have been proud to take over editorship of *Lecture Notes in Tropical Medicine*. The fifth edition has necessarily continued the process of multi-authorship, but all contributors are either staff of the Liverpool School, or teachers on the Liverpool DTM&H course. We have tried to ensure that style and presentation are uniform, whilst allowing authors freedom of content.

Though a clinically based approach has been continued, there have also been major changes. The chapters on malaria, tuberculosis and AIDS have been revised and expanded, to reflect their major importance in current tropical medical practice. There are some completely new chapters (for example on non-communicable dis-eases, and on refugee health), in recognition of the emergence of new health challenges in the tropics. We have tried to include relevant basic science and evidence-based treatment where appropriate, but have concentrated on clear therapeutic advice for the practising tropical doctor. Thus, clinical algorithms and flow charts have been included, and the bibliographies have been expanded, and now include a variety of websites.

We have reluctantly followed the decision of the British National Formulary to adopt American spellings of some drugs. The old and new versions of these are listed at the end of the List of Abbreviations.

Royalty profits from this book are donated to a fund to assist Liverpool medical students to travel overseas for their elective studies. All the authors of this edition are pleased to continue this tradition, established with the first edition by Dion Bell.

We hope the new edition is useful to students and practitioners of tropical medicine. As always, we welcome any comments or criticisms.

Geoff Gill
Nick Beeching
Liverpool, UK, 2003

List of Abbreviations

AAFB	acid- and alcohol-fast bacilli	CMI	cell-mediated immunity
Ab	antibody	CMR	crude mortality rate
ACE	angiotensin-converting enzyme	CMV	cytomegalovirus
ACR	adequate clinical response	CNS	central nervous system
AFB	acid-fast bacilli	COPD	chronic obstructive pulmonary
Ag	antigen		disease
AgB	antigen B	CRP	C-reactive protein
AIDP	acute inflammatory demyelinating	CSF	cerebrospinal fluid
	polyneuropathy	CT	computerized tomography
AIDS	acquired immune deficiency	CVP	central venous pressure
	syndrome	DAT	direct agglutination test
ALA	amoebic liver abscess	DCL	diffuse cutaneous leishmaniasis
AMAN	acute motor axonal neuropathy	DD5	double diffusion test for arc 5
ARC	AIDS-related complex	DDT	dichlorodiphenyl-trichloroethane
ARI	annual risk of infection	DDS	4,4-diaminodiphenylsulphone
BB	borderline leprosy	DEC	diethylcarbamazine citrate
BCG	bacille Calmette–Guérin	DF	dengue fever
b.d.	twice daily	DHF	dengue haemorrhagic fever
BI	bacterial index	DHFR	dihydrofolate reductase
BL	borderline lepromatous leprosy	DHPS	dihydropteroate synthetase
BMI	body mass index	DIC	disseminated intravascular
BP	blood pressure		coagulation
b.p.m.	beats per minute	DKA	diabetic ketoacidosis
BT	borderline tuberculoid leprosy	DOT	direct observation of therapy
cAMP	cyclic adenosine monophosphate	DSS	dengue shock syndrome
CATT	card agglutination test for	DTH	delayed-type hypersensitivity
	trypanosomes	DTP	diptheria, tetanus and pertussis
CCF	congestive cardiac failure	EBV	Epstein–Barr virus
CCHF	Crimean–Congo haemorrhagic	ECG	electrocardiogram
	fever	ECHO	enteric cytopathogenic human
CFT	complement fixation test		orphan (virus)
CHD	congestive heart disease	EEG	electroencephalography
CHE	complex humanitarian	EIA	enzyme immunoassay
	emergency	EITB	enzyme-linked
CIATT	card indirect agglutination test for		immunoelectrotransfer blot
	trypanosomes	ELISA	enzyme-linked immunoabsorbent
CL	cutaneous leishmaniasis		assay
CM	cryptococcal meningitis	ENL	erythema nodosum leprosum

EPI	Extended Programme of Immunization	IL	interleukin
ERCP	endoscopic retrograde cholangiopancreatography	i.m.	intramuscular
		IMCI	Integrated Management of Childhood Illness strategy
ESR	erythrocyte sedimentation rate	INR	international normalized ratio
ETF	early treatment failure	IPT	intermittent presumptive therapy
FAR	fever–arthralgia–rash	i.v.	intravenous
FCPD	fibrocalculous pancreatic diabetes	IVDU	intravenous drug user
FES	fasciola excretory–secretory	KS	Kaposi's sarcoma
FEV$_1$	forced expiratory volume in 1 second	LBRF	louse-borne relapsing fever
		LL	lepromatous leprosy
FGT	formol gel test	LP	lumbar puncture
FVC	forced vital capacity	LR	leishmaniasis recidivans
G6PD	glucose-6-phosphate dehydrogenase	LRTI	lower respiratory tract infection
GABA	γ-aminobutyric acid	LTF	late treatment failure
GCS	Glasgow coma score	LVF	left ventricular failure
GPA	Global Programme on AIDS	MAEC	minianion exchange column technique
GTT	glucose tolerance test		
HAART	highly active anti-retroviral therapy	MAT	microscopic agglutination test
		MCH	mean corpuscular haemoglobin
HAV	hepatitis A virus	MCHC	mean corpuscular haemoglobin concentration
HbA	adult haemoglobin		
HbF	fetal haemoglobin	MCL	mucocutaneous leishmaniasis
HbS	sickle haemoglobin	MCV	mean corpuscular volume
HBIg	hepatitis B immunoglobulin	MDR	multidrug resistant
HBV	hepatitis B virus	MHCT	microhaematocrit
HCC	hepatocellular carcinoma	ML	mucosal leishmaniasis
HCV	hepatitis C virus	MMDM	malnutrition-modulated diabetes mellitus
HDCV	human diploid cell vaccine		
HDV	hepatitis D virus	MOTT	mycobacteria other than tuberculosis
HEV	hepatitis E virus		
HFRS	haemorrhagic fever with renal syndrome	MRDM	malnutrition-related diabetes mellitus
Hib	*Haemophilus influenzae* type b	MRI	magnetic resonance imaging
HIV	human immunodeficiency virus	MSF	Médecins sans frontières
HLA	human leucocyte antigen	MTB	*Mycobacterium tuberculosis*
HNK	hyperosmolar non-ketotic coma	MUAC	mid-upper arm circumference
HPV	human papilloma virus	NCD	non-communicable disease
HTLV-1	human T lymphotrophic virus type 1	NGO	non-governmental organizations
		NK	natural killer
HUS	haemolytic uraemic syndrome	NNN	Novy, MacNeal and Nicolle's medium
ICT	immunochromatographic card test		
		NSAID	non-steroidal anti-inflammatory drug
IDP	internally displaced person		
IFAT	indirect fluorescent antibody test	NTS	non-typhi *Salmonella*
IFN	interferon	OLM	ocular larva migrans
Ig	immunoglobulin	PA	postero-anterior

PAS	periodic acid–Schiff	SMB	suckling mouse brain
PCECV	purified primary chick embryo cell vaccine	SP	sulfadoxine–pyrimethamine
		STD	sexually transmitted disease
PCP	*Pneumocystis carinii* pneumonia	STI	sexually transmitted infection
PCR	polymerase chain reaction	t.d.s.	three times daily
PCV	packed cell volume	TB	tuberculosis
PDEV	purified duck embryo vaccine	TBRF	tick-borne relapsing fever
PE	pre-erythrocytic	TCBS	thiosulphate–citrate–bile salt–sucrose
PEG	pegylated		
PF	peak flow	TIF	thiomersal, iodine and formol
PGL	persistent generalized lymphadenopathy	TNF	tumour necrosis factor
		TPE	tropical pulmonary eosinophilia
PHC	primary health clinic	TT	tuberculoid leprosy
PKDL	post-kala-azar dermal leishmaniasis	UFM	under-fives mortality
		UN	United Nations
PPD	purified protein derivatives	UNHCR	United Nations High Commission for Refugees
PUO	pyrexia of unknown origin		
PVCV	purified vero cell vaccine	UTI	urinary tract infection
q.d.s.	four times a day	VHF	viral haemorrhagic fever
QBC	quantitative buffy coat	VL	visceral leishmaniasis
RIG	rabies immune globulin	VLM	visceral larva migrans
r.p.m.	revolutions per minute	VTC	voluntary testing and counselling
RVF	Rift Valley fever	W/H	weight-for-height index
SAT	standard agglutination test	WBC	white blood cell count
SFP	selective feeding programme	WBCT20	20-min whole blood clotting test
SLE	systemic lupus erythematosus	WHO	World Health Organization

New drug names

New	Old
aciclovir	acyclovir
amoxicillin	amoxycillin
anthelmintic	antihelminthic
beclometasone	beclomethasone
chlorphenamine	chlorphenyramine
lidocaine	lignocaine
nonoxinol '9'	non-oxynol 9
phenobarbital	phenobarbitone
sulfamethoxazole	sulphamethoxazole
tiabendazole	thiabendazole
thioacetazone	thiacetazone

PART I

A General Approach to Syndromes/Symptom Complexes

CHAPTER 1

Gastrointestinal Presentations

Dysphagia, 3

Abdominal pain, 3

Diarrhoea, 4

Haematemesis, 3

Malabsorption, 4

Further reading, 10

The most important gastrointestinal presentation in the tropics is diarrhoea, and the majority of this chapter is devoted to this problem. However, other presentations of gastrointestinal disease are first discussed.

Dysphagia

Significant recent-onset dysphagia should always raise the possibility of oesophageal carcinoma. This malignancy is particularly common in certain parts of the tropics, e.g. some areas of Central and East Africa. Oesophageal candidiasis (AIDS-related) is also a common cause of tropical dysphagia. In South America, the mega-oesophagus of Chagas' disease should be considered. Finally, peptic strictures, corrosive chemical ingestion and foreign bodies (fish bones especially in some areas) may also be important causes of impaired swallowing.

Haematemesis

In all areas of the world, an upper gastrointestinal haemorrhage can be caused by peptic ulceration, gastritis, oesophagitis, and gastric or oesophageal carcinoma. Gastritis, gastric erosions and gastric ulcers may be drug related; e.g. corticosteroids and non-steroidal anti-inflammatory drugs (NSAIDs). *Helicobacter pylori* is recognized globally as a major cause of

gastric and duodenal inflammation and/or ulceration. Oesophageal varices may be a particularly common cause of haematemesis in many tropical areas — at least 25% of all cases in some series. The underlying liver disease can be the late result of chronic viral hepatitis, or schistosomal hepatic fibrosis.

Abdominal pain

In 'western' populations, severe abdominal pain can result from appendicitis, mesenteric adenitis, perforated peptic ulcers, biliary colic, cholecystitis and intestinal obstruction (commonly because of adhesions or malignancy). This list is far from exhaustive, but serves to demonstrate that the spectrum of causes in the tropics is much wider. The following 'exotic' causes of acute severe abdominal pain may need to be considered.
- Abdominal tuberculosis
- Typhoid (including typhoid perforation)
- Hydatid cyst rupture
- Amoebic colitis (including perforation)
- Amoebic liver abscess (which may rupture)
- Intestinal obstruction caused by *Ascaris lumbricoides*
- Ectopic ascariasis (e.g. biliary and/or pancreatic obstruction)
- Sickle cell crisis
- Splenic rupture
- Hyperinfection syndrome of strongyloidiasis.

Malabsorption

Malabsorption can be a feature of infection with *Giardia lamblia*, *Strongyloides stercoralis*, intestinal tuberculosis (TB) infection, as well as AIDS. Perhaps the most common cause, however, is the temporary lactase-deficient situation that may occur after any significant acute infective diarrhoeal illness. Milk and milk products may need to be avoided, although yoghurt is usually tolerated, because of its high bacterial lactase content.

Tropical sprue

A particularly well-described form of tropical malabsorption is 'tropical sprue'. This occurs predominantly in India and South East Asia, as well as the Caribbean and Central America. Patients develop non-bloody diarrhoea (sometimes steatorrhoea) often with abdominal bloating and significant weight loss. There may be a history of initial acute diarrhoeal illness, which is thought to be the precipitant (although the exact mechanism is unknown). As well as biochemical features of malabsorption, duodenal biopsy typically shows partial villous atrophy. The illness can be prolonged and debilitating. Traditional treatment with tetracycline (for associated bacterial small bowel overgrowth) and folic acid is often highly effective.

Diarrhoea

Diarrhoeal illness is one of the most important causes of morbidity and mortality in the tropics, causing over 6 million deaths per year, and is clearly linked with poor hygiene and contamination of water and food. A wide variety of viral, bacterial and parasitic pathogens have been implicated in the pathogenesis of diarrhoea but it is impossible and unnecessary to test for all of these in individual cases. Systematic review of epidemiological, clinical and host factors usually enables a sensible working aetiological diagnosis to be established. The working diagnosis can be used to decide whether specific investigation should be performed, or to direct empirical antimicrobial therapy in the minority of cases in which it is required. The mainstay of management of diarrhoeal illness is the assessment and maintenance of adequate hydration and electrolyte balance, irrespective of the aetiology, as well as the introduction of control measures in an epidemic setting to prevent further cases.

History

It is essential to establish that both the doctor and the patient are talking about the same thing, especially if interpreters are being used to take the clinical history. A useful working definition of diarrhoea is the passage of three or more loose or watery bowel motions in 24h. The distinction between soft or loose diarrhoea is more difficult, but bowel motions can be described as diarrhoeal when they assume the shape of the collecting container. This definition works with acute diarrhoeal illness but is less satisfactory with chronic diarrhoeal illness related to malabsorption, in which bulky sticky soft bowel motions are abnormal but may not be fluid enough to move around in the container. Key features in the history are the presence or absence of visible blood in the stool (dysentery), the presence and degree of abdominal pain, the presence of tenesmus and the presence of fever. The duration of illness is important — chronic diarrhoea can usefully be defined as diarrhoea lasting more than 14 days, although a more precise definition (especially in the context of an immunocompromised host) is the passage of three or more loose or watery stools a day for 21 days or more.

In the historical assessment of fluid balance, the volume and frequency of faecal loss should be estimated, together with the frequency and approximate volume of any vomiting. The amount of fluid intake should be checked, as should the frequency of urinary output during the last 24h.

The epidemiological setting is important. Illness in close family contacts should be ascertained, and enquiry made about whether the patient has attended any functions or eaten unusual foods in the preceding 48–72h. If so, have any other guests had similar illness? Point source outbreaks can be caused by toxin-mediated food poisoning, in which case vomiting is often a predominant feature and incubation periods are usually shorter than 24 h. Unusual systemic pathogens (e.g. anthrax of the gut) or non-infectious poisoning caused by adulterated or contaminated food products must always be considered. Bacterial pathogens causing small or large bowel diarrhoea usually have intermediate incubation periods of 12–72 h. More detailed food histories are not otherwise very helpful, except in the case of ex-patriates who have unwisely overindulged in very spicy foods ('tasting the chilli twice'), or who have recently arrived in the tropics (traveller's diarrhoea). Diarrhoea developing in patients who are already hospitalized suggests a nosocomial or antibiotic-associated cause, while outbreaks of diarrhoeal illness in a refugee or camp setting imply specific infections such as shigellosis or cholera (see below).

Other illness

Diarrhoea can be a prominent feature of many systemic illnesses, including malaria, pneumonia and enteric fever, especially in children, and evaluation of the patient should exclude these as potential causes. Surgical and other intra-abdominal conditions may mimic gastroenteritis, as can inflammatory bowel disease. In older or immobile patients, constipation with overflow diarrhoea must be excluded. Alcohol and drugs frequently cause diarrhoea with or without nausea and vomiting.

Host factors

Conditions that cause hypochlorhydria (e.g. gastric surgery, H_2 antagonists and proton pump inhibitors) reduce the gastric acid barrier to many bacterial pathogens, so a smaller infective dose is required. Patients with established cardiovascular or renal disease are less likely to tolerate dehydration, as are those on diuretics and patients with poorly controlled diabetes. Pre-existing large bowel problems such as inflammatory bowel disease predispose to complications of dysenteric infections such as toxic mega-

colon, signs of which may be partly masked by concurrent steroid therapy. Bowel tumours can produce diarrhoea with or without blood or weight loss. Small bowel problems, including lymphoma, can cause prolonged diarrhoea. Immunosuppression of the patient, particularly by HIV, predisposes to increased invasiveness (local and systemic) of bacterial pathogens such as non-typhoidal *Salmonella*, increased recurrence of such pathogens, and chronic diarrhoea caused by a variety of protozoa.

Examination

General examination must include assessment of the state of hydration. This is more difficult to quantify clinically in adults than in children but key features are summarized in Table 1.1. Measurement of any postural drop in blood pressure is particularly useful. Rectal examination should be performed, except in obvious cases of cholera for example, and is particularly important in older patients who are more likely to have non-infectious bowel problems. Systemic causes of diarrhoea and signs of immunosuppression (e.g. zoster scars and oral candidiasis) should be sought.

Clinical syndromes of diarrhoea

Apart from acute toxin-mediated food poisoning, diarrhoeal illness can be broadly classified into small bowel secretory diarrhoea, small bowel malabsorption and large bowel inflammatory diarrhoea. Each of these groups may be acute or chronic, and there is considerable overlap (Table 1.2).

Small bowel secretory diarrhoea is exemplified by cholera and non-invasive *Escherichia coli* infections, in which toxins specifically promote secretion of water and electrolytes into the bowel lumen and inhibit their reabsorption. Such secretion can be competitively overcome by a steady intake of balanced electrolyte solutions containing adequate amounts of glucose, but not too much to produce an osmotic diarrhoea. This is the scientific basis for the success of oral rehydration therapy, in which the correct quantities of salts and glucose are added to sterile water for rehydration.

DEHYDRATION SEVERITY

	Mild	Moderate	Severe
Subjective			
General state	Alert, active, up and about	Weak, lethargic, able to sit and walk	Dull, inactive, unable to sit or walk
Ability to perform daily activities	Able to perform daily activities without difficulty	Able to perform daily activities with some difficulty, e.g. stays away from work, needs support	Unable to perform daily activities, stays in bed or needs hospitalization
Thirst	Not increased	Increased thirst	Feels very thirsty
Objective			
Pulse	Normal	Tachycardia	Tachycardia
Blood pressure	Normal	Normal or decrease, 10–20 mmHg systolic	Decrease >20 mmHg systolic
Postural hypotension	No	Yes or no	Yes
Jugular venous pressure	Normal	Normal or slightly flat	Flat
Dry mucosa (mouth, tongue)	No	Slight	Severe
Skin turgor	Good	Fair	Poor
Sunken eye balls	No	Minimal	Sunken
Body weight loss	<5%	5–10%	>10%

Table 1.1 Clinical classification of severity of dehydration in adults. Adapted with permission from Manatsathit *et al.* (2002).

Malabsorption is a common complication of infectious diarrhoea in the tropics, as many races have relatively low disaccharidase activity in the small bowel enterocytes. Disruption of 'normal' bowel activity readily leads to failure to break down sugars and a moderately prolonged lactose intolerance. This is particularly common after infections that cause flattening of the small bowel mucosa (such as giardiasis and cryptosporidiosis). Large bowel diarrhoea is usually caused by direct invasion of the bowel by pathogens such as *Entamoeba histolytica*, bacteria such as *Campylobacter* species, or *Clostridium difficile* after antibiotic therapy. Other parasites such as *Schistosoma mansoni* can also cause prolonged large bowel diarrhoea. In heavy *Trichuris*

trichiura infections, oedema of the rectal mucosa together with continued efforts to defaecate resulting from tenesmus can lead to rectal prolapse. A summary of the major pathogens in inflammatory and non-inflammatory diarrhoea is shown in Table 1.3.

Investigations

A useful algorithmic approach to individual patient diagnosis and management is summarized in Fig. 1.1. In most tropical settings, microbiological investigation proves impossible or very limited. Microscopy of faeces for leucocytes, suggestive of invasive pathogens in the large bowel, is commonly advocated but is of questionable time-effectiveness compared with macroscopic inspection of faeces for blood (and smell) when resources are limited. However, cholera vibrios may be observed with their characteristic 'shooting star' motility even with-

CLINICAL FEATURES OF DIARRHOEA

Non-inflammatory	Inflammatory
Symptoms Nausea, vomiting; abdominal pain and fever not major features	Abdominal pain, tenesmus, fever
Stool Voluminous, watery	Frequent, small volume; blood-stained, pus cells present, mucus
Site Proximal small intestine	Distal ileum, colon
Mechanism Osmotic or secretory	Invasion of enterocytes leading to mucosal cell death and inflammatory response

Table 1.2 Clinical features of inflammatory and non-inflammatory diarrhoea. From Hart (2003) with permission from Elsevier Science.

out dark ground facilities, and this is very useful when culture is not available. Investigations for faecal parasites should be limited to specific settings (e.g. chronic diarrhoea complicating HIV), and are almost never indicated in nosocomial diarrhoea. Fresh stool microscopy for active trophozoites should only be requested when amoebic dysentery is truly suspected. Blanket requests for faecal microscopy for 'ova, cysts and parasites' on all patients are a waste of time in most settings. Such requesting patterns overload laboratories, demoralize their staff and lead to reports of questionable quality with little effect on clinical management decisions.

In an outbreak setting, full microbiological identification of the pathogen and assessment of the antimicrobial resistance patterns is very helpful, and should be pursued even if outside assistance is required. In sporadic cases, detailed microbiological tests may be inappropriate, but clinicians need to be aware of the local antibiotic sensitivities of organisms such as *Shigella*, *Salmonella* and *Campylobacter* if they are to use empirical antimicrobial therapy in a responsible and effective manner. Other investigations, such as serum electrolytes, peripheral white cell count and blood cultures, are per-

formed in a hospital setting but again may not be available routinely.

Management

Detailed management of individual pathogens is beyond the scope of this brief chapter. The key is the correction of fluid and electrolyte imbalance. Severely dehydrated patients need rapid intravenous replacement of fluid loss, preferably using a physiologically balanced electrolyte solution such as Hartmann's fluid (see also p. 164). Large volumes of dextrose solution can be dangerous. Intravenous fluid can be supplemented and rapidly replaced by oral rehydration, which is more successful if small volumes of fluid are taken steadily rather than large volumes at a time. Specific World Health Organization (WHO) oral rehydration solution is ideal, but the water in which it is dissolved must be sterile. Alternative oral rehydration therapy mixtures can also be used for adults and food, including milk products, is usually reintroduced as early as possible after initial resuscitation of children. Fluid balance should be carefully monitored and a cholera bed is useful for less mobile patients with profuse diarrhoea. The fluid faeces can then be collected through a hole in the middle of the bed directly into a measuring bucket. If a large-bore disposable Foley's urinary catheter is available, this can be inserted into the rectum when diarrhoea is profuse and watery (e.g. in

PATHOGENS IN DIARRHOEA

Inflammatory	Non-inflammatory
Viruses	
Nil	Rotavirus
	Adenovirus 40/41
	Astrovirus
	Norovirus (Norwalk agent)
	Calicivirus
	Small round structureless virus
	Coronavirus
	Torovirus
	Bredavirus
	Picobirnavirus
Bacteria	
Enteroinvasive *Escherichia coli* (EIEC)	Enterotoxigenic *E. coli* (ETEC)
Enterohaemorrhagic *E. coli* (EHEC), e.g. 0157	Enteropathogenic *E. coli* (EPEC)
Enteroaggregative *E. coli* (EAggEC)	*Vibrio cholerae*
Aeromonas hydrophila	*Vibrio parahaemolyticus*
Campylobacter spp.	*Campylobacter* spp.
Salmonella spp.	*Salmonella* spp.
Shigella spp.	*Plesiomonas shigelloides*
Yersinia enterocolitica	*Bacillus cereus*
Clostridium difficile	*Clostridium perfringens*
Protozoa	
Entamoeba histolytica	*Cryptosporidium* spp.
Balantidium coli	*Giardia intestinalis*
	Cyclospora cayetanensis
	Isospora belli
	Microsporidia (e.g. *Enterocytozoon bieneusi*)
Helminths	
Schistosoma spp.	*Strongyloides stercoralis*

Table 1.3 Pathogens in inflammatory and non-inflammatory diarrhoea. From Hart (2003) with permission from Elsevier Science.

cholera), removing the need for frequent evacuation, and allowing accurate measurement of faecal losses by volume.

Antidiarrhoeal agents such as codeine or loperamide should be avoided in patients with acute invasive or large bowel disease, and should not be used in young children. Antiemetics should be used sparingly and again avoided in young children. Empirical or specific antimicro-bial treatment should be reserved for specific situations such as proven amoebiasis, prolonged severe infection in a vulnerable host, or in outbreak settings, e.g. cholera or shigellosis. Chronic diarrhoea presents a different challenge and patients with HIV-related diarrhoea often progress through successive therapeutic trials of co-trimoxazole, metronidazole and albendazole. Such patients may need 'hospital at home' support including provision of adequate antidiarrhoeal medications.

In a refugee camp outbreak setting, logistical support must be requested at an early stage for

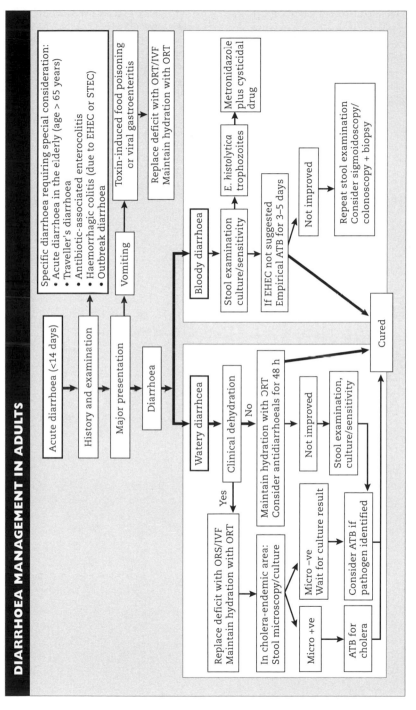

Fig. 1.1 Algorithm for the management of diarrhoea in adults. (Adapted from Manatsathit et al. (2002) with permission.) Stool examination and culture depends on local availability, affordability and practice. In suspected cholera, dark field microscopy is ideal (or, if not available, a search for 'shooting star' bacteria on light microscopy will do). In epidemic situations, a clinical diagnosis is sufficient. When antibiotics are used, the choice either depends on culture and sensitivity results, or local experience. If available, ciprofloxacin is a good choice. ATB, antibiotic; EHEC, enterohaemorrhagic *Escherichia coli*; IVF, intravenous fluids; ORT, oral rehydration therapy.

detailed epidemiological investigation, triage and treatment facilities; as well as provision of an adequate water supply, rehydration solutions and latrines (Chapter 60).

Further reading

Hart CA. Introduction to acute infective diarrhoea. In: Cook GC, Zumla A, eds. *Manson's Tropical Diseases*, 21st edn. London: Elsevier Science, 2003: 907–13. [Good overview with references of both adult and paediatric diarrhoea causes and effects.]

Manatsathit S, DuPont H, Farthing M *et al.* Guideline for the management of acute diarrhoea in adults. *J Gastroenterol Hepatol* 2002; **17** (Suppl): S54–S71. [Superb working party report produced by acknowledged experts from Thailand, India and Africa as well as 'western' authorities. Detailed definitions, practical approaches and many references.]

World Health Organization. Management of the child with a serious infection or severe malnutrition: guidelines for care at the first referral level in developing countries. Chapter 4 (diarrhoea). WHO/FCH/00.1. Geneva: World Health Organization, 2000: 45–55. [*Essential practical guide to Integrated Management of Childhood Illness* for those managing children in the tropics. Chapter 4 has details of assessment and correction of dehydration.]

CHAPTER 2

Respiratory Presentations

Assessment, 11
Investigation of respiratory
disease, 13

Common
presentations, 14

Respiratory disease in the
HIV-infected, 16

Disorders of the respiratory tract are the most important cause of ill health in human populations around the world. The normal physiological functioning of the respiratory tract exposes it to the most prolonged and intimate contact with the external environment of any organ system, leading to a steady exposure to airborne pollutants and pathogens with disease-causing potential.

Infectious diseases dominate respiratory illness in the tropics in all age groups; acute viral and bacterial infections in childhood, and tuberculosis and bacterial pneumonia in adults. Increasing urbanization and access to cigarettes has already led to a rise in smoking-related pulmonary disease and asthma, and this is likely to be a growing trend (Chapter 59). This will not only increase the burden of respiratory illness and incapacity but will complicate the assessment and management of pulmonary conditions.

Assessment

History
Symptoms of respiratory illness are few and can be summarized as breathlessness, cough and chest pain. An initial assessment of respiratory illness can be usefully undertaken by stratifying presentation by age group and by symptom duration (Table 2.1). Further definition of each symptom may provide useful diagnostic information.

Breathlessness (dyspnoea)
• Positional dyspnoea suggests a cardiac cause or a structural abnormality of the thoracic cage.
• Nocturnal dyspnoea is a feature of cardiac breathlessness and obstructive airways disease.
• The effort required to precipitate breathlessness also provides a good gauge of the level of respiratory impairment. Breathless at rest or inability of a young child to feed indicate severe restriction.
• Association with tightness and wheeze is a key indicator of bronchospasm but must not be confused with stridor, which derives from upper airways pathology.

Cough
• A productive cough is the hallmark of many conditions. The quantity of sputum produced may provide diagnostic information about chronic obstructive pulmonary disease or bronchiectasis. The expectoration of mucopurulent material is an indicator of neutrophil activity and infection.
• Haemoptysis is often an indicator of serious underlying pathology, but it is important to establish that blood is being coughed and not coming from the upper airway or enteric tract. Tuberculosis, bronchiectasis and neoplasia are primary concerns.
• Extreme paroxysms of coughing in a child, particularly in association with the characteristic whoop, indicates a diagnosis of whooping cough. A 'barking' cough with inspiratory stri-

CLASSIFICATION BY SYMPTOM DURATION

Presentation

Sudden	Acute	Subacute	Chronic
Cough and dyspnoea			
Foreign body	Pneumonia	CCF	Tuberculosis
Pneumothorax	Bronchitis	Bronchial obstruction	Bronchiectasis
Pulmonary embolus	Asthma*	Lung abscess	COPD
	Pulmonary eosinophilia*	Neoplasia	Paragonimiasis
	Croup	Kaposi's sarcoma	Silicosis/asbestosis
	Acute LVF	CHD	Deformity of thoracic cage or content
	Malaria		Pulmonary hypertension
	Helminths		Mycotic infections
	Allergic alveolitis*		
Chest pain			
Pneumothorax	Pneumonia/pleurisy	Empyema	Neoplasia
Pulmonary embolus			

Abbreviations: CCF, congestive cardiac failure; CHD, congestive heart disease; COPD, chronic obstructive pulmonary disease; LVF, left ventricular failure.
*The hallmark of these conditions is the recurrent nature of the attacks.

Table 2.1 Relationship between duration of presenting symptoms and diagnostic categories. Sudden indicates onset over a period of minutes, acute over a period of several hours, subacute over a period of several days and chronic usually several weeks to months.

dor is the hallmark of laryngotracheobronchitis —croup.

Chest pain
• Complaints of chest pain should be assessed for their association with breathing and coughing. Pain derived from the pleura will be noticeable on breathing and is lateralized. Tracheal pain has a tearing or burning quality and is felt retrosternally, particularly on coughing. Occasionally, a dull continuous central chest pain may be caused by a bronchial neoplasm.

In addition to the presenting symptoms it is important to enquire about smoking, occupation and to identify clues to underlying disease, in particular features of HIV. In children, a family history of tuberculosis contact is important when investigating chronic cough or failure to thrive.

Non-respiratory illness presenting with predominantly respiratory symptoms should also be considered. Dyspnoea and tachypnoea are features of metabolic acidosis. Systemic illness—in particular diabetic ketoacidosis—can be easily confused with acute pneumonia as can any systemic acidosis (e.g. poisoning, severe sepsis, renal failure). Chronic breathing difficulties are a feature of anaemia and thyroid disease. Alteration of breathing pattern and breathlessness can occur with neurological injury, during the early stages of tetanus and botulism and following envenomation.

Examination
Examination should confirm a diagnosis formulated from the history, and assess the severity of respiratory illness.
• The general condition of an individual provides clues to a diagnosis. In particular, cachexia,

or failure to thrive in a young child, will indicate malnutrition or chronic underlying illness. Oral thrush, skin rashes and old herpes zoster scars are highly suspicious of HIV infection, heightening the possibility of pneumococcal pneumonia or tuberculosis.

• Tachypnoea can be a feature of any respiratory illness. In the context of an acute presentation, rates in adults above 30/min suggest severe disease particularly in association with diastolic blood pressure below 60 mmHg and/or a tachycardia in excess of 120 b.p.m. The criteria for tachypnoea in childhood are very different from adults. It is diagnosed only if the respiratory rate is over 60/min before the age of 2 months, over 50/min from 2–12 months, over 40/min from 1–5 years and over 30/min (as for adults) over the age of 5 years.

• Cyanosis should be looked for in the oral mucosa. When present it indicates at least 10% desaturation of haemoglobin and the need for supplemental oxygen. However, cyanosis is difficult to detect when there is significant mucosal pigmentation. In addition, severe anaemia obscures cyanosis as does carbon monoxide poisoning. Thus, the absence of cyanosis should be treated cautiously in the presence of other severity markers.

• Altered consciousness and confusion usually indicate severe acute disease and can necessitate specific management to protect the airway. Meningism can be found with severe pneumonia, although meningitis is infrequently present.

• Examination of the chest will usually readily identify a significant pleural effusion. Lobar pneumonia or tuberculosis can be diagnosed on the basis of bronchial breathing over dense lobar consolidation or apical cavitation; however, findings are often more subtle and lack textbook descriptions. A few focal inspiratory crackles may be the only findings and a decision to use limited diagnostic facilities or a therapeutic trial must be made. The increasingly atypical presentation of tuberculosis in the era of HIV makes this a constant diagnostic possibility.

Investigation of respiratory disease

The limited access to investigational facilities throughout the tropics encourages a symptom-related approach to management. This is often successful; however, there are specific conditions where investigations are prudent or essential for good management:

• suspicion or confirmation of tuberculosis;
• failure of therapy; or
• severe illness.

In addition, when non-pulmonary disease is suspected, specific investigations are required (e.g. haemoglobin, blood glucose). Investigations should be undertaken in a systematic manner, starting with the simplest required to establish a diagnosis. Indeed, many intensive investigations (blood oximetry, blood gas measurements, pulmonary function tests, bronchoscopy and CT scanning) are rarely available outside of specialized centres.

Sputum/respiratory secretions

Sputum examination is a valuable investigation because it is simple to perform and can provide a clear aetiological diagnosis. Ideally, sputum examination in the laboratory is best performed in a safety cabinet to minimize infective risk.

Sputum must be from the lower respiratory tract and instructions should be given to the patient to ensure expectoration of an appropriate sample. Samples are best collected outside to limit the hazard of cross-infection. When sputum cannot be produced, placing the patient in a head down position or injecting normal saline into the trachea through the cricoid membrane can induce expectoration. Failing this, bronchoscopy or trans-tracheal aspiration (placing a catheter into the lower respiratory tract via needle puncture of the cricoid membrane) is required. In the presence of dense consolidation, lung aspiration can be considered: a needle and syringe primed with 1 mL normal saline or sterile water is passed into the consolidated tissue through the thoracic wall and aspirated.

The aspirated material can be smeared onto slides for examination and injected into liquid culture media. In young children when sputum is difficult to collect, gastric washings may be considered for the investigation of possible tuberculosis (mycobacteria are gastric acid resistant).

Microscopy of respiratory secretions requires access to simple laboratory equipment and reagents. Gram staining for bacteria and Ziehl–Neelsen staining for mycobacteria should be performed routinely. Occasionally, an unstained wet preparation of sputum examined under low power may be useful for identifying *Strongyloides*, paragonimiasis or fungal elements. Cytology for malignant cells can also be performed on sputum but requires a skilled pathologist.

Culture of respiratory secretions requires good laboratory facilities. However, it is rarely possible to culture all submitted samples and this is best reserved for identifying smear-negative cases of tuberculosis or in complicated or severe pneumonia to establish antibiotic sensitivities to guide therapy.

Although a variety of other techniques are available for identifying pathogens in respiratory secretions, they do not have a place for routine care in the tropics. These include direct immunofluorescence for viruses and *Pneumocystis carinii*, antigen detection for pneumococci, and molecular techniques for several organisms including tuberculosis.

Blood

Blood cultures are a valuable investigation in all febrile conditions. Recovery of a pathogen provides a clear aetiological diagnosis. In addition, blood cultures are frequently the investigation by which an unusual cause of pneumonia is established, e.g. *Salmonella typhi*, melioidosis, *Rhodococcus equi*.

Pleural fluid

Sampling of pleural fluid is simple to perform and should be considered for most effusions. Fluid can be aspirated by use of a needle and syringe. Occasionally, this fails because pleural fluid is loculated, has formed a thick empyema or the chest findings result from chronic pleural scarring. Use of an Abrams pleural biopsy needle allows a piece of tissue to be subjected to histological review, although this is more invasive and requires some practice on the part of the operator.

Pleural fluid should be subjected to the same investigations as outlined for sputum. Protein measurements may be helpful in confirming an effusion to be a transudate—a protein level below 30 g/dL and an absence of inflammatory cells.

Chest X-ray

The postero-anterior (PA) chest X-ray is fundamental to the management of respiratory disease, but limited resources usually necessitate careful selection of patients. Those with severe acute respiratory illness, those who fail to respond to therapy and smear-negative cases of chronic cough are the ones most frequently requiring a chest X-ray. Details of radiographical abnormalities can be found in the chapters on specific diseases.

Common presentations

In general, respiratory presentations in the general medical clinic tend to fall into a small number of syndromes.

Acute breathlessness and fever in the small child

Lower respiratory tract infection (LRTI) is a leading killer of children. Consequently, the early assessment and management of this syndrome is a core component of the Integrated Management of Childhood Illness (IMCI) strategy promoted by the World Health Organization (WHO). Simple assessment at the primary care level using features of rapid breathing and subcostal recession (chest indrawing) of a child with fever and cough is used to distinguish children with an LRTI who require antibiotics and possible hospital admission, from those with an upper respiratory tract infection. Early

CAUSES OF PLEURAL EFFUSIONS

Common	Infrequent	Rare
Parapneumonic	Constrictive pericarditis	Thoracic duct damage
Primary tuberculosis	Amoebic liver abscess*	Pancreatitis
Post-primary tuberculosis	Neoplasia	Haemorrhagic fevers
Empyema	–Lung carcinoma	Filariasis
–Pneumococcal	–Burkitt's lymphoma	Paraneoplastic (e.g. Meigs' syndrome)
–Staphylococcal	–Kaposi's sarcoma	Collagen vascular diseases
–Anaerobic	–Mesothelioma	Hypothyroidism
Pulmonary embolus	Nephrotic syndrome	
Heart failure	Liver cirrhosis	

*Right-sided effusion only.

Table 2.2 Causes of pleural effusions.

initiation of therapy is essential for a good outcome.

Although many LRTIs are initiated by viral infections (amongst which respiratory syncitial virus, parainfluenza, adenovirus and measles are important), super-added bacterial infections are frequent. *Streptococcus pneumoniae* and *Haemophilus influenzae* Pittman type b predominate so antimicrobial therapy must include agents effective against these two bacteria. The presentation of tuberculosis in infants is often occult.

Acute breathlessness, cough and fever in adults

Acute bacterial pneumonia is the principal diagnostic consideration and the diagnosis and management of this is covered in Chapter 28. Non-infectious causes become more prevalent in older adults when uncontrolled hypertensive disease or ischaemic heart disease may present with acute pulmonary oedema and low-grade fever.

Chronic cough and malaise

Most chronic respiratory problems present in this way. As a consequence, establishing a clear diagnosis can be challenging. It is important to exclude or confirm tuberculosis, which repre-

sents a serious public health threat but is readily treatable (Chapter 12). A small number of conditions are specific to the tropics and may need to be considered under the right epidemiological circumstances: paragonimiasis in South East Asia and restricted areas of West Africa (Chapter 29); endemic mycoses in South and Central America (Chapter 54); and pulmonary complications of schistosomiasis in endemic regions (Chapter 17).

Breathlessness and wheeze

Asthma is an increasingly important problem in the tropics, particularly in urban centres. The expiratory wheeze or whistling associated with lower airways obstruction must be differentiated from inspiratory phase stridor, which indicates upper airway obstruction. The presence of paroxysmal or diurnal cough, breathlessness and wheeze preferably supported by variation in peak flow measurements reliably indicates airways obstruction. The principal differential diagnosis of asthma is tropical pulmonary eosinophilia (Chapter 30). Although the symptoms are identical, a high peripheral eosinophil count above 1×10^9/L, demonstration of microfilaria in blood and clinical response to filaricides support the diagnosis.

Pleural effusion

Symptoms associated with pleural effusions can be of short or long duration depending on the

PROBLEMS COMPLICATING HIV INFECTION

Common	Infrequent
Bacterial pneumonia	Pneumocystis carinii pneumonia*
Tuberculosis	Rhodococcus equi infection
Acute bronchitis	Nocardiasis
Sinusitis	Lymphoid interstitial pneumonitis†
Bronchiectasis	Lymphoma
Pulmonary cryptococcosis	Pulmonary hypertension
Pulmonary Kaposi's sarcoma	Penicillinosis‡
	Melioidosis‡
	Invasive mycoses§

* Common in children under 1 year old.
† Common in children.
‡ South East Asia.
§ South and Central America.

Table 2.3 Respiratory problems complicating HIV infection.

nature of the underlying problems, but the identification of an effusion is usually quickly made on examination. The cause of an effusion may be suggested from the history, but it is rare that a clear diagnosis can be established on history and examination alone and sampling of the pleural fluid is essential. Parapneumonic effusions or empyema and tuberculous effusions head the differential diagnosis. Malignant effusions must also be considered when an infective aetiology is not readily apparent. Effusions —more so than other respiratory presentations—may indicate extrapulmonary or systemic problems (Table 2.2).

Respiratory disease in the HIV-infected

Respiratory problems head the list of conditions leading to hospital admission of HIV-infected adults (Table 2.3). Bacterial (particularly pneumococcal) pneumonia is strongly associated with HIV infection. It has a similar predictive value for HIV infection in adults to herpes zoster—around 90% in eastern and southern Africa. HIV infection is the principal factor driving the tuberculosis epidemic in Africa. Thus, HIV coinfection varies between 50 and 70%. HIV testing should be considered in all cases of pneumonia and tuberculosis.

Cough is a frequently reported symptom in individuals with advanced immunosuppression. Although tuberculosis always requires exclusion, hypostatic pneumonia, bronchial mucus hypersecretion, pharyngeal thrush, acute bronchitis and postnasal drip from sinus disease all contribute to the causes of cough. Management is especially difficult; the patient appears ill and multiple courses of antibiotics are often prescribed. Therapeutic trials of antituberculosis therapy are frequently required, but are often disappointing.

CHAPTER 3

Neurological Presentations

Rapid assessment of patient with coma in the tropics, 19

Classification and further investigation of patients with coma, 23

Indications and contraindications for lumbar puncture in suspected CNS infections, 24

Further reading, 25

Neurological presentations are more common in the tropics than in the developed industrial world. Infectious diseases make a major contribution, but non-infectious causes are also important (Table 3.1). A complex mixture of socioeconomic and environmental factors contribute to the increased incidence.

Reasons for increased incidence of neurological disorders in the tropics

• Non-infectious neurological disorders — trauma is more common in the tropics, especially road traffic accidents. Patterns of vascular disease are catching up with those in the developed world, but the usage of drugs to control them lags behind.

• Infectious neurological diseases — the climate supports transmission of insect-borne pathogens (malaria, trypanosomiasis, arthropod-borne viruses). Environmental factors include the close proximity of homes to zoonotic infections. Vaccine-preventable diseases are more common (e.g. measles, tetanus, diphtheria, polio). There is also unregulated use of over-the-counter antibiotics, leading to the partial pretreatment of central nervous system (CNS) infections, which hampers diagnosis and therapy, and promotes the development of antibiotic resistance.

• Poverty, overcrowding, poor sanitation and lack of education about disease risk factors and prevention are important. These may lead, for example, to cysticercosis and typhoid.

• Immunosuppression, particularly as a result of HIV, allows many other infections such as cryptococcal and tuberculous meningitis.

Neurological syndromes

Neurological diseases — particularly infections — can present with a range of syndromes.

• *Encephalopathy* — a reduced level of consciousness from any cause (infectious, metabolic, vascular, traumatic).

• *Meningism* — clinical signs of meningeal irritation (headache, neck stiffness, Kernig's sign; see below).

• *Paralysis* — weakness of one or more limb, respiratory or bulbar muscles, which may be a result of damaged upper motor neurones, lower motor neurones, peripheral nerves, or muscles.

• *Chronic neurological presentations* — insidious presentation over weeks or months, often with changes in personality, behaviour or other psychiatric illness. Fever may not be prominent, even with an infectious cause (Table 3.2).

• *Headache* — may be the only symptom (e.g. in cryptococcal meningitis).

• *Other focal neurological signs* — including hemispheric signs, brainstem signs, seizures and movement disorders.

CAUSES OF NEUROLOGICAL DISEASE

Vascular
Ischaemia/infarct
Subarachnoid/subdural/extradural/
    intracerebral haemorrhage
Hypertension/hypotension

Infectious
Direct effect on CNS
Bacteria
    Meningocccccus, streptococci, *Haemophilus
    influenzae*, tuberculosis, leprosy
Viruses
    Arboviruses, herpes viruses, enteroviruses,
    rabies
Parasites
    Protozoans
        malaria *(Plasmodium falciparum)*
        African trypanosomiasis *(Trypanosoma
        gambiense* and T. *rhodesiense)*
        toxoplasmosis *(Toxoplasma gondii)*
        amoebiasis *(Entamoeba histolytica)*
    Trematodes (flukes)
        paragonimiasis
        schistosomiasis (especially *Schistosoma
        japonicum)*
    Cestodes (tapeworms)
        cysticercosis *(Taenia solium)*
        hydatidosis *(Echinococcus granulosus)*
    Nematodes (roundworms)
        ascariasis *(Ascaris lumbricoides)*
        parastrongyliasis *(Parastrongylus
        cantonensis)*
        gnathostomiasis
        trichinosis *(Trichinella spiralis)*
    Spirochetes
        Neurosyphilis *(Treponema pallidum)*
        Lyme disease *(Borrelia burgdorferi)*
        Leptospirosis *(Leptospira* species)
        Louse-borne/epidemic relapsing fever *(B.
        recurrentis)*
        Tick-borne/endemic relasing fever *(B. duttonii)*
    Rickettsiae
        Epidemic/louse-borne typhus *(Rickettsia
        prowazekii)*

Endemic/murine/flea-borne typhus *(R.
    typhi/R. mooseri)*
Scrub typhus *(O. tsutsugamushi)*
Rocky Mountain spotted fever *(R. rickettsii)*
Fungi
    Cryptococcosis
    Histoplasmosis
    Aspergillosis
    Coccidioidomycosis
    Candidiasis
    Paracoccidiomycosis
    Blastomycosis
    Nocardiasis*

Indirect effect of infection
Toxin-mediated infectious diseases (tetanus,
    diphtheria, shigellosis)
Immune-mediated postinfectious
    inflammatory (GBS, acute disseminated
    encephalomyelitis)

Metabolic
Hypoglycaemia
Diabetic ketoacidosis
Hepatic encephalopathy
Uraemia
Hyponatraemia
Hypothyroidism/hyperthyroidism
Addison's disease

Tumours/trauma/toxins
Alcohol
Drugs (medical, recreational, traditional)
Pesticides
Poisons

Other
Hydrocephalus
Epilepsy
Psychiatric disease (hysteria)
Inflammatory
Nutritional
Degenerative

Abbreviations: CNS, central nervous system; GBS, Guillain–Barré syndrome.
Nocardia are actinomycete bacteria which are grouped with fungi because of their morphology and
behaviour.

Table 3.1 Causes of neurological disease (VIMTO).

CHRONIC NEUROLOGICAL PRESENTATIONS

Infectious	Other
Sleeping sickness (especially *Trypanosoma rhodesiense*, *T. gambiense*)	Tumours
Tuberculous meningitis	Chronic subdural haemorrhages
HIV encephalopathy	Lead, other heavy metal poisoning
Toxoplasma gondii and other parasitic space-occupying lesions	Dementia
Bacterial abscesses	Vitamin deficiencies
Partially treated bacterial meningitis	Drugs
Neurosyphilis	Toxins
Cryptococcal meningitis and other fungi	
Subacute sclerosing panencephalitis	

Table 3.2 Causes of chronic neurological presentations in the tropics.

Pathological processes

These neurological syndromes are explained by a range of pathological processes.

• *Encephalitis*—inflammation of the brain substance, usually in response to viral infection, but also in response to other pathogens.

• *Meningitis*—inflammation of the meningeal membranes covering the brain, in response to bacterial, viral or fungal infection.

• *Myelitis*—inflammation of the spinal cord. This may occur across the whole cord (causing transverse myelitis, which is often postinfectious) or be confined to the anterior horn cells.

• *Neuropathy*—damage to peripheral nerves (e.g. Guillain–Barré syndrome, diphtheria, leprosy, rabies, vitamin deficiencies).

• *Space-occupying lesions* (Table 3.3)—these cause pathology in the brain or spinal cord directly (by interrupting neuronal pathways), and indirectly (by causing localized swelling, raised intracranial pressure and brainstem herniation syndromes). Typically, they present with focal signs or a chronic insidious deterioration.

Rapid assessment of patient with coma in the tropics

1 **Stabilize the patient,** and treat any immediately life-threatening conditions.

CNS SPACE-OCCUPYING LESIONS

Tumours and metastases
Haemorrhage
Bacterial abscesses
Tuberculomas
Parasites
   Protozoa (toxoplasmosis, amoebiasis)
   Trematodes (paragonimiasis, schistosomiasis)
   Cestodes (cysticercosis, hydatidosis)
   Nematodes (ascariasis)
Fungi
   Aspergillosis, blastomycosis, nocardiasis

Table 3.3 Causes of central nervous system space-occupying lesions in the tropics.

• Airways.
• Breathing—give oxygen; intubate if breathing is inadequate or gag reflex impaired.
• Circulation—establish venous access.
Obtain blood for immediate bedside blood glucose test (hypoglycaemia?).
Malaria film (look for parasites and pigment of partially treated malaria).
FBC, U&E, blood cultures, arterial blood gases.
• Disability.
Give intravenous (i.v.) glucose (e.g. 10% glucose 50 mL in adults, 5 mL/kg in children), irrespective of blood glucose.
Give adults 100 mg thiamine i.v., especially if alcohol abuse is suspected.

Immobilize cervical spinal cord if neck trauma is suspected.
• Rapidly assess AVPU scale (alert, responds to voice, to pain, or unresponsive).
If patient responds to pain or is unresponsive, examine the pupils, eye movements, respiratory pattern, tone and posture for signs of cerebral herniation (see below).
If herniation is suspected start treatment for this.
• If purpuric rash is present give penicillin or chloramphenicol (or third generation cephalosporin) for presumed meningococcal meningitis (after taking blood cultures).
• Look for and treat generalized seizures, focal seizures and subtle motor seizures (mouth or finger twitching, or tonic eye deviation).

2 Take a history, while preliminary assessment and resuscitation proceeds. This is the single most useful tool in determining the cause of coma. In particular:
• Duration of onset of coma.
  Rapid onset (minutes–hours) suggests a vascular cause, especially brainstem cerebrovascular accidents or subarachnoid haemorrhage. If preceded by hemispheric signs, then consider intracerebral haemorrhage. Coma caused by some infections (e.g. malaria, encephalitis) can also develop rapidly, especially when precipitated by convulsions.
  Intermediate onset (hours–days) suggests diffuse encephalopathy (metabolic or, if febrile, infectious).
  Prolonged onset (days–weeks) suggests tumours, abscess or chronic subdural haematoma (see Table 3.2).
• Any drugs?
• Any trauma?
• Important past medical history (e.g. hypertension)?
• Family history (e.g. tuberculosis)?
• Known epidemic area (e.g. viral encephalitis)?

3 Perform a rapid general medical examination, and in particular:
• Check pockets for drugs.

• Note temperature (febrile or hypothermia) and blood pressure (hypo- or hypertensive).
• Examine for signs of trauma (check ears and nose for blood or cerebrospinal fluid (CSF) leak).
• Smell the breath for alcohol or ketones (diabetes?).
• Examine the skin for:
  rash (meningococcal rash, dengue or other haemorrhagic fever, typhus, relapsing fever);
  needle marks of drug abuse;
  recent tick bite or eschar (tick-borne encephalitis, tick paralysis, tick-borne typhus or relapsing fever);
  chancre, with or without circinate rash (trypanosomiasis, especially *Trypanosoma rhodesiense*);
  healed dog bite (rabies); or
  snake bite.
• Examine for lymphadenopathy (e.g. Winterbottom's sign of posterior cervical lymphadenopathy in African trypanosomiasis).
• Examine the fundi for papilloedema (long-standing raised intracranial pressure) or signs of hypertension.

4 Determine the coma score to allow subsequent changes to be accurately monitored. The scale in Box 3.1 is for adults and children over 5 years of age and in Box 3.2 for young children.

5 Neurological examination. A detailed description of the neurological examination is beyond the scope of this chapter. For most practical purposes the ability to recognize the following four clinical patterns (and combinations of them) will allow appropriate classification and subsequent investigation and treatment.
• *Meningism*—with or without encephalopathy.
• *Diffuse encephalopathies*—usually metabolic or infectious.
• *Supratentorial focal damage* (above the cerebellar tentorium)—usually manifests as hemispheric signs.
• *Damage in the diencephalon or brainstem* (midbrain, pons or medulla)—may indicate a syndrome of cerebral herniation through the

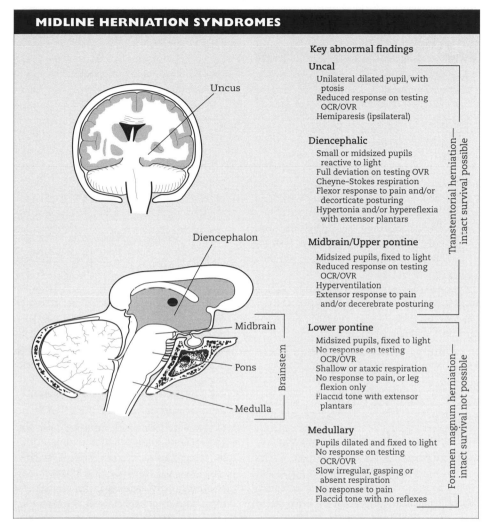

MIDLINE HERNIATION SYNDROMES

Key abnormal findings

Uncal
Unilateral dilated pupil, with ptosis
Reduced response on testing OCR/OVR
Hemiparesis (ipsilateral)

Diencephalic
Small or midsized pupils reactive to light
Full deviation on testing OVR
Cheyne–Stokes respiration
Flexor response to pain and/or decorticate posturing
Hypertonia and/or hyperreflexia with extensor plantars

Midbrain/Upper pontine
Midsized pupils, fixed to light
Reduced response on testing OCR/OVR
Hyperventilation
Extensor response to pain and/or decerebrate posturing

Lower pontine
Midsized pupils, fixed to light
No response on testing OCR/OVR
Shallow or ataxic respiration
No response to pain, or leg flexion only
Flaccid tone with extensor plantars

Medullary
Pupils dilated and fixed to light
No response on testing OCR/OVR
Slow irregular, gasping or absent respiration
No response to pain
Flaccid tone with no reflexes

Uncus

Diencephalon

Midbrain

Pons

Medulla

Brainstem

Transtentorial herniation— intact survival possible

Foramen magnum herniation— intact survival not possible

Fig. 3.1 Sagittal section of brain showing anatomy and key abnormal findings of midline herniation syndromes, and (above) coronal section showing herniation of the uncus of the temporal lobe—this compresses the ipsilateral third nerve (to cause a palsy of CNIII), and the contralateral cerebral peduncle (to cause an ipsilateral hemiparesis).

tentorial hiatus or the foramen magnum (Fig. 3.1). The importance of these syndromes is being increasingly recognized in non-traumatic coma (particularly that caused by infections). Although the level of brainstem damage is given in brackets below (and in Fig. 3.1), recognizing the presence or absence of brainstem signs, and in particular early signs of reversible damage, is usually more important than determining their exact localization.

Assessment of the following five points allows most patients to be classified.

1 Check for neck stiffness (if no trauma), and Kernig's sign (extension of knee when hip is already flexed causes pain).

COMA SCALE FOR ADULTS AND CHILDREN OVER 5 YEARS

Best motor response
6 Obeys command
5 Localizes supraorbital pain
4 Withdraws from pain on nail bed
3 Abnormal flexion response
2 Abnormal extension response
1 None

Best verbal response
5 Oriented
4 Confused
3 Inappropriate words
2 Incomprehensible sounds
1 None

Eye opening
4 Spontaneous
3 To voice
2 Pain
1 None

Box 3.1 Modified Glasgow coma scale for adults and children over 5 years. Total score is the sum of best score in each of the three categories (maximum score 15). 'Unrousable coma' reflects a score <9

BLANTYRE COMA SCALE FOR YOUNG CHILDREN

Best motor response
2 Localizes painful stimulus
1 Withdraws limb from pain
0 Non-specific or absent response

Best verbal response
2 Appropriate cry
1 Moan or inappropriate cry
0 None

Eye movements
1 Directed (e.g. follows mother's face)
0 Not directed

Box 3.2 Blantyre coma scale for young children. Total score is sum of best score in each of three categories (maximum score 5). 'Unrousable coma' reflects score <2

2 Examine pupil reaction to light (Fig. 3.1). A normal reaction (constriction) is seen in a diffuse encephalopathy. A unilateral large pupil is seen in herniation of the uncus of the temporal lobe. The pupils are reactive (small or mid-sized) in the diencephalic syndrome. Unreactive pupils occur in brainstem lesions (mid-sized in midbrain or pontine lesions; large in medullary lesions). Pinpoint pupils occur following opiate or organophosphate overdose, or in isolated pontine lesion. Other drugs can cause large unreactive pupils.

3 Assess eye movements (holding eyelids open if necessary).
 • *Spontaneous eye movements* — eyes spontaneously roving or eyes following indicates the brainstem is intact (a diencephalic syndrome or a diffuse encephalopathy).
 • *Oculocephalic (doll's eye) reflex* — when rotating the head, the eyes normally deviate away from the direction of rotation. A normal

response indicates that the brainstem is intact (diffuse encephalopathy). Reduced or absent responses occur in uncal herniation, brainstem damage or, rarely, deep metabolic coma.
 • *Oculovestibular reflex* — caloric response to water should be tested if the result of the oculocephalic reflex is unclear. Check that the eardrum is not perforated, then irrigate by injecting 20 mL ice-cold water. Nystagmus is the normal response and indicates 'psychogenic coma'. Both eyes deviate towards the irrigated ear in coma with the brainstem intact. A reduced or absent response indicates an uncal syndrome or a damaged brainstem.

4 Assess breathing pattern. A normal pattern occurs in diffuse encephalopathy. Cheyne–Stokes breathing and hyperventilation occur in reversible herniation syndromes. Shallow, ataxic or apnoeic respiration occurs in more severe syndromes (Fig. 3.1). Hyperventilation also occurs in acidosis or may be caused by aspiration pneumonia, which is common in coma.

5 Assess response to pain by applying painful stimulus to the supraorbital ridge and nail bed of each limb.
 • *Hemiparesis* — most often indicates supra-

tentorial hemispheric focal pathology (other signs include asymmetry of tone and focal seizures), but also occurs in uncal herniation.

• 'Decorticate posturing'—flexion of arms with extension of legs, indicating damage in the diencephalon, and 'decerebrate posturing' (extension of arms and legs caused by midbrain/upper pontine damage) may both be reversible. No response, or leg flexion only, are more severe.

Symmetrical posturing (decorticate or decerebrate) and hemiparetic focal signs are also occasionally seen in metabolic encephalopathies (e.g. hypoglycaemia; hepatic, uraemic or hypoxic coma; sedative drugs), cerebal malaria, and intra- or postictally. Other pointers to metabolic disease include asterixis, tremor and myoclonus preceding the onset of coma.

Classification and further investigation of patients with coma

At this stage, if the history, general examination and preliminary investigation have not made one diagnosis extremely likely, most comatose patients will fall into one of three categories, based on the presence or absence of meningism, supratentorial and brainstem signs.

1 **Coma only** (no hemispheric signs, brainstem signs or meningism — 'sleeping beauties').

• If patient is febrile (or has a history of fever), suspect CNS infection (especially cerebral malaria) or metabolic coma plus secondary aspiration pneumonia.

• If afebrile, coma is likely to be metabolic (hypoglycaemia, drugs, alcohol, diabetic ketoacidosis, toxins), psychogenic (test caloric response to water—causes nystagmus), or, occasionally, resulting from subarachnoid haemorrhage or other cerebrovascular accident.

2 **Coma with meningism,** but no focal signs.

• If febrile, CNS infection (especially bacterial meningitis) is likely.

• If afebrile, subarachnoid haemorrhage is likely.

3 **Coma with focal signs** (with or without meningism). Decide if the signs are 'hemispheric signs', 'brainstem signs', or both.

• Hemispheric signs only. If febrile, consider CNS infection (especially encephalitis, bacterial meningitis, abscess, etc.). If afebrile, consider space-occupying lesion (Table 3.3), cerebrovascular accident or trauma.

• Brainstem signs only may be caused by either focal pathology within the brainstem (e.g. encephalitis) especially if markedly asymmetrical signs, or by herniation of the brainstem (through the foramen magnum) secondary to a diffuse process (e.g. diabetic ketoacidosis, or late bacterial meningitis) causing raised intracranial pressure.

• Hemispheric and brainstem signs may be a result of either a supratentorial lesion causing hemispheric signs and sufficient swelling to precipitate brainstem herniation (e.g. cerebral bleed, abscess) or patchy focal pathology in the hemispheres and brainstem (e.g. toxoplasmosis, viral encephalitis).

LUMBAR PUNCTURE GUIDELINES

All patients with suspected CNS infection should have a lumbar puncture, except those with the following contraindications:

• Obtunded state with poor peripheral perfusion or hypotension

• Deteriorating level of consciousness, or deep coma (responsive only to pain, GCS < 8)

• Focal neurological signs present:
Unequal, dilated or poorly responsive pupils
Hemiparesis/monoparesis (in patients with coma)
Decerebrate or decorticate posturing
Absent 'doll's eye' movements
Papilloedema

• Hypertension and relative bradycardia

• Within 30 min of a short convulsive seizure

• Following a prolonged convulsive seizure or tonic seizure

Abbreviation: GCS, Glasgow coma score.

Table 3.4 Guidelines for lumbar puncture in patients with suspected central nervous system (CNS) infections.

CSF FINDINGS

	Acute bacterial meningitis	Viral meningo-encephalitis	Tuberculous meningitis	Fungal	Normal
Opening pressure	Increased	Normal/increased	Increased	Increased	10–20 cm*
Colour	Cloudy	'Gin' clear	Cloudy/yellow	Clear/cloudy	Clear
Cells/mm³	High–very high	Normal–high	Slightly increased	Normal–high	
	1000–50000	0–1000	25–500	0–1000	<5
Differential	Neutrophils	Lymphocytes	Lymphocytes	Lymphocytes	Lymphocytes
CSF:plasma glucose ratio	Low	Normal	Low–very low (e.g. <30%)	Normal–low	66%
Protein (g/L)	High	Normal–high	High–very high	Normal–high	
	>1	0.5–1	1–5	0.2–5.0	<0.5

Normal values

Normal CSF opening pressure is <20 cm water for adults, <10 cm for children below age 8.
A normal CSF glucose is usually quoted to be 66% that of the plasma glucose, but in many tropical settings a cut-off of 40% is found to be more useful.
A bloody tap will falsely elevate the CSF white cell count and protein. To correct for a bloody tap, subtract 1 white cell for every 700 red blood cells/mm³ in the CSF, and 0.1 g/dL of protein for every 1000 red blood cells.

Some important exceptions

In patients with acute bacterial meningitis that has been partially pretreated with antibiotics (or patients <1 year old) the CSF cell count may not be very high and may be mostly lymphocytes.
In viral CNS infections, an early lumbar puncture may give predominantly neutrophils, or there may be no cells in early or late lumbar punctures.
Tuberculous meningitis may have predominant CSF polymorphs early on.
Listeriosis can give a similar CSF picture to tuberculous meningitis, but the history is shorter.
CSF findings in bacterial abscesses range from near normal to purulent, depending on location of the abscess and whether there is associated meningitis or rupture.
An Indian ink test, and if negative a cryptococcal antigen test, should be performed on the CSF of all patients in whom cryptococcosis is possible.

Table 3.5 Cerebrospinal fluid (CSF) findings in central nervous system infections.

Indications and contraindications for lumbar puncture in suspected CNS infections (Table 3.4)

For many years lumbar puncture was performed in all patients with suspected CNS infections, in both the tropics and western in-dustrialized nations. It has gone out of fashion in the latter, following concerns that it was being performed on patients with contraindications, and may have precipitated herniation.

In patients with a contraindication, treatment should be started and then a lumbar puncture reconsidered later. In many tropical settings in Africa and Asia, where CNS infections are very common, lumbar puncture is still considered an essential investigation. Here the benefits of accurate diagnosis and appropriate treatment may outweigh the theoretical risk of herniation, and even patients with relative contraindications often receive lumbar punctures with no apparent harm.

Cerebrospinal fluid findings in CNS infections

Although most patients with CNS infections will have findings that are straightforward to interpret, there may be considerable overlap (Table 3.5). Ideally, the decision about starting antibiotics should await the result of the lumbar puncture (if it is available quickly). However, antibiotics should be started immediately for patients with a typical meningococcal rash, because of the speed with which meningococcal septicaemia can become fatal. In such patients, if it is certain that the rash is meningococcal it has been argued that the lumbar puncture is not necessary, because the diagnosis is already made, though others advocate always doing a lumbar puncture.

Further reading

Kirkham FJ. Non-traumatic coma in children. *Arch Dis Child* 2001; **85:** 303–12. [An excellent review of pathophysiology and management.]

Kneen R, Solomon T, Appleton RA. The role of lumbar punctures in CNS infections. *Arch Dis Child* 2001; **87:** 181–3. [Detailed discussion of the use of lumbar puncture.]

CHAPTER 4

Febrile Presentations

Pathogenesis and symptomatic treatment of fever, 26

Clinical approach to the patient with fever, 26

Acute fevers with a negative malarial blood film, 28

Treatment of common causes of fever lasting <2 weeks, 28

Common causes of fever lasting >2 weeks, 29

Fever in HIV-infected patients, 29

Common clinical problems with febrile patients, 30

Pathogenesis and symptomatic treatment of fever

Fever is a physiological response to infection, but there are other non-infective causes of inflammation and fever to be considered. The principal cytokines initiating fever are interleukin 1 and interleukin 6; these alter thermoregulation in the hypothalamus mediated by prostaglandins. Antipyretics in common use act by inhibition of pyrogenic prostaglandin production: these are either non-steroidal anti-inflammatory drugs or paracetamol (acetaminophen). Although antipyretics are widely used and have beneficial properties in terms of analgesic effects and reducing discomfort, they have never been proved to improve the outcome of infection or to reduce the complications of pyrexia. Paracetamol is the antipyretic of choice because it is free of side-effects at normal dosage and, unlike aspirin, it is not associated with Reye's syndrome in children.

Clinical approach to the patient with fever

History

Patients often complain of symptoms, such as generalized myalgia and arthralgia, that suggest the presence of fever, but are actually afebrile (temperature <37.5°C). Such patients rarely have significant underlying pathology and are best managed conservatively. A history of rigors or night sweats is much more suggestive of fever and, even if initially afebrile, they are best managed as if a fever were present. Similarly, significant weight loss is also indicative of underlying organic disease. Localizing symptoms should be sought, but these carry far more weight if they are volunteered at the outset. Important symptoms are headache, photophobia, cough, sputum, pleurisy, localized pain, diarrhoea (especially if bloody) and urinary symptoms. Coryza and upper respiratory symptoms generally suggest a viral illness. Prior treatment with antibiotics may make diagnosis difficult. Freshwater exposure suggests schistosomiasis (Katayama fever). The pattern of fever is rarely helpful in making a diagnosis in practice, but duration of fever is useful. Older children will give a history in the same way as adults. For infants and babies, enquiry should be made about feeding, weight gain (often charted), general activity and the health of the parents.

Examination

A temperature >37.5°C is clinically significant. If there is a convincing history of fever but no significant pyrexia on presentation, and if the patient is not sick enough to be admitted, a self-recorded temperature chart usually resolves the matter. A pulse rate >125 or systolic blood pressure <100 mmHg in an adult suggests the patient is seriously ill and in need of empirical treatment. Spontaneous haemorrhage suggests a viral haemorrhagic fever. The eyes should be inspected for anaemia, jaundice, conjunctival injection (measles and leptospirosis) and the fundi examined if lumbar puncture is likely to be needed or bacterial endocarditis is possible. The mouth should be examined for candidiasis (HIV infection), Koplick's spots (measles) and pharyngitis. The tympanic membranes of all young children should be inspected, but only in adults if there are relevant symptoms. Cervical and axillary lymphadenopathy should be sought (pharyngitis, HIV, cytomegalovirus [CMV], Epstein–Barr virus [EBV], tuberculosis, lymphoma, toxoplasmosis, syphilis) and also occipital lymphadenopathy (rubella, trypanosomiasis). The skin should be carefully inspected for rash (viral exanthems, non-blanching meningococcal petechiae), an eschar (tick-borne rickettsial infection) or anaesthetic patches with pigmentary change (leprosy). Skin sepsis and cellulitis are common causes of fever (streptococcal or staphylococcal). Conscious level, orientation and neck stiffness need to be assessed. Psychosis may be a manifestation of typhoid.

A more detailed neurological examination is not required unless the history suggests a neurological problem, a lumbar puncture is likely to be needed or leprosy is possible. The chest and heart require examination for signs of consolidation (pneumonia often fails to give respiratory symptoms), pleural or pericardial effusion (tuberculosis, HIV, empyema) and heart murmurs (bacterial endocarditis, rheumatic heart disease). In infants and babies a raised respiratory rate may be the only evidence of pneumonia. Abdominal tenderness and peritonism should be sought (appendicitis, peritonitis, pelvic inflammatory disease). Localized right lower intercostal tenderness suggests amoebic liver abscess. Hepatomegaly (malaria, tuberculosis, hepatitis, schistosomiasis, hepatoma, amoebic liver abscess) and splenomegaly (malaria, typhoid, leishmaniasis, HIV, infectious mononucleosis, lymphoma and leukaemia, portal hypertension, brucellosis, disseminated tuberculosis) are important signs. Demonstrable ascites requires a diagnostic tap. Any detectable joint effusion should also be tapped. Urinalysis should, of course, be part of the examination.

Initial investigation

Laboratory tests that are useful to discriminate those who require further investigation and treatment from those who require symptomatic management only are malaria films and full blood count. A malaria film should always be performed if there has been a visit to a malarious area and there is fever or symptoms suggestive of fever, irrespective of the patient's presentation. It is useful triage to arrange a malaria film as soon as a febrile patient presents, even before seeing a doctor. Where facilities are available, measurement of urea and electrolytes (renal failure in septicaemic shock, severe malaria, leptospirosis, haemolytic uraemic syndrome; hyponatraemia in tuberculosis), liver function tests (viral hepatitis) and C-reactive protein (CRP) are helpful. If the CRP is <5 mg/L, significant underlying pathology is unlikely, unless the erythrocyte sedimentation rate (ESR) is raised: this may suggest a rare case of systemic lupus erythematosus (SLE). The ESR is simple to perform and may be helpful as a non-specific marker of inflammation (bearing in mind that the ESR may be raised in the elderly and in the general population in the tropics). It is good practice to save a specimen of acute serum, if you have facilities to analyse acute and convalescent viral titres, to make a retrospective diagnosis. A chest X-ray should be performed if no obvious cause of fever is present. Appropriate bacterial cultures are important and, where available, should always be performed prior to starting antibiotic treatment. These include: blood cultures, especially if typhoid or paratyphoid is possible; urine culture, where symptoms or urinalysis suggest urinary tract infection; and stool microscopy and culture, if bloody diarrhoea is present. Sputum culture is generally not helpful, except for tuberculosis, but this is often unavailable in resource-poor settings. Sputum microscopy for acid-fast bacilli should be available in most settings, and is important if pulmonary tuberculosis is suspected (chronic cough, weight loss, night sweats). Lumbar puncture is necessary if symptoms and signs suggest meningitis. Infants and babies may give no specific clues of meningitis, so lumbar puncture should be performed if they are significantly unwell with no other identifiable cause of fever and a negative malaria film. An adult patient who has severe headache but no meningism, and in whom HIV is known or suspected, requires lumbar puncture in order to exclude cryptococcal meningitis. Usually, evidence of raised intracranial pressure is a contraindication to lumbar puncture. Genital swabs for microscopy and culture should be taken if sexually transmitted infection is suspected; syphilis serology is also relevant, especially with a rash extending to the palms and soles. Viral tests are usually restricted to serology, generally for HIV and hepatitis B, in resource-poor settings. Counselling is necessary prior to HIV testing; I have found refusal common in West Africa.

Acute fevers with a negative malarial blood film

The white blood cell count (WBC) divides this group into two, as shown in Box 4.1.

Treatment of common causes of fever lasting < 2 weeks

It is important to remember that in general in at least 50% of adults and older children with genuine fever, no cause will be identified and the fever will resolve spontaneously in a few days. These patients come to no harm from their presumed infection, and the important thing is to keep diagnostic procedures and therapeutic intervention to a minimum. In young children and babies many infections are viral, and if the cause is not immediately apparent, it may become so in a few days. Such infections require sympto-

ACUTE FEVERS WITH A NEGATIVE MALARIAL BLOOD FILM	
Polymorphonuclear leucocytosis?	
Yes	No
Pyogenic infection	Viral infections
Leptospiral infection	Rickettsial infections
Relapsing fevers	Typhoid
Amoebic liver abscess	
Gout	

Box 4.1   Acute fevers with a negative malarial blood film

matic treatment only, with the exception of suspected measles (requiring high-dose vitamin A supplements). In adults, even when there is likely to be an underlying cause for infection, it is preferable to delay anti-infective treatment until the diagnosis is established, and certainly cultures should always be taken first where possible. In children with neutrophilia who appear unwell there may be a case for giving antibiotics. There are instances where empirical treatment should be started immediately, but treatment has to be guided by the diagnostic and therapeutic options available. For example, it is bad practice to blindly give treatment for malaria unless there are no diagnostic facilities whatsoever. Semi-immune people in endemic areas may have a few detectable malaria parasites circulating harmlessly, with their fever caused by something entirely different. Malaria does not cause a raised neutrophil count, but thrombocytopenia is very common and requires no intervention unless very low with spontaneous bleeding. Empirical treatment should be started immediately for meningitis if the patient is ill (with appropriate symptoms and signs) or if there is going to be a delay in obtaining the results of lumbar puncture. In practice, this usually means treatment with chloramphenicol; in well-resourced settings a third generation cephalosporin is preferred. Similarly, a shocked septicaemic patient will require empirical antibiotics immediately after blood cultures have been taken. Rickettsial infection is often a clinical diagnosis; the patient should be treated with tetracycline once investigations have been performed to exclude malaria and typhoid. Dengue fever should be suspected if there is general body pain and severe retro-orbital headache with generalized blanching erythema and a negative malaria film. Treatment is supportive only, and the diagnosis is only established retrospectively when serology is available. The fever should not last longer than 2 weeks and classically has a saddleback pattern. Treatment for pneumonia should be started on clinical grounds, although chest X-ray is certainly helpful. A low threshold for treating severe pharyngitis with penicillin V is reasonable in the tropics, partly because post-streptococcal complications are seen much more frequently, and because the rare but severe Lemierre's syndrome has increased in frequency as antibiotic treatment for sore throat has become unfashionable.

Common causes of fever lasting >2 weeks

The following list contains the most common causes of prolonged fever, simply subdivided according to the most usual white blood cell picture.

1 **Chronic fever with neutrophilia:**
  • deep sepsis;
  • amoebic liver abscess;
  • erythema nodosum leprosum;
  • cholangitis;
  • relapsing fever.
2 **Chronic fever with eosinophilia:**
• invasive (toxaemic) Schistosoma mansoni and S. japonicum infections;
  • invasive Fasciola hepatica infection;
  • acute lymphangitic exacerbations of Wuchereria bancrofti and Brugia malayi infections;
  • gross visceral larva migrans caused by Toxocara canis.
3 **Chronic fever with neutropenia:**
  • malaria;

• disseminated tuberculosis;
• visceral leishmaniasis;
• brucellosis.
4 **Chronic fever with normal WBC:**
  • HIV related (see below);
  • localized tuberculosis;
  • brucellosis;
  • secondary syphilis;
  • trypanosomiasis;
  • toxoplasmosis;
  • bacterial endocarditis;
  • SLE;
  • chronic meningococcal septicaemia.
5 **Chronic fever with a variable WBC picture:**
  • tumours;
  • lymphomas;
  • drug reactions.
Omitted from the list are those conditions where the localizing signs are so obvious that they could not be overlooked, such as pyogenic arthritis.

Fever in HIV-infected patients

The likely cause of fever in HIV infection changes according to the CD4 count, but this piece of information and even HIV serostatus are often unavailable. HIV and/or AIDS in infants and young children presents with a different spectrum of febrile infections compared with adults; often they are simply common childhood infections (e.g. acute respiratory infection, measles, respiratory syncytial virus, chickenpox) but with greater severity and duration. In a recent study in Malawi, it was found that the mortality of perinatally acquired HIV infection reached 89% by 3 years of age in the absence of antiretroviral treatment. Symptoms suggestive of HIV infection in a child are cough, ear discharge, oropharyngeal ulcers, fever and skin rash (when present for >2 weeks). Signs suggestive of HIV infection are malnutrition (wasting and stunting), oral thrush, oropharyngeal ulcers, lymphadenopathy and evidence of pulmonary infection. As in adults, tuberculosis is more common, especially extrapulmonary forms. Lymphocytic interstitial pneumonitis is rarely reported in sub-Saharan Africa compared with the UK.

Symptoms and signs suggestive of HIV infection in adults are chronic diarrhoea and weight loss, dysphagia (especially when associated with oral thrush), symptomatic sexually transmitted infection, chronic cough and night sweats, herpes zoster (vesicles or scars in dermatomal distribution), oropharyngeal candidiasis, generalized lymphadenopathy and skin lesions suggestive of Kaposi's sarcoma or persistent herpes simplex. Symptomatic HIV seroconversion (a febrile illness reminiscent of infectious mononucleosis, sometimes accompanied by aseptic meningitis) is rarely identified in sub-Saharan Africa. Apart from this initial illness, HIV does not directly cause fever. Fever in early HIV infection may signify underlying pneumococcal pneumonia, tuberculosis or the onset of herpes zoster (although these infections can still occur in late-stage HIV disease). Once the CD4 count drops below 200×10^6/L, fever may signify opportunistic infections such as cryptococcosis and toxoplasmosis. Non-typhoid *Salmonella* septicaemia is particularly common. In South East Asia, *Penicillium marneffii* septicaemia may be found. Tuberculosis is a common finding at postmortem, but *Mycobacterium avium-intercellulare* is rarely reported from sub-Saharan Africa. Similarly, *Pneumocystis carinii* pneumonia is a rare finding, although it has been reported from Zimbabwe. In areas endemic for Burkitt's lymphoma, a disseminated disease characterized by generalized lymphadenopathy and fever may occur. Malaria is statistically associated with a higher parasitaemia in HIV infection. In contrast, disseminated leishmaniasis (chronic fever, splenomegaly, pancytopenia, amastigotes in bone marrow and/or splenic aspirates) is a feature of late-stage HIV infection, but even so is not diagnostic of underlying HIV.

Common clinical problems with febrile patients

Managing a febrile patient with no localizing symptoms or signs and little or no laboratory or radiological back-up is a realistic problem. If the patient is unwell, empirical treatment has to be given. In a malarious area, the first line should be antimalarials, and a significant fall in fever would be expected after 3–4 days. If there is no response, then the next pathogen of importance is typhoid. It used to be standard practice to treat with chloramphenicol, but resistance is widespread so ciprofloxacin is to be preferred if available. Typhoid fever should respond to appropriate treatment in 4–5 days, but it can take 7–10 days if there is low-grade ciprofloxacin resistance. If there is still no response, empirical antituberculous therapy may be indicated, although regimens containing rifampicin will treat other infections as well as tuberculosis.

Another difficult scenario is a fever which has lasted >3 weeks and has not resolved after at least 1 week of investigation; i.e. pyrexia of unknown origin (PUO). Such patients should be clinically assessed at regular intervals in case new signs or symptoms come to light. There is very little information regarding underlying causes of PUO in the tropics. When PUO in the developed world is investigated, then infection accounts for about one-third of cases (mainly intra-abdominal abscess, tuberculosis, infective endocarditis and complications of HIV infection), neoplasia for 20% (especially lymphoma and occasionally renal cell carcinoma), autoimmune disorders for 10% (e.g. adult Still's disease, temporal arteritis, Wegener's granulomatosis and polyarteritis nodosa), miscellaneous causes for 15% (e.g. drug fever, non-infective granulomatous disorders, haematomas, e.g. subdural), and cause is unknown in 25%. Travel and exposure history, symptoms and signs should guide investigation and laboratory results. If available, an ultrasound is non-invasive and helpful for demonstrating hepatic disease (e.g. tumour, abscess, schistosomiasis, fascioliasis), splenomegaly, ascites and renal tract disorders. A computerized tomography (CT) scan of the thorax, abdomen and pelvis is relatively non-invasive and has a high diagnostic yield for abscesses, tumours and lymphadenopathy, but is not widely available in the tropics. Other potentially useful investigations include lymph node biopsy (if enlarged), bone marrow aspirate, trephine and culture, and liver biopsy. In the tropics, tuberculosis, osteomyelitis, dental sepsis, hepatoma and SLE appear more common than in adults with PUO in Europe. There is very little written about PUO affecting babies and children in the tropics, but it is safe to say that infections secondary to HIV-induced immunosuppression are increasing in importance, and neoplasia is less common than in adults. PUO in HIV-infected adults in the tropics presents a different spectrum of disease, and is usually associated with a CD4 count $< 200 \times 10^6$/L. A recent series of patients investigated in Brazil found (in descending order) tuberculosis, *Pneumocystis carinii*, *Mycobacterium avium* complex, non-Hodgkin's lymphoma, cryptococcal meningitis, sinusitis, salmonellosis, histoplasmosis, neurosyphilis and isosporiasis.

Finally, there are many other exotic and uncommon infective causes of fever. Some, like trypanosomiasis, *Borrelia recurrentis* and babesiosis will show up unexpectedly on examination of the blood film. Others have to be thought of and deliberately sought, such as bartonellosis, *Borrelia burgdorferi* and Q fever.

CHAPTER 5

Dermatological Presentations

Skin ulcers, 32

Skin itching, 32

Creeping eruptions, 34

Papules, 34

Skin nodules, 34

Changes in pigmentation, 34

Urticaria, 35

Bullae, 35

Petechial rashes, 35

Further reading, 35

Skin disease is ubiquitous among the poor of underdeveloped countries. Bacterial infections often secondary to insect bites or scabies are particularly likely in childhood and superficial fungal infections, especially pityriasis versicolor, are present in most adults. Many outpatient clinics are besieged with children suffering from tropical ulcers or skin infections and suitable plans for effective low-cost treatment are essential. Several of the major tropical diseases also have skin manifestations and it is essential that the clinician should not miss the diagnosis of leprosy.

Whenever possible take a systematic history to include details of any travel, contact with insects or sensitizing agents, use of drugs or skin applications, similar rashes in family or contacts and the evolution of the skin problem. Examine the entire skin surface together with the scalp and mucous membranes in a good light and note the character and distribution of all skin lesions, distinguishing the original lesions from modifications caused by scratching or secondary infection.

Skin ulcers

Trauma and insect bites account for many acute skin ulcers but these usually heal quickly. All the causes of chronic ulceration seen in temperate climates, including diabetic ulcers, venous and arterial ulcers, are seen but their frequency is much less in most young tropical populations. Some of the more important causes of chronic ulcers are given in Table 5.1.

Skin itching

A very wide range of dermatological conditions and health problems may cause itching but in the tropics the most common causes include the following.

- *Scabies* — typical distribution but burrows often masked by secondary infection.
- *Insect bites* — papular urticaria on exposed surfaces.
- *Superficial fungal infection* — examine scrapings for hyphae after clearing in 10% potassium hydroxide.
- *Eczema* — often a personal or family history of allergy or recent exposure to drugs or topical sensitizer.
- *Onchocerciasis* — geographical distribution, examine for nodules and take skin snips.

Itching can be a prominent symptom in African trypanosomiasis, during the initial invasion with schistosomes or hookworms, with anaemia, Hodgkin's disease, jaundice, diabetes and in the elderly with dry skin. Itching is also a feature of the creeping eruptions.

CHRONIC ULCERS

Type of ulcer	Main characteristics	How diagnosis is established
Tropical ulcer	Painful; rapid onset, usually lower leg	Heals on non-specific regimen and does not recur in same place
Buruli ulcer	Extensive; mainly painless; very deep undermined edges	Finding acid-fast bacilli in edge of ulcer
Cutaneous leishmaniasis	Single or multiple; often with infiltrated edges; not undermined	Amastigotes in edges of lesions. Culture or PCR
Desert sore, veld sore (cutaneous diphtheria)	Usually single, painful onset with vesicle; adherent slough, undermined, paralysis from toxin	Culture. The combination of sore + slough + palsy is diagnostic
Tertiary syphilis (gumma)	Chronic, usually painless ulcers on extremity	Serological tests for syphilis. Spirochaetes cannot be found
Tuberculous ulcer	Frank ulcers often follow subcutaneous TB; there may be adjacent cold abscess; evidence of TB elsewhere	Microscopy for AAFB. Culture
Sickle cell disease in adults	Rare in Africa where few adults with the disease survive. Ulcers often symmetrical on lower legs	Patient is obviously anaemic; sickling test is positive; more than 90% of the Hb is HbS on electrophoresis
Dracontiasis (guinea worm)	The pearly prolapsed uterus is seen early	By identifying the worm, or larvae expelled after exposure to water
Trophic ulcer of leprosy	Painless; may be deeply penetrating on the sole	Associated evidence of nerve damage: thick nerves, loss of sensation
Diabetic ulcer	In the tropics usually neuropathic. On soles of feet over bony prominences. Usually painless	Evidence of neuropathy. Known or newly diagnosed diabetes
Malignant ulcers	Squamous cell carcinoma on lip or old ulcer proliferating edge	Biopsy and histology
Mycoses (subcutaneous or deep)	Usually ulcerates within a preformed granulomatous nodule	Biopsy, culture and histology. Microscopy for hyphae and spores

Abbreviations: AAFB, acid- and alcohol-fast bacilli; Hb, haemoglobin; HbS, haemoglobin S; PCR, polymerase chain reaction; TB, tuberculosis.

Table 5.1 Chronic ulcers.

Creeping eruptions

• Larva migrans from dog hookworms form a slowly extending, persistent, itching track most often on the foot or lower leg. Multiple infections cause severe itching.
• Track-like lesions from the larvae of some species of *Paragonimus* from *Gnathostoma spinigerum* or from some fly larvae are less common.
• Larva currens is the name given to rapidly moving tracks caused by migrating *Strongyloides stercoralis* larvae. These urticaria-like tracks are found between the neck and the knees and last for hours or a day or two only.

Papules

• Milia.
• Onchocerciasis.
• HIV related.
• Scabies.
• Insect bites.
• Acne.
• Cercarial dermatitis.
• Tungiasis.

Skin nodules

• Furuncle.
• Furuncular myiasis.
• Leprosy. The nodules are frequently over the ears, eyebrows and face. The diagnosis is readily confirmed by slit skin smears for acid-fast bacilli.
• Erythema nodosum resulting from leprosy is sometimes widespread. Tuberculosis, streptococcal infection and sarcoidosis are other causes.
• Leishmaniasis. Single or multiple nodules may take months to ulcerate; they are predominantly on exposed surfaces but spread along lymphatics occurs. Diffuse cutaneous leishmaniasis and nodular post-kala-azar dermal leishmaniasis can resemble nodular leprosy.

• Kaposi's sarcoma. Chiefly affecting limbs of older persons in endemic areas, but any skin surface and mucosae often with lymph node enlargement in AIDS.
• Fungal infections including chromoblastomycosis, sporotrichosis, *Histoplasma dubosii* and *H. capsulatum*, *Paracoccidioides brasiliensis*, *Penicillium marneffei*. Some are particularly common as secondary infections in the immunosuppressed.
• Subcutaneous nodules are a feature of onchocerciasis.
• Cysticercosis.
• Juxta-articular nodules are found in late yaws.
   Remember also 'non-tropical' causes such as rheumatoid nodules and gout.

Changes in pigmentation

Hypopigmented macules
• Postinflammation and scarring.
• Pityriasis versicolor. 'Raindrop' patches over trunk with slight scaling.
• Leprosy. Loss of sensation to heat and light touch, enlarged nerves.
• Onchocerciasis. Patches chiefly over shins; nodules and positive skin snips.
• Yaws. Chiefly affects palms and soles; positive syphilis serology.
• Vitiligo. Usually white sometimes hyperpigmented borders; symmetrical; association with autoimmune disease.
• Post-kala-azar dermal leishmaniasis.

Hyperpigmentation
• Pellagra. Affects sun-exposed skin and may be associated with diarrhoea and dementia.
• Pregnancy.
• Chronic arsenic poisoning. Slaty-grey colour of trunk with small areas of normal skin. Hyperkeratosis of palms and soles.
• Addison's disease. Look for pigment in oral mucosa.
• Cachexia of malignant disease.
• Hypertrophic lichen planus. Warty patches typically involving calves, forearms and lower back.
• Kaposi's sarcoma.

Urticaria

Acute urticaria may follow jellyfish stings or other envenomation, contact with plants, arthropods or drugs such as penicillin.

The diagnosis of the cause of chronic urticaria can be very difficult because such a wide variety of both internal and external causes may be responsible. Some common causes acquired in the tropics are as follow.
• Papular urticaria from insect bites.
• Katayama syndrome in schistosomiasis. Follows freshwater exposure by days to months and associated with cough, wheeze and marked eosinophilia.
• Intestinal helminths including roundworms and hookworms and migrating larvae of *Strongyloides stercoralis*.
• Filarial infection.
• Drugs.
• Food additives.

Bullae

Large fluid-filled blisters may result from a variety of causes; some of the more common ones in the tropics include the following.
• Bullous impetigo.
• Insect bites.
• Sunburn.
• Burns and scalds.
• Drug eruptions.
• Snake bite and other causes of envenoming.
• Larva migrans.

• Pemphigus including Brazilian pemphigus foliaceous.
• Porphyria associated with sun exposure.

Petechial rashes

• Meningococcal septicaemia.
• Dengue and dengue haemorrhagic fever.
• Viral haemorrhagic fevers including Lassa, Ebola and Crimean–Congo.
• Vasculitis including Henoch–Schönlein purpura.
• Disseminated intravascular coagulation.
• Infective endocarditis.

Further reading

Canizares O. *A Manual of Dermatology for Developing Countries*, 2nd edn. Oxford: Oxford University Press, 1993. [A useful textbook that includes the skin distribution diagrams of common dermatoses.]

Caumes E, Carriere J, Guermonprez G et al. Dermatoses associated with travel to tropical countries: a prospective study of the diagnosis and management of 269 patients presenting to a tropical disease unit. *Clin Infect Dis* 1995; **20:** 542–8. [A description of travellers with dermatological conditions acquired abroad.]

Saw S-M, Koh D, Adjani MR et al. A population-based survey of skin diseases in adolescents and adults in rural Sumatra, Indonesia, 1999. *Trans R Soc Trop Med Hyg* 2001; **95:** 384–8. [Gives comparative prevalence data for skin disease in various developing countries.]

CHAPTER 6

The Patient with Anaemia

Clinical diagnosis of anaemia, 36
Laboratory investigations, 36

Management of anaemia in the absence of a laboratory, 38
Blood transfusion in developing countries, 39

Further reading, 39

Anaemia is the most common medical condition worldwide, but it is not a diagnosis in itself. Whenever possible, the underlying cause should be determined and alleviated to prevent recurrence. It is important to have a knowledge of the local causes of ill health as this can help to prioritize investigations and treatment, especially when diagnostic resources are limited. Where schistosomiasis is common this may be a frequent local cause of anaemia, whereas amongst rural farming communities it may be chronic hookworm infestation. In poorer countries the aetiology of anaemia is often multifactorial and exacerbated by poor nutrition and high levels of infections and infestations.

Clinical diagnosis of anaemia

Examination of the degree of pallor of tongue, nails and conjunctiva can provide a reasonable indication of anaemia when it is moderate or severe but is not helpful for detecting mild anaemia. Clinical examination alone is 66% sensitive and 68% specific for haemoglobin levels of 5–8 g/100 mL in Malawian children and 82% specific and 65% specific for haemoglobin levels of less than 7 g/100 mL in pregnant women in Kenya. A good clinical history and examination are essential in determining the cause of anaemia (Box 6.1).

Laboratory investigations

At smaller hospitals, the laboratory may provide tests such as haemoglobin, blood film examination and malaria slide microscopy. It may also offer a transfusion service. Accurate haemoglobin estimation and a good blood film examination will enable the diagnosis and cause of anaemia to be established in the majority of cases. Features on the blood film can indicate the presence of iron or folate deficiency, haemoglobinopathies or enzymopathies and malignant or proliferative haematological disorders. Well-resourced hospitals may be able to estimate iron, folate and vitamin B_{12} levels, examine bone marrow samples and carry out a range of other tests including endoscopy, to determine the cause of anaemia.

Measurement of haemoglobin

Although this is the most commonly performed laboratory test it is difficult to get an accurate result using manual techniques under conditions prevailing in most developing countries. The World Health Organization (WHO) recommended reference method for haemoglobin, the haemoglobincyanide method (WHO 1986, Cheesbrough 1998), requires a spectrophotometer and a well-supervised and qualified technician. Any method that depends on

HISTORY AND EXAMINATION OF THE ANAEMIC PATIENT

Enquire about:
- Symptoms of hypotension/heart failure/anaemia
- Diet (specifically iron and folate content)
- Blood loss (from gastrointestinal, urogenital or gynaecological systems)
- Childhood or family history of haemoglobinopathy
- Chronic inflammatory process
- Tendency to bleed or excessive infections (suggesting bone marrow dysfunction)

Examine for:
- Heart failure/postural hypotension
- Pallor
- Jaundice
- Fever
- Spoon-shaped nails
- Skeletal abnormalities (suggesting haemoglobinopathy)
- Splenomegaly
- Tuberculosis and other chronic disorders
- Petichiae, lymphadenopathy, gum infiltration (suggesting leukaemia)

Box 6.1 History and examination of the anaemic patient

manual dilution of the sample (e.g. Sahli, Lovibond) requires careful pipette technique to maintain accuracy. It is essential that clinicians satisfy themselves that their laboratory's results are reliable before using haemoglobin measurements to guide patient management. This can best be done by enrolling the laboratory in an external quality monitoring scheme or setting one up between a group of local laboratories. Other simple ways of maintaining the quality of results include:
- repeat testing of the same sample with each batch of tests;
- compare packed cell volume (PCV) with haemoglobin results (PCV should be approximately three times the haemoglobin value);
- plot weekly cumulative averages of haemo-globin results to determine any 'drifting' of results; and
- ensure that reagents are within their shelf life, technical staff are qualified and regularly supervised, and that the method in use is appropriate for the level of health care and local infrastructure.

Examination of peripheral blood film

If the cause of anaemia remains elusive after basic investigations, determining whether the anaemia is microcytic, macrocytic or normocytic will narrow down the possibilities.

Microcytic anaemia

The most common cause of microcytic anaemia is iron deficiency. To elucidate the cause, specific questions should be asked relating to blood loss and dietary insufficiency, and may require stool examination for parasites and occult blood, and endoscopic examination of the gastrointestinal tract to exclude occult malignancy. The thalassaemias may produce similar morphological changes to iron deficiency but the clinical setting and further investigations such as haemoglobin electrophoresis and measurement of HbA_2 and HbF may help to confirm the diagnosis. α-Thalassaemia trait may be particularly difficult to diagnose and referral to a specialist centre may be necessary. If features of both microcytic and macrocytic anaemia are present, nutritional deficiency of both iron and folate is likely.

Macrocytic anaemia

A high mean cell volume (MCV) with oval macrocytes and hypersegmented neutrophils suggests folate or vitamin B_{12} deficiency. A high MCV may also indicate the presence of early red cells. These may be produced in response to blood loss or destruction and can be detected as polychromatic cells on a peripheral blood film. Specific stains can be used to confirm that these cells are reticulocytes. A high MCV can also be associated with alcohol excess and liver disease, or drugs such as hydroxyurea. A combination of red cell fragments, thrombocytopenia

SUGGESTED MANAGEMENT OF ANAEMIA IF NO LABORATORY TESTS ARE AVAILABLE

• Give iron, folate and antihelmintics and monitor response (clinically detectable improvement should occur within 4 weeks as the haemoglobin should rise at rate of 0.5–0 g/100 mL each week.
• Continue iron for at least 3 months after normal haemoglobin is achieved.
• If acute life-threatening haemolysis is suspected (anaemia with jaundice and dark urine) and there is no obvious cause or underlying infection, a trial of folate and prednisolone 0.5–1 mg/kg may be worthwhile until transfer to a higher level facility can be arranged.
• If still no response, refer to specialist centre.

Box 6.2 Suggested management of anaemia if no laboratory tests are available

and polychromasia on the blood film indicates microangiopathic haemolytic anaemia and the need for further tests such as coagulation studies, assessment of renal function and a search for infection or neoplastic disease.

Normocytic anaemia

Normochromic normocytic anaemia is usually caused by an underlying chronic non-haematological disease. Investigations should include screening for renal disease, infections, autoimmune diseases and neoplasia. In the presence of anaemia, a lack of polychromasia and reticulocytes suggests primary failure of erythropoiesis, or lack of haematinics. A combination of fever, anaemia and thrombocytopenia may indicate acute leukaemia and a blood film should be examined urgently for the presence of blasts. Examination of the bone marrow may be helpful to detect aplastic anaemia or early myelodysplastic syndrome and immunophenotyping assists in confirming leukaemia subtypes.

PRESCRIBING BLOOD: A CHECKLIST FOR CLINICIANS

Always ask yourself the following questions before prescribing blood or blood products for a patient.
1 What improvement in the patient's clinical condition am I aiming to achieve?
2 Can I minimize blood loss to reduce this patient's need for transfusion?
3 Are there any other treatments I should give before making the decision to transfuse, such as intravenous replacement fluids or oxygen?
4 What are the specific clinical or laboratory indications for transfusion in this patient?
5 What are the risks of transmitting HIV, hepatitis, syphilis or other infectious agents through the blood products that are available for this patient?
6 Do the benefits of transfusion outweigh the risks for this particular patient?
7 What other options are there if no blood is available in time?
8 Will a trained person monitor this patient and respond immediately if any acute transfusion reactions occur?
9 Have I recorded my decision and reasons for transfusion on the patient's chart and the blood request form?
Finally, if in doubt, ask yourself the following question.
   If this blood was for myself or my child, would I accept the transfusion in these circumstances?

Box 6.3 Prescribing blood: a checklist for clinicians

Management of anaemia in the absence of a laboratory

At health centres without laboratory facilities the management of anaemia will depend critically on knowledge of the locally prevalent diseases. Good clinical skills are essential to elicit relevant information and to detect subtle signs of pallor. Mild anaemia is likely to be missed if haemoglobin cannot be measured (for management of anaemia see Box 6.2).

In children under 5 years in malaria endemic regions, local policy often involves treating all fevers as malaria. The majority of these children are also anaemic and their consultation at the health centre should be used as an opportunity to detect and treat any other conditions and also to provide relevant education for the child's carer.

Blood transfusion in developing countries

Blood transfusions should only be given for specific clinical indications and in accordance with local or international guidelines (Box 6.3; European Commission 1995, WHO 1999). Transfusions carry serious risks of transmitting infections such as HIV and hepatitis B, and can also be responsible for acute reactions. These risks are increased in situations where there is no laboratory quality assurance programme. Contrary to the recommendations of the WHO, many blood transfusions in developing countries are donated by coerced or remunerated donors rather than voluntary donors. Hospitals therefore often have no, or very little, stock of blood for emergency use. The problem has been worsened by the increasing prevalence of HIV infection amongst potential blood donors. Patients with severe anaemia, usually children with malaria, need blood transfusion within a few hours of admission as an emer-gency life-saving measure. Before prescribing blood transfusions, clinicians need to balance carefully the risks and benefits. They should satisfy themselves that all other options, such as intravenous fluids and haematinics, have been excluded. Transfusions should only be used as a last resort (WHO 1999).

Further reading

Cheesbrough M. *District Laboratory Practice in Tropical Countries*. Cambridge, UK: Tropical Health Technology, 1998. [Widely recognized as a leading standard laboratory manual for developing countries. Clearly written and well illustrated with basic explanations of rationale and principles of tests.]

Gerard C, Sondag-Thull D, Watson-Williams EJ, Fransen L, eds. *Safe Blood in Developing Countries*. Brussels: European Commission, 1995. [General overview of practical issues around the systems needed to ensure that the blood supply is safe in developing countries.]

World Health Organization. *Methods Recommended for Essential Clinical Chemistry and Haematological Tests for Intermediate Hospital Laboratories*. WHO/LAB/86.3, 1986. [Provides information about recommended validated methods and standards for tests used in diagnosis and management of common conditions including anaemia.]

World Health Organization. *The Clinical Use of Blood*. WHO/BTS/99.2, 1999. [Provides prescribers of blood with information to assist them to make appropriate decisions about the use of blood and to avoid unnecessary transfusions. A pocket handbook is available to accompany this publication.]

CHAPTER 7

A Syndromic Approach to Sexually Transmitted Infections

The need for a public health
  approach, 40
Syndromic management, 40
Local adaptations, 42
How to use the flow charts, 42

The 'four C's' of syndromic
  management, 43
Advantages of syndromic
  management, 45

Disadvantages of syndromic
  management, 47
Further reading, 48

The global prevalence of chronic viral sexually transmitted infections (STIs) is likely to be over a billion. In some populations, almost every adult has either active or latent infection with viruses such as genital herpes virus, genital human papilloma virus (HPV), hepatitis B or HIV. In addition, the World Health Organization (WHO) estimated the incidence of curable bacterial STIs such as gonorrhoea, chlamydia and syphilis was 333 million annually in 1995. In view of this it is hardly surprising that the management of individuals with STIs and the consequences thereof constitutes the bulk of the outpatient workload in high-prevalence settings such as sub-Saharan Africa.

The need for a public health approach

The rapid spread of the HIV epidemic has renewed interest in the early detection and treatment of curable STIs. From early on in the HIV epidemic it was clear that STIs have a role in the spread of HIV. Individuals with recurrent presentations of genital ulcer or discharge were shown to be at a higher risk of acquiring HIV. In addition, HIV-infected individuals with an ulcer or discharge were found to have higher rates of viral shedding of HIV and therefore to be more infectious to partners. Because rates of shed-ding revert to lower levels after treatment, the appropriate management of STIs reduces the risk of HIV transmission. It makes sense therefore to include STI management as an HIV prevention strategy. Currently, however, the public health benefits of this approach have not been realized on a large scale.

Syndromic management

In 1991, the WHO developed a system of syndromic management for STIs. The aim was effective management of STIs in resource-poor countries with high prevalence rates of STIs. The emphasis was on an integrated approach at primary health care centres, with no requirement for specialist clinics, highly trained personnel or laboratory facilities. Syndromic management is based on the identification of consistent groups of symptoms and easily recognized signs (syndromes) (Table 7.1). A patient with an STI syndrome presents to a health facility for care and the practitioner has merely to tell the difference between a genital ulcer and a discharge. Once identified, a step-by-step flow chart guides the practitioner. A single course of treatment is provided at the first clinic visit, which deals with the majority of the organisms responsible for producing each syndrome in a given area.

COMMON SYNDROMES AND AETIOLOGIES

Syndrome	Infectious causes	Aetiological agent*	Non-infectious causes
Urethral discharge	Gonococcal urethritis Non-gonococcal (NSU)	*Neisseria gonorrhoeae* No aetiology (25%) *Chlamydia trachomatis* *(most of the rest)* *Ureaplasma urealyticum* *Trichomonas vaginalis* *Candida albicans*	Physiological Trauma
	Intra-urethral ulcers or warts	Herpes simplex HPV	
Vaginal discharge	Vaginal infections	*C. albicans* *Gardnerella vaginalis* *T. vaginalis*	Physiological, trauma, retained products, tampons
	Cervical infections	*C. trachomatis* *N. gonorrhoeae* Herpes simplex HPV *Treponema pallidum*	Carcinoma of cervix (linked with HPV infection)
Genital ulcer			
Multiple and painful	Herpetic Chancroid Scabies	Herpes simplex *Haemophilus ducreyi* *Sarcoptes scabiei*	Behçet's disease Stevens–Johnson syndrome
Single and painful	TB Superinfection of painless ulcers	*Mycobacterium* *tuberculosis* *Staphylococcus aureus*	Carcinoma
Multiple and painless	Secondary syphilis	*T. pallidum*	
Single and painless	Primary syphilis LGV	*T. pallidum* *C. trachomatis* (LGV strains)	
	Granuloma inguinale	*Calymmatobacterium* *granulomatis*	
Bubo			
Genital ulcer visible	Chancroid	*H. ducreyi*	
No genital ulcer visible	LGV	*C. trachomatis*	
Non-sexually transmitted infections	Plague Filariasis Infections of the lower limb		

Abbreviations: HPV, human papilloma virus; LGV, lymphogranuloma venereum; NSU, non-specific urethritis; TB, tuberculosis.
* Bold italic indicates infections targeted by syndromic management.

Table 7.1 Some common syndromes and their aetiologies.

Local adaptations

Local data on aetiology and bacterial sensitivity patterns must be taken into consideration when designing a flow chart. For example, penicillin resistance amongst *Neisseria gonorrhoea* is on the increase globally. In some countries as much as 50% of gonorrhoea cases are resistant to penicillin. This represents a major threat to the cheap and effective treatment of urethral and cervical discharge syndromes. Moreover, as in the treatment of pneumococcal disease, penicillin resistance is associated with resistance to multiple antibiotics including macrolides. The inevitable consequence has been therapeutic failures and higher priced therapies such as ciprofloxacin being included in syndromic management.

The successful implementation of syndromic management programmes requires that the appropriate drugs be accessible, available and affordable. Unfortunately, many of the antibiotics used for the treatment of STIs in the West (such as quinolones) are too expensive for STI control programmes or individuals to afford in resource-poor countries. The choice of drugs in most countries is therefore often a compromise between what is affordable and what is therapeutically required. Drugs are generally dispensed at health centre level where other more urgent conditions may put demands on supplies originally intended for STI treatment only. Ensuring that the supply of STI drugs is used for the treatment of STIs and guarding it against drug theft are additional challenges on a strained system.

Each country therefore has a slightly different flow chart validated in its own setting according to local prevalence and incidence rates of STIs. WHO flow charts in use that manage the most common clinical situations include the following:
• urethral discharge syndrome in men (Fig. 7.1);
• vaginal discharge (Fig. 7.2);
• lower abdominal pain in women (Fig. 7.3); and
• genital ulcer disease (Fig. 7.4).

How to use the flow charts

1 Symptoms determine which flow chart to select. Patients should be specifically asked about the onset of symptoms and whether the condition is associated with pain.

2 Signs indicate the likelihood of pathology. Patients must be examined for the presence of ulcers, in males the urethra milked for discharge and in females a bimanual examination should be performed. Speculum examination of the cervix should be performed if available.

3 Investigations are limited in many settings. If a microscope is available, a Gram stain provides a sensitive indicator of gonococcal infection in urethral discharge. A wet mount from the vaginal specimen will reveal *Candida albicans*, *Trichomonas vaginalis* and clue cells of bacterial vaginosis. However, investigation with microscopy does not improve the sensitivity and specificity of the flow charts for lower abdominal pain and vaginal discharge in women. To identify women at greater risk of cervical infection, questions about local risk factors have been added. Laboratory-assisted diagnosis is rarely helpful in genital ulcer disease as mixed infections are common.

4 Management should be guided by the local guidelines. Alternatives are recommended on the locally produced flow charts for patients with allergies. Special circumstances such as pregnancy are also covered. In general, quinolones are recommended to treat gonorrhoea and tetracyclines to cover chlamydial infection. Concurrent therapy for *Chlamydia* and gonorrhoea should be given to all patients with gonorrhoea, as dual infection is common. While the treatment of choice for syphilis remains intramuscular benzylpenicillin, the penicillins and tetracyclines have no place in the treatment of chancroid owing to widespread resistance in all geographical areas. To cover chancroid a quinolone or macrolide must be added.

5 Some patients fail to respond to treatment. The likeliest causes are reinfection from partner, poor compliance or drug resistance. In the case of persistent urethral discharge, *Tri-*

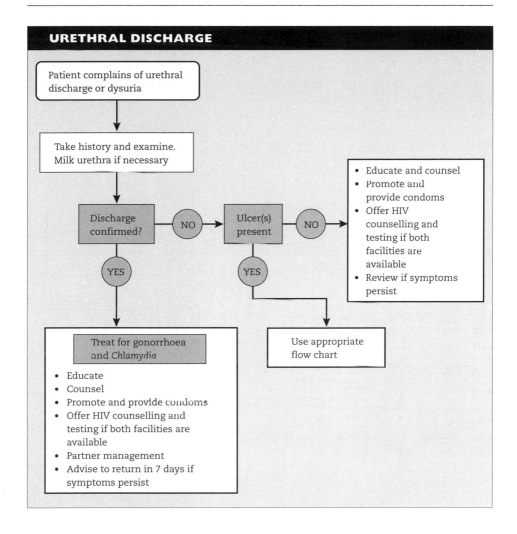

Fig. 7.1 Urethral discharge in men.

chomonas vaginalis should be considered before referral, as there are high prevalence rates in some geographical settings. In the case of genital ulcer disease, if symptoms persist after adequate treatment of the index case and partner, the patient should be referred to rule out other causes, including chronic viral STIs, coinfection with HIV, carcinoma or a non-sexually transmitted disease. Where prevalence rates of HIV infection are high, a large number of genital ulcers are likely to be brought about by atypical herpes simplex virus infection.

The 'four C's' of syndromic management

One of the greatest barriers to the successful implementation of syndromic management has been the attitude of health care workers. Training on the use of flow charts therefore always includes training on basic counselling skills. This is the first 'C'. Emphasis in training is also put on Contact tracing, Compliance and Condoms.

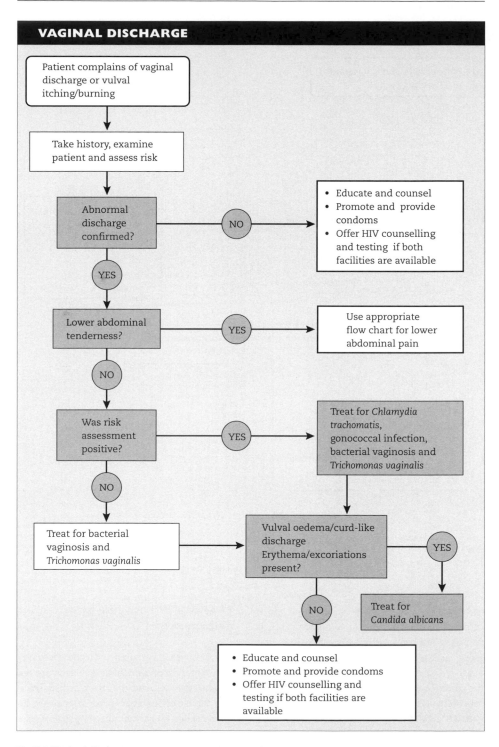

Fig. 7.2 Vaginal discharge.

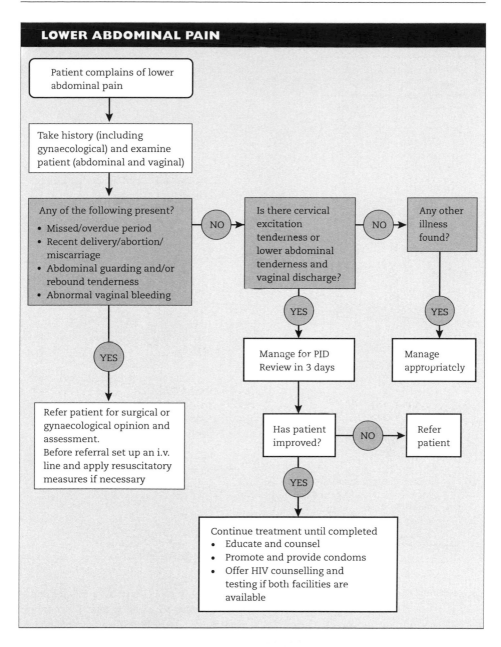

Fig. 7.3 Lower abdominal pain in women.

Advantages of syndromic management

The syndromic approach to STI management is simple and problem-orientated and can be integrated into existing health facilities without requiring specialist clinics, doctors or nurses. It allows for rapid diagnosis and treatment of the individual at the first visit, thus saving resources for the client and the provider and improving surveillance. Flow charts developed for the

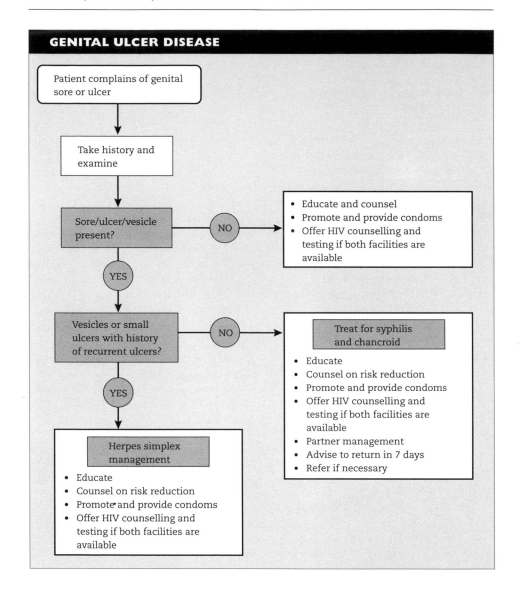

GENITAL ULCER DISEASE

Patient complains of genital sore or ulcer

Take history and examine

Sore/ulcer/vesicle present? — **NO** →
- Educate and counsel
- Promote and provide condoms
- Offer HIV counselling and testing if both facilities are available

YES

Vesicles or small ulcers with history of recurrent ulcers? — **NO** →

Treat for syphilis and chancroid
- Educate
- Counsel on risk reduction
- Promote and provide condoms
- Offer HIV counselling and testing if both facilities are available
- Partner management
- Advise to return in 7 days
- Refer if necessary

YES

Herpes simplex management
- Educate
- Counsel on risk reduction
- Promote and provide condoms
- Offer HIV counselling and testing if both facilities are available

Fig. 7.4 Genital ulcer disease.

treatment of urethral discharge in men are robust and well validated in numerous settings. Cases of genital ulcer disease resulting from chancroid and syphilis have dropped dramatically in Nairobi, Kenya as a result of intensive syndromic management in high-risk cohorts. Studies of a population-wide approach to syndromic management in Mwanza, Tanzania showed a 42% reduction in rural HIV incidence rates in those receiving access to syndromic management of STIs. However, reduction of HIV incidence was not demonstrated in a study conducted in neighbouring Rakai district in Uganda where a similar cohort was enrolled in a mass STI treatment campaign. A closer analysis of the two data sets reveals different baseline HIV and gonococcal prevalence rates as well as different risk behaviour levels. Syndromic management is therefore most likely to influence

HIV incidence in areas with high levels of risk behaviour and low prevalence of HIV.

Disadvantages of syndromic management

The focus on syndromic management has public health limitations. The main disadvantage is that the uninfected are not targeted and therefore asymptomatic infections are not detected and there is no provision for screening. In addition, there is insufficient emphasis on partner notification and a lost opportunity to promote condom use and provide information on STIs. The approach relies entirely on the self-presentation of those who perceive themselves to have symptoms. In the majority of settings the flow charts have a low sensitivity and specificity for cervical gonococcal and chlamydial infections in symptomatic women. This frequently leads to overtreatment of the individual and possible increased rates of antibiotic resistance. A lack of consideration of the differential diagnoses is commonplace.

The design of STI control programmes in the tropics has been much enhanced by the widespread use of the syndromic management approach. However, programmes need to combine the simple management of syndromes with interventions that target the general population, promote condoms and educate on other prevention methods. Levels of appropriate health-seeking behaviour are low (Fig. 7.5). Work in the community and with health care workers themselves is therefore needed to challenge stigma, gender roles and myths surrounding STIs. Partner notification and treatment are essential to interrupt the chain of transmission and prevent reinfection and programmes should be accompanied by access to other services such as counselling and testing. Efforts to improve partner notification should be voluntary and ensure the confidentiality of patients and their partners, as fear of rejection and of domestic violence are very real concerns that underlie poor rates of partner notification recorded in many programmes. Good STI management challenges traditional views of medicine as clinicians work closely with public health

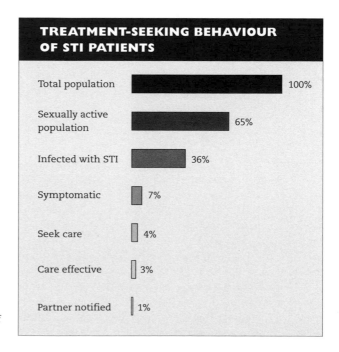

TREATMENT-SEEKING BEHAVIOUR OF STI PATIENTS

Total population	100%
Sexually active population	65%
Infected with STI	36%
Symptomatic	7%
Seek care	4%
Care effective	3%
Partner notified	1%

Fig. 7.5 Operational model of treatment-seeking behaviour of STI patients.

specialists and all sectors of the community for it to be a success.

Further reading

Dallabetta G, Laga M, Lamptey P et al. *Control of Sexually Transmitted Diseases: A Handbook for the Design and Management of Programmes*. Arlington: AIDSCAP/Family Health International, 2001. [This manual was developed primarily for programme managers in resource poor settings.]

Dallabetta GA, Gerbase AL, Holmes KK. Problems, solutions and challenges in syndromic management of sexually transmitted diseases. *Sex Transm Infect* 1998; **74**: (suppl 1): S1–11. [This supplement to the journal deals with experiences and viewpoints on syndromic management.]

Grosskurth H, Mosha F, Todd J et al. Impact of improved treatment of sexually transmitted diseases on HIV infection in Tanzania: randomized, controlled trial. *Lancet* 1995; **346**: 530–6. [Showed a 42% reduction in HIV incidence following effective syndromic management.]

Wawer M, Sewankambo N, Serwadda D et al. Control of sexually transmitted diseases for AIDS prevention in Uganda: a randomized community trial. *Lancet* 1999; **353**: 525–35. [Showed a reduction in some STIs, but no significant reduction in HIV incidence despite mass treatment.]

Wisdom A. *A Colour Atlas of Sexually Transmitted Disease*. London: ELBS edition, 1992. [Available at low cost.]

www.who.int/Reproductive_health [Useful source for the guidelines for the management of sexually transmitted infections with several flow charts that can be downloaded.]

CHAPTER 8

Splenomegaly in the Tropics

Reasons for enlarged
  spleens, 49
Massive tropical
  splenomegaly, 49

Hyper-reactive malarial
  splenomegaly, 50

Splenectomy in the tropics, 50
Further reading, 51

Enlarged spleens are common in tropical practice. The disorders that cause splenomegaly in temperate regions are also present in the tropics but, in addition, there is a heavy burden of infections and parasitic infestations. The spleen responds to this by augmenting its major physiological functions of phagocytosis and antibody production. The subsequent splenic enlargement is particularly common in children living in areas of high malaria transmission where it is used as a population indicator of transmission intensity. The 'spleen rate' may reach 100% in children and then decline to less than 10% in adults as they acquire clinical malarial immunity. In areas with stable endemic malaria, adults' spleens are about twice as large as those in non-malarious areas.

Reasons for enlarged spleens

Spleens can enlarge:
• in response to a need for excess physiological activity (e.g. phagocytosis of abnormal red cells as in haemoglobinopathies, or antibody production to combat infection); and
• because of a structural abnormality (e.g. portal hypertension or infiltration by malignant cells).

The degree to which the spleen enlarges depends on the underlying cause but in most cases the spleen size rarely exceeds 10 cm when measured from the left costal margin. Acutely en-

larged spleens are often tender and soft on examination, and are associated with a higher risk of rupture than chronically enlarged spleens which tend to be firmer and more fibrous. Conditions in which the spleen may be moderately (< 10 cm) enlarged include chronic haemolysis (e.g. recurrent malaria, haemoglobinopathies, spherocytosis), portal hypertension and haematological malignancies such as chronic lymphocytic leukaemia, lymphomas, acute leukaemias and polycythaemia vera.

Massive tropical splenomegaly

The most common causes of massive splenomegaly in the tropics are hyper-reactive malarial splenomegaly (formerly called tropical splenomegaly syndrome), lymphomas, schistosomiasis, visceral leishmaniasis, haemoglobinopathies, chronic myeloid leukaemia, myelofibrosis and miscellaneous disorders such as splenic cysts, tumours and lipid storage diseases (Fig. 8.1). Although massive splenomegaly has been reported to be common in many tropical African countries there are few data available on prevalence. Published rates in Africa vary from 1–2% in Nigeria to 0.4–1.2% in Gambia. The highest prevalence of massive splenomegaly is in Papua New Guinea where up to 80% of some ethnic groups are affected. Almost all of this is caused by hyper-reactive malarial splenomegaly. The diagnosis and

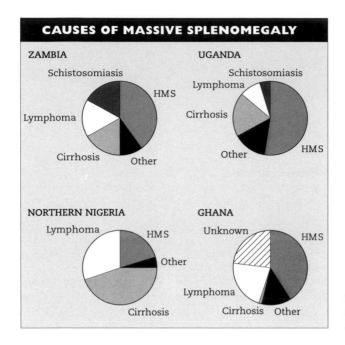

Fig. 8.1 Causes of massive splenomegaly from various African countries.

management of most of the conditions associated with massive splenomegaly are discussed elsewhere.

Hyper-reactive malarial splenomegaly

This is caused by an abnormal response to repeated malaria infections that results in overproduction of immunoglobulin M (IgM). The consequent immune complexes are removed by the spleen, which can enlarge to huge proportions (Fig. 8.2, see colour plate facing p. 146). The disorder is more common in women and predominantly affects those aged between 20 and 40 years. Patients are surprisingly asymptomatic but eventually develop symptoms of anaemia, malaise and abdominal discomfort. Pregnant women with hyper-reactive malarial splenomegaly commonly experience sudden episodes of haemolysis which may be life-threatening. A mild reduction in platelets and white cells secondary to hypersplenism is common in hyper-reactive malarial splenomegaly

and almost all patients have anaemia and hepatomegaly.

Criteria for diagnosis include splenomegaly over 10 cm from the left costal margin and a sustained reduction in spleen size of at least 40% on antimalarial treatment. Differentiation from splenic lymphoma may be difficult without molecular techniques but a diagnosis of hyper-reactive malarial splenomegaly is more likely if the patient is under 40 years of age and has a peripheral blood lymphocyte count of less than 10×10^9/L. Life-long treatment is necessary as the condition will recur if treatment is stopped. Proguanil 100 mg/day is the drug of choice as it is safe for continued use over many years. On this treatment the haemoglobin increases and the spleen slowly shrinks although it may never become impalpable.

Splenectomy in the tropics

The indications for splenectomy in tropical practice are similar to those in temperate regions but the balance between risk and benefit may be altered by the lack of intensive care and other support services in poorer countries. Elective splenectomy for patients with enlarged spleens in Uganda had an early postoperative mortality of 4.8%. Particular risks of splenectomy in tropical countries include the following.

• Larger spleens make the procedure technically difficult.

• Hypersplenism leads to reduced platelet counts and potentially increased risk of bleeding. The risk of morbidity from haemorrhage is exacerbated by the lack of emergency blood supplies and platelet transfusions.

• Increased susceptibility to bacterial infections, especially encapsulated organisms. Septicaemia with these organisms can be associated with disseminated intravascular coagulation, which has a mortality of 50–80% in established cases. Vaccination for *Haemophilus influenza* type b, *Neisseria meningitidis* and *Streptococcus pneumoniae* may not be available.

• Very few data are available about the risk of malaria infection postsplenectomy. It is likely that the risk is less if the individual acquired malarial immunity at an early age and has been a long-term resident in a malarious area.

To reduce the risk of postsplenectomy infections, a partial rather than total splenectomy should be performed, leaving a portion of spleen with arterial structures *in situ*.

Further reading

Bedu Addo G, Bates I (2002) Causes of massive tropical splenomegaly in Ghana. *Lancet* 2002; **360**: 449–54. [Review of causes of massive splenomegaly in tropical, malaria-endemic country determined using current technology.]

Crane G. Tropical splenomegaly. Part 2. Oceania. In: Luzzatto L, ed. *Clinics in Haematology*. Philadelphia: WB Saunders, 1981; and

Fakunle Y. Tropical splenomegaly. Part 1. Tropical Africa. In: Luzzatto L, ed. *Clinics in Haematology*. Philadelphia: WB Saunders, 1981. [These two references summarize the historical evidence about hyper-reactive malarial splenomegaly from the perspective of different continents.]

Fakunle Y. Splenomegaly In: Parry E, ed. *Principles of Medicine in Africa*, 2nd edn. Oxford, UK: Oxford University Press, 1984: 886–92. [Overview of the causes of splenomegaly from several African countries demonstrating the geographical variation.]

PART 2

Major Tropical Infections

CHAPTER 9

Malaria

Importance and
 distribution, 55
Life cycle, 55
Clinical features, 57
Classical stages of fever, 57
Progress of the untreated
 attack, 58

Peculiarities of *Plasmodium*
 falciparum infection, 58
Malaria in pregnancy, 61
Immunity in malaria, 61
Diagnosis, 62
Treatment, 63

The problem of relapse, 68
Chemoprophylaxis, 68
Epidemiology, 69
Global malaria eradication, 71
Further reading, 71

Importance and distribution

Malaria is the most important of all tropical parasitic diseases, causing many deaths and much morbidity. It is widely distributed in the tropical and subtropical zones. There are four parasite species that cause human malaria, all of which belong to the genus *Plasmodium*. In order of their prevalence worldwide these are *Plasmodium falciparum* (the cause of nearly all of the deaths caused by malaria), *P. vivax*, *P. malariae* and *P. ovale*.

A malarial infection can occasionally be transferred directly from one person to another by blood transfusion, accidental inoculation or across the placenta. However, transmission usually depends on an insect vector, in which the parasite spends several weeks undergoing the sexual part of its life cycle.

Life cycle (Fig. 9.1)

Malaria is usually transmitted by the bite of an infected female anopheline mosquito. The infecting agent is the sporozoite, a microscopic spindle-shaped cell which is in the mosquito's saliva. Thousands of sporozoites may be injected in a single bite. The sporozoites disappear from the blood within 8 h, and the successful ones enter liver cells. The process by which the malaria parasites multiply asexually is called schizogony, whether it takes place in a hepatocyte or in an erythrocyte.

Inside the liver cell, the sporozoite divides by asexual fission to form a cyst-like structure called a pre-erythrocytic (PE) schizont, which contains tens of thousands of merozoites. Each merozoite consists of a small mass of nuclear chromatin within a tiny sphere of cytoplasm. When the PE schizont is mature, it ruptures and liberates its contained merozoites. These now enter the bloodstream to penetrate red cells.

The time between the bite of the infecting mosquito and the appearance of parasites in the blood is the prepatent period. It is 7–30 days in *P. falciparum* (usually around 10 days), and longer in the other species. It may be very long; in the case of *P. vivax* and *P. ovale* many months or even more than a year. This dormant stage of the parasite is the hypnozoite.

Merozoites, released into the bloodstream from hepatic PE schizonts, attach themselves to red cells by means of surface receptors. The parasite then penetrates the red cell and resides in a vacuole with a lining derived from the red cell surface. Here it begins the process of blood schizogony. Schizogony occurs in the circulating blood in the cases of *P. vivax*, *P. ovale* and *P. malariae*, so in all these infections schizonts are

MALARIA LIFE CYCLE

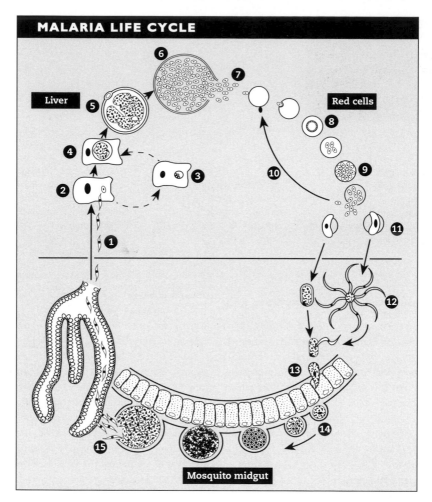

Liver

Red cells

Mosquito midgut

Fig. 9.1 Malaria life cycle. 1, Sporozoites, injected through the skin by female anopheline mosquito; 2, sporozoites infect hepatocytes; 3, some sporozoites develop into 'hypnozoites' (*Plasmodium vivax* and *P. ovale* only); 4, liver-stage parasite develops; 5–6, tissue schizogony; 7, merozoites are released into the circulation; 8, ring-stage trophozoites in red cells; 9, erythrocytic schizogony; 10, merozoites invade other red cells; 11, some parasites develop into female (macro-) or male (micro-) gametocytes, taken up by mosquito; 12, mature macrogametocyte and exflagellating microgametes; 13, ookinete penetrates gut wall; 14, development of oocyst; 15, sporozoites penetrate salivary glands. (Reproduced with permission from Zaman V. *Atlas of Medical Parasitology*. Balgawlah, NSW, Australia: ADIS Press, 1978.)

commonly seen in the peripheral blood films from infected patients. In *P. falciparum*, schizogony only occurs in capillaries deep within the body. At the stage of the maturing trophozoite, parasite antigens are expressed on the surface of the red cell. Some of these antigens are capable of linking to receptors expressed on the endothelial cells lining capillaries in various organs and tissues of the body. The resulting cytoadherence of parasitized erythrocytes to endothelial surfaces leads to the gathering or sequestration of large numbers of mature parasites in deep tissues.

The periodicity of schizogony characteristically coincides with paroxysms of fever and this led to the traditional names of the different types of human malaria.

• *Tertian malaria* — fever every third day, if the first day is given the number 1 : *P. vivax* and *P. ovale*.
• *Subtertian malaria* — fever slightly more often than every third day: *P. falciparum*.
• *Quartan malaria* — fever every fourth day if the first day is given the number 1 : *P. malariae*.

Plasmodium falciparum malaria was sometimes called malignant tertian malaria, because of its much greater lethal potential than the other tertian malarias. These antique names for malaria are best avoided, not only because they can be confusing, but because the periodicity they imply often fails to develop. Many patients have lost their lives from *P. falciparum* malaria because they never developed the periodic fever that their doctors wrongly believed to be invariable.

Some of the merozoites entering red cells do not develop into schizonts, but develop more slowly into solid-looking parasites called gametocytes. These may persist in the circulation for many weeks without destroying the red cells containing them, and they are the forms infective to the mosquito. In each species of malaria, the gametocytes are differentiated into male and female. When the female mosquito swallows the male and female gametocytes in her blood meal, they develop further in her stomach. The male gametocytes rapidly develop to produce spermatozoon-like microgametes, and the female gametocyte becomes the egg-like macrogamete.

Clinical features

There are no recognized symptoms associated with the liver stage of malaria infections, or (as far as we know) with rupture of tissue schizonts. The development of a blood stage infection is necessary for malarial illness.

Infections with each of the four different malaria species have many clinical features in common. These result from the release, when red cell schizonts rupture, of 'malaria toxins' or pyrogens. The common features are as follows.

• *Fever* — often irregular. Fever is believed to be mediated by host cytokines, which are secreted by leucocytes and other cells in response to the released pyrogens. The pattern of regularly periodic fever often does not occur until the illness has continued for a week or more. It depends on synchronized schizogony. Why schizogony should ever become synchronized is unknown, but an intriguing explanation has been suggested. High temperatures slow the growth of mature more than of young parasites. Fever itself may therefore allow young parasites to 'catch up' with older ones, leading to increasing synchrony with successive cycles.
• *Anaemia*. This is caused by a combination of haemolysis and bone marrow suppression. Haemolysis is usually most severe in *P. falciparum* malaria because cells of all ages can be invaded in this infection.
• *Splenomegaly*. The spleen enlarges early in the acute attack in all types of malaria. When a patient has had many attacks, the spleen may be of enormous size and lead to secondary hypersplenism.
• *Jaundice*. A mild jaundice caused by haemolysis may occur in all types of malaria. Severe jaundice only occurs in *P. falciparum* infection, and results from specific liver involvement.

Classical stages of fever

In a paroxysm of malaria, the patient may notice the following stages.

1 *Cold stage* — the patient shivers or has a frank rigor; the temperature rises sharply.

2 *Hot stage* — the patient is flushed, has a rapid full pulse and a high temperature is sustained for a few hours.

3 *Sweating stage* — the patient sweats freely, or is even drenched, and the temperature falls rapidly.

These stages are most often recognized in *P. vivax* infection. For the clinician or epidemiologist, the important point is that an individual with a symptomatic malarial infection, or even with severe disease, may be afebrile at one particular time; a history of recent febrile symptoms and/or repeated measurements are important. In rare cases, a patient may be persistently afebrile in the presence of a very severe *P. falciparum* infection. Hyperpyrexia may complicate malaria, especially in attacks of *P. falciparum*.

Progress of the untreated attack

The natural history of untreated malaria differs with each species.

Plasmodium falciparum

Following a single exposure to infection, the patient will either die in the acute attack (a common event) or survive with the development of some immunity and residual anaemia. Attacks may recur over the course of the next year (a phenomenon called recrudescence, caused by the persistence of blood forms in small numbers between attacks) but then die out spontaneously in the absence of reinfection.

Plasmodium malariae

Following a single exposure to infection, and an incubation period that may extend to many weeks, the patient develops a recurrent fever that occurs at increasing intervals. There may be considerable anaemia, and enlargement of the liver and spleen. If no treatment is given to clear the blood forms of the parasite, recrudescences may occur from time to time for more than 30 years. The severity of the attacks tends to diminish as time goes by, until bouts of fever last only a few days.

Plasmodium vivax and P. ovale

Plasmodium vivax and *P. ovale* malaria cause very similar illnesses, with bouts of fever that relapse periodically but irregularly over a period of up to 5 years. These are true relapses and not simple recrudescences, because they may occur despite treatment with drugs that entirely eliminates the parasites from the blood. The relapses are caused by reinvasion of the blood by merozoites produced when hypnozoites awake from dormancy and develop into PE schizonts.

Peculiarities of *Plasmodium falciparum* infection

The important difference between *P. falciparum* and the other plasmodia that infect humans is the capacity of *P. falciparum* to cause severe (or complicated) disease. Nearly all of the million or more malaria deaths that occur each year result from *P. falciparum* infections. In endemic areas, most of the clinical impact of *P. falciparum* infection falls on young children. Nevertheless, the majority of infections cause only a self-limiting febrile illness or, as immunity increases, no illness at all. For reasons that are still not understood, some infections progress to severe disease, and some of these are fatal. In areas with limited or unstable transmission, adults (including tourists) with *P. falciparum* infection may develop severe or complicated disease, especially if diagnosis is neglected or delayed.

Complicated *P. falciparum* malaria, also known as 'severe' malaria, may take a number of clinical forms, which are listed in Table 9.1. In young children in endemic areas, who suffer the greatest malaria mortality, five clinical syndromes predominate: prostration, severe anaemia, cerebral malaria, acidosis and hypoglycaemia. A child may suffer from just one of these complications, or from any combination of them. Other complications seen in adults are unusual in children in endemic areas. Non-immune adults may develop any combination of these or of the other syndromes listed in Table 9.1.

Specific syndromes caused by *Plasmodium falciparum*

Altered consciousness and coma

In a patient with *P. falciparum* malaria and coma,

MICROCIRCULATORY ARREST

Organs most affected	Main symptoms or signs	Typical misdiagnosis
Stomach and intestines	Vomiting and diarrhoea	Gastric flu; cholera (diarrhoea is *not* bloody)
Brain	Delirium Disorientation Stupor Coma Convulsions Focal neurological signs	Encephalitis, meningoencephalitis (there may be misleading CSF abnormalities)
Kidneys	Renal failure, with or without oliguria or haemoglobinuria	Nephritis
Liver	Jaundice and fever	Hepatitis
Lungs	Pulmonary oedema	Pneumonia, heart failure

Table 9.1 Microcirculatory arrest in *P. falciparum* infection.

several possible causes of altered consciousness must be considered.
• A metabolic explanation, such as *hypoglycaemia* (especially in a young child or pregnant woman) or *acidosis*. Correction of either of these may restore consciousness.
• The patient may be having a *seizure*. This is sometimes manifested by only very minor twitching movements or none at all, but an anticonvulsant drug may restore consciousness.
• The patient may be *postictal* after a recent seizure, when recovery is likely within a few minutes or hours.
• *Very severe anaemia* may also impair consciousness.

If none of these complications accounts for the coma, or if coma persists despite finding and correcting these, then the patient may have 'cerebral malaria'. When a diagnosis of cerebral malaria is made, it is important not to overlook other possible explanations of the illness (e.g. meningitis, encephalitis, severe pneumonia or head injury) in an individual who happens to be parasitaemic.

Cerebral malaria

This common complication is one of the important causes of malaria deaths. It is a diffuse disturbance of cerebral function, characterized by altered consciousness, commonly accompanied by convulsions. The onset may be gradual or sudden, usually within hours or days of the first febrile symptoms of malaria. It is not uncommon for a child to become comatose without any preceding fever or other symptoms. Coma may be accompanied by flaccidity of limbs, or by any combination of hypertonicity, posturing and opisthotonos. Any repetitive muscular movement, even of a minor degree, may reflect underlying seizure activity.

Recently, a characteristic retinopathy has been observed in children and adults with cerebral malaria. A short-acting mydriatic can be used to dilate the pupils, and the retinopathy can then be seen with a direct ophthalmoscope and with practice. The changes consist of areas of whitening in the macula and extramacular optic fundus, with patchy whitening of small vessels. These features are sufficiently distinctive to be diagnostically helpful. Most children with cerebral malaria also have white-centred retinal haemorrhages, and about 10% have some degree of papilloedema, neither of these being distinctive of malaria.

For research purposes, a strict definition of cerebral malaria requires that the parasitaemic patient be unable to localize a painful stimulus (Glasgow coma score ≤8; Blantyre coma score ≤2), that coma persists despite correction of metabolic defects and seizures, and that no other explanation for the coma can be found. If effective antimalarial drugs are given, together with supportive care, about 80% of patients with cerebral malaria recover. Coma usually persists for 1–3 days after the start of treatment, on average somewhat longer in adults than in children. If recovery does occur, a minority of patients (5–10%) are left with a neurological deficit, such as hemiparesis, cerebellar ataxia, amnesia, diffuse spasticity or epilepsy. Clinically obvious sequelae may resolve over a period of months, but some are permanent. We do not know how many individuals may suffer more subtle impairment (e.g. of memory or intelligence) after cerebral malaria.

The pathogenesis of cerebral malaria remains unclear. The usual histopathological finding in fatal cases is the presence of large numbers of erythrocytes containing mature parasites in the capillaries and venules of many organs, including the brain. Because irreversible brain damage is unusual in those who recover, it seems unlikely that the microcirculation is totally obstructed by these sequestered cells. The highly active, developing and dividing parasites may consume essential nutrients such as oxygen and glucose and release toxic products, including lactate, with detriment to surrounding tissues. As the schizont ruptures the red cell, substances are released that are known to stimulate the release of cytokines from host cells: these in excessive local concentrations may contribute to coma and other complications of P. falciparum malaria (Figs 9.2 & 9.3, see colour plate facing p. 146).

Perivascular 'ring haemorrhages' are commonly found in the brain at autopsy (Fig. 9.4, see colour plate facing p. 146). Their numbers correlate with the number of retinal haemorrhages visible by ophthalmoscopy during life, but their pathogenetic significance is not known. Raised intracranial pressure is usual in children with cerebral malaria, but there is no firm evidence that the raised pressure itself contributes to mortality. Factors contributing to raised intracranial pressure include cerebral oedema and, probably to a lesser extent, the mass of sequestered parasitized erythrocytes.

Severe anaemia

This complication is most common in children between 6 and 24 months of age in areas of intense transmission of P. falciparum. There is direct red cell destruction when schizonts rupture, and further haemolysis, both of parasitized and unparasitized erythrocytes, occurs through autoimmune mechanisms.

Red cell destruction is not, however, the whole story. One would expect a brisk reticulocytosis in a haemolytic anaemia, but this is often absent in acute malaria, reflecting impaired bone marrow function. Like fever, this is believed to be mediated by a host cytokine response to the infection. Marrow aspirates often show evidence of dyserythropoiesis, including phagocytosis of parasitized red cells by macrophages, and phagocytosis of apparently uninfected red cells.

Severe anaemia may be found by chance when a patient attends for some unrelated problem, or it may lead to breathlessness, weakness and, occasionally, impaired consciousness. Dyspnoea in a child with severe malarial anaemia is most commonly a manifestation of acidosis; occasionally, dyspnoea is a result of heart failure.

Acidosis

Tissue anoxia leads to anaerobic metabolism and the release of lactic acid. The resulting acidosis is initially compensated for by deep breathing, which eventually may be insufficient to prevent the arterial blood pH from falling. Factors contributing to tissue anoxia include: the sequestration of parasitized red cells, which may impair tissue perfusion; anaemia; hypovolaemia; and hypotension. Rapid fluid volume replacement, with whole blood if necessary, may be life-saving.

Hypoglycaemia

Hypoglycaemia is a common complication of untreated *P. falciparum* malaria in children, as it is of many other infections in children. Hypoglycaemia may also occur in adults with malaria and pregnant women are particularly susceptible. The principal mechanism of malarial hypoglycaemia is probably cytokine-induced impairment of hepatic gluconeogenesis, although the consumption of glucose by millions of parasites may also contribute. Hypoglycaemia sometimes develops as a complication of quinine or quinidine therapy, probably because these drugs stimulate the pancreas to secrete insulin; again, pregnant women are particularly susceptible.

'Blackwater fever'

This obsolete term used to be applied to the syndrome that sometimes occurs in *P. falciparum* malaria when severe intravascular haemolysis is associated with haemoglobinuria and renal failure. The syndrome still occurs, especially in non-immune adults with severe *P. falciparum* infection. In children in the endemic areas of sub-Saharan Africa, haemoglobinuria sometimes occurs in *P. falciparum* malaria, but it is rarely accompanied by renal failure. In some cases, haemoglobinuria is precipitated by a drug or dietary factor in an individual with glucose-6-phosphate dehydrogenase (G6PD) deficiency. Haemolysis in this condition usually only affects the older cells, so ceases when the haemoglobin has dropped to about 6 g/dL.

Bleeding disorder

A minor degree of disseminated intravascular coagulation (DIC) is common in *P. falciparum* malaria, and DIC severe enough to cause bleeding is an occasional complication in adults.

Malaria in pregnancy

All types of malarial infection can lead to abortion. In *P. falciparum* infection, even in women normally immune, pregnancy is associated with an increased likelihood of developing parasitaemia and with higher parasite densities, especially in the first pregnancy. Anaemia is a common consequence, and many women enter labour with a low haemoglobin concentration, making peripartum blood loss more dangerous. Organ complications such as coma and renal failure are rare in pregnant women living in endemic areas but, among the non-immune, pregnant women are liable to the same complications as other adults.

Plasmodium falciparum in endemic areas is an important cause of low birth weight, especially in first-pregnancy babies, who are then at increased risk of dying in infancy from any of a variety of causes. Low birth weight because of maternal malaria presumably results from the fact that the placenta becomes packed both with late-stage parasites and host mononuclear cells, especially in the first pregnancy.

In endemic areas, it is common to find malaria parasites in umbilical venous blood; it is less common to find them in the neonate's peripheral blood, and these usually disappear within the first 2 days of life. Illness brought about by congenital infection is rare in endemic areas, but may develop in infants born to non-immune mothers. *Plasmodium vivax* is a more common cause of congenital malaria than *P. falciparum*; the illness presents within a few days or weeks of birth with fever, haemolytic anaemia and failure to thrive.

Immunity in malaria

Immunity in malaria is most pronounced in *P. falciparum* infection. In areas of very high transmission, if a child survives to the age of 5 or 6 years, he or she is likely to have achieved a high degree of immunity to the lethal effects of the infection. This immunity has two main components: an ability to limit parasitaemia by the development of specific protective immunoglobulin (IgG) and cell-mediated immunity (antiparasitic immunity), and a physiological tolerance such that parasitaemia produces little or no fever or subjective illness (antitoxic immunity). In order to maintain this immunity, frequent re-exposure to infection is required. If re-exposure does not occur, the immunity wanes over a period of a few years. Although West African students living in the UK gradually lose their protective antibodies over a 5-year period, they rapidly regain immunity on re-exposure to infection, but the price may well be two or more attacks of malaria on first returning home. The development of a high degree of immunity in an entire population exposed to high levels of *P. falciparum* infection has an extremely important effect on the epidemiology of the infection.

As the CD4 count falls, an individual with HIV infection becomes increasingly susceptible to *P. falciparum* infection, and is likely to develop a greater density of parasites in the peripheral blood than more immunocompetent people. It has not yet been shown that such individuals are susceptible to more severe malaria disease. HIV-positive pregnant women are more likely than others to have malaria in pregnancy and to fail to clear it with standard treatment. Conversely, malaria increases the plasma viral load in HIV-infected people, and there is suggestive evidence that placental malaria may enhance mother–child transmission of HIV.

Non-immune protective factors in malaria

There are several non-immune factors that affect susceptibility to malaria. *P. vivax* is unable to infect red cells lacking the Duffy blood-group antigen. This is believed to account for the natural resistance of people of West African origin (who lack the Duffy antigen) to infection with this parasite.

Individuals whose haemoglobin genotype is AS are resistant to the lethal effects of *P. falciparum* infection, but no more resistant to infection itself than those with normal (AA) haemoglobin. This is because the sickle trait prevents the development of high parasitaemia, probably partly as a result of parasitized red cells sickling in the circulation and being removed by the spleen before they can develop into schizonts.

Evidence that G6PD deficiency has a similar protective effect does exist, but is less striking. Sickle-cell anaemia itself is not protective, for malarial infection is disastrous in such patients.

There is now good evidence that the beta-thalassaemia trait confers protection against *P. falciparum*. Malnutrition was once thought to protect against the lethal effects of *P. falciparum* infection, but recent case–control studies have failed to confirm this.

Immune disorders in malaria

Some complications of malaria are related to immune effects. *Malarial nephrosis* is an occasional complication of *P. malariae* infection in children. Antigen–antibody complex is bound firmly to the glomerular basement membrane. An intractable nephrotic syndrome results, with non-selective proteinuria and a bad prognosis. Neither treatment with corticosteroids nor eradication of the malaria seems to influence the outcome. The intractability of the condition seems to be determined by the permanence of the complex-binding mechanism.

A more tractable condition is hyper-reactive malarial splenomegaly (formerly known as tropical splenomegaly syndrome), in which marked splenomegaly in *P. falciparum* infection is associated with infiltration of the hepatic sinusoids with lymphocytes, with or without features of secondary hypersplenism. Serum IgM levels are very high. This condition usually resolves within a few months if the patient is given continuous effective chemoprophylaxis.

There is good evidence that an acute attack of malaria has general immunosuppressive effects. The effects of chronic malaria are less well-defined. Interventions against malaria (e.g. bed-nets) have sometimes led to a fall in mortality from other causes, including respiratory infections, suggesting that malarial immunosuppression may increase susceptibility to other common pathogens. The immunosuppressive effects of malaria may account for the tendency of the Epstein–Barr virus to produce Burkitt's lymphoma in malaria-endemic areas.

Diagnosis

Direct diagnosis

The specific diagnosis of malaria is made by examining the blood, by making a film, drying and staining it. The thin blood film shows the undistorted parasites within the red cells. It is of most use in the detailed study of parasite morphology and species identification. Its disadvantage is that it requires a very prolonged search to detect a low parasitaemia, so its sensitivity is low. A patient may have a fever resulting from *P. falciparum* and yet have no parasites detected by searching the thin film.

The thick film, in which cells are piled upon each other 10–20 deep and lysed and stained at the same time, allows far more red cells to be examined at a time, but it has the disadvantage that the parasites in the lysed cells are distorted. Although readily recognizable as malaria parasites, their specific features of identification may be ambiguous or entirely lost. However, in experienced hands, the thick film is the best method to use for answering the question, 'Does the patient have malaria?'

Serodiagnosis

Serodiagnosis of malaria is of no use for diagnosis of the acute attack. It depends on finding specific antibodies, and most methods in common use are incapable of distinguishing between antibodies to the different species of parasite. Antibodies may be detectable for several years after the last attack of malaria. The main use of serodiagnosis is in excluding malaria in a patient suffering from recurrent bouts of fever who does not present during a bout. Serology may also be used in surveys as an approximate measure of exposure of a population to malaria. The most frequently used serological technique is the indirect fluorescent antibody test (IFAT).

New methods of diagnosis

Many new techniques for identifying malaria parasites are being developed. The quantitative buffy coat (QBC) technique makes use of the fact that parasitized erythrocytes have a different specific gravity from unparasitized red cells and can therefore be looked for in a particular segment of the blood in a centrifuged capillary tube. Several dipstick methods are now available, by which parasite antigens are detected by placing a drop of blood on a dipstick impregnated with antibody. Polymerase chain reaction (PCR) can be used to detect parasite DNA. While these methods are useful in research studies, they have not replaced thick and thin films for routine clinical diagnosis.

Treatment

The treatment of a patient with malaria is supportive and specific. Supportive treatment may include the following.

1 Reducing the temperature if hyperpyrexia is present—especially common with *P. falciparum* infection. Oral or rectal paracetamol is the method of choice. If this is unavailable, tepid sponging and fanning offers temporary benefit.

2 Rehydration, especially when vomiting and diarrhoea have been prominent, and in the patient with deep breathing suggestive of acidosis.

3 Monitoring renal output and taking corrective measures if necessary (first, rapidly correct any hypovolaemia; if oliguria persists, maintain careful fluid balance).

4 Monitoring the need for blood tranfusion, which may be life-saving. However, blood should only be transfused when there are strong clinical indications; e.g. when the haemoglobin concentration is < 4 g/dL (haematocrit < 12%) or when higher levels are accompanied by coma, acidosis or hyperparasitaemia. In most patients, the haemoglobin concentration rises rapidly when the attack has been terminated by specific chemotherapy.

5 Terminating convulsions with appropriate drugs—rectal diazepam or lorazepam, rectal or intramuscular paraldehyde, and intramuscular phenobarbital are some options. These should be used in a sequence, proceeding to the second or third option only when the others have failed.

6 Monitoring of blood glucose and correction of hypoglycaemia where necessary (10–40 mL of 50% dextrose diluted×3 with saline and infused over 5–10 min).

7 Reducing acidaemia. Rehydration, blood transfusion (when appropriate) and antimalarial therapy are usually sufficient for this purpose. The use of bicarbonate infusion is not of proven benefit, but may be attempted with care in severe acidosis.

8 Treating DIC if this complication is severe enough to cause bleeding; fresh whole blood, platelet-rich plasma and fresh frozen plasma may be given according to availability.

9 Giving antibiotics is likely to be helpful:
- if the diagnosis of malaria is in doubt;
- in an unconscious patient in whom lumbar puncture is deferred; or
- in patient groups known to have a high risk of bacteraemia accompanying severe malaria.

This may vary geographically: at-risk groups identified have been children with cerebral malaria in coastal Kenya, and young children with severe anaemia in southern Malawi.

Specific chemotherapy

Specific treatment is directed to terminating the parasitaemia as rapidly as possible. The drug of choice depends on national policy in the particular country, and on the likely place of origin of the patient's parasites. Drug resistance is an increasing problem throughout the world, and the picture changes with time. Many endemic countries now have a national programme that sets policy for first-line treatment of uncomplicated malaria, with other drugs for treatment of failures or of severe disease. In some countries, multidrug resistance threatens to make malaria untreatable, and new additions to the armamentarium of drugs are urgently needed. In general, national policy should be followed. Treat non-severe malaria with oral drugs if the patient can take them. Complicated *P. falciparum* malaria requires parenteral antimalarial drugs, at least until there is clinical improvement and the patient can swallow.

Drugs that prevent the development of the blood stages which are causing the illness are traditionally called schizonticides. Some of them also act against the gametocytes of some species, but this has no relevance to the clinical situation. Some of the schizonticides also have useful anti-inflammatory effects. The most widely used schizonticide has until recently been chloroquine, but the spread of parasite chloroquine resistance has limited the use of this drug in recent years. Chloroquine remains the first-line treatment for non-severe *P.*

falciparum malaria in some semi-immune populations in parts of Africa, and it is the drug of choice for all non-*falciparum* malarias (*P. vivax* resistance to chloroquine is increasingly common but, as the disease is not life-threatening, it is reasonable to try chloroquine first). Drug doses given below are for average-weight adults in a tropical setting.

Chloroquine

This is a synthetic compound of the 4-aminoquinoline group. It is a bitter white powder. As a base, it forms salts with acids. Those in common use are diphosphates (Aralen, Resochin, Avloclor) and sulphate (Nivaquine). It is a powerful schizonticide; it also has anti-inflammatory action and so helps to reduce the non-specific symptoms of malaria (malaise, headache, myalgia).

Chloroquine is relatively non-toxic if properly administered in the correct dosage. The drug should be given orally when possible. Intravenous infusion must be over 2–4 h, and intramuscular or subcutaneous chloroquine should be divided into frequent small doses rather than given as a single large injection. The drug is taken up by the liver, so higher blood levels follow parenteral rather than oral administration, allowing larger, less frequent doses by mouth.

Main toxic effects include: gastrointestinal effects (nausea, vomiting, etc.); a fall in blood pressure; generalized itching (a common complaint in black-skinned people only); the hair may turn white with chronic overdosage; the vision may also be affected with prolonged use (acute effects, corneal crystal deposition; chronic effects, retinopathy).

Dosages are as follows, and all doses are expressed as dose of base as this is the active part of the drug.

- **Adults**
  Oral—600 mg initially, 300 mg 6 h later, then 300 mg/day for 2 days (total dosage 1.5 g).
  Intravenous—5 mg/kg (maximum 300 mg) infused over 3 h in saline, repeated every 8 h to total of 25 mg/kg.
  Intramuscular or subcutaneous—2.5 mg/kg (maximum 150 mg) every 4 h to total 25 mg/kg.

When giving chloroquine parenterally, change to oral treatment as soon as the patient can take it.

- **Children**

Oral—The dose should be in proportion to the body weight (using the full dose at 60 kg).

Amodiaquine

This is a 4-aminoquinoline with a molecule that has some resemblance to chloroquine and some to quinine. Preparations include tablets of amodiaquine hydrochloride (Camoquin (PD)) containing 200 mg of base. The dose is the same (in terms of base) as for chloroquine. Toxic effects are also similar to chloroquine, but agranulocytosis has been reported.

Amodiaquine is effective against some strains of chloroquine-resistant P. falciparum both in vivo and in vitro; it is no longer recommended for prophylaxis because of the incidence of agranulocytosis when used for this purpose.

Quinine

The main use of quinine is for treatment of severe P. falciparum malaria in areas of chloroquine resistance. Its isomer, quinidine, is equally effective. Quinine is a natural alkaloid derived from cinchona bark, a bitter crystalline powder practically insoluble in water. It forms salts of varying solubility: the sulphate and bisulphate are used for oral preparations and the dihydrochloride or chloride for injection. It is a powerful schizonticide; it also has an anti-inflammatory action.

Toxic effects may vary. The symptom complex of tinnitus, deafness, dizziness, nausea and vomiting, which is known as cinchonism, is almost inevitable with normal doses of quinine; therapy does not need to be stopped or changed on account of such symptoms unless they are severe. Others include hypoglycaemia (especially in pregnancy); hypotension (if excessive dose or if given too fast intravenously); thrombocytopenia (a rare idiosyncratic reaction); and erythematous rash. Overdose of quinine can cause deafness, blindness and severe hypotension. Quinine has a stimulatory effect on uterine muscle, and overdose can cause abortion. However, this is not a reason to avoid quinine in pregnancy, because the benefit of curing malaria greatly outweighs the risk of uterine excitation from therapeutic doses of the drug.

Doses are given below, and are expressed as dose of salt.

- **Adults**

Oral—600 mg 8-hourly for 7–14 days (usually as sulphate).

Parenteral—The intravenous route is preferred. First (loading) dose of 20 mg/kg (maximum dose 1400 mg) quinine dihydrochloride infused over 4 h in an isotonic glucose–electrolyte fluid (e.g. half-strength Darrow's–5% dextrose); subsequent doses of 10 mg/kg similarly infused over 2 h, at 12-hourly intervals for children, or infused over 4 h, at 8-hourly intervals for adults. Change to oral therapy as soon as the patient can take it.

Intramuscular—Injection of quinine dihydrochloride may be given if intravenous infusion is impossible. It is usually well-tolerated if given deep, with aseptic precautions. The provided ampoule (300 mg/mL) should be diluted fivefold with sterile water to reduce the pain of the injection. As with intravenous infusion, the first dose should be a loading dose of 20 mg/kg: to reduce the volume of this large dose, it may be divided between the two thighs, and further divided by giving half at time zero and the remainder 4 h later.

Mefloquine

This is a 4-quinoline methanol drug chemically related to quinine. It is bound to plasma, has a half-life of 21 days and is effective as a single adult oral dose of 750–1250 mg. It has the major disadvantage that no parenteral preparation is available, and naturally occurring RI resistance has been reported. It is available in 250 mg tablets as Lariam. Minor toxic effects include headache, dizziness and disturbances of sleep. Occasional severe toxic effects are fits, psychomotor disturbances and psychoses. Individuals with a history of convulsions or neuropsychiatric disease are therefore advised not to use mefloquine. There have been many hundreds of well-observed cases in which the drug has been used in pregnancy without adverse effect on mother or fetus; nevertheless it is wise if possible to avoid the use of mefloquine in pregnancy on general grounds. Patients on cardiosuppressant drugs or beta-blockers should not take mefloquine because of its additional effects on the myocardium.

Sulfadoxine–pyrimethamine (Fansidar)

The two components of this widely used therapy inhibit enzymes required by the parasite for folic acid synthesis. Sulfadoxine (like other sulphonamides) competitively inhibits the enzyme dihydropteroate synthetase (DHPS), while pyrimethamine (like the biguanide proguanil) inhibits dihydrofolate reductase (DHFR). Sulfadoxine–pyrimethamine (SP) has been widely used as an alternative to chloroquine in the treatment of uncomplicated P. falciparum malaria. It has two major benefits as a first-line treatment in impoverished countries: it is very cheap, and it is used as a single-dose treatment.

When introduced in South East Asia as therapy for malaria in the wake of chloroquine resistance, SP resistance rapidly appeared until SP was useless. Its use in Africa has had mixed effects, with increase of resistance being rapid in some areas and much slower in others. Resistance appears to be the result of mutations in the parasites' DHFR and DHPS genes, and the capacity to detect these mutations by PCR may provide a useful means of monitoring and predicting the spread of SP resistance in different populations.

Atovaquone–proguanil (Malarone)

This is a recently licensed combination therapy for the treatment of uncomplicated P. falciparum malaria. The standard regimen is four tablets daily for 3 days.

Halofantrine

This is an effective antimalarial that is used in some parts of Africa. However, it is cardiotoxic, prolonging the QT interval and leading to arrhythmias. It is therefore no longer recommended for the treatment of malaria.

The artemisinin drugs

These compounds are derived from the plant Artemisia annua, which has been used for thousands of years as a herbal remedy for fevers in China. The plant's active components, arteether, artemether and artesunate are highly effective antimalarials, active against chloroquine-resistant and multidrug-resistant P. falciparum and useful in the treatment of severe and complicated malaria as well as uncomplicated disease. Parasites are cleared from the circulation faster by artemisinins than by quinine or chloroquine. When used for severe disease, artemisinins are as effective as quinine, less toxic and more convenient to use, although their more rapid antiparasitic action has not yet been shown to translate into improved survival.

If used alone for treatment of P. falciparum infection, an artemisinin drug must be given for at least 5 and preferably 7 days; shorter courses are followed by recrudescence of parasites in over half of cases. Artemisinins are therefore usually used in combination with other antimalarial drugs.

Oral, intramuscular, intravenous and rectal formulations of various artemisinins are available.
- Artesunate is available as tablets, suppositories and a powder for preparing an intravenous solution. Rectal artesunate is being studied as a possible first-line drug for patients with severe malaria in remote areas where injections cannot be given, pending the patient's transfer to a hospital.
- Artemether is an oil-soluble derivative suitable for intramuscular injection. In several studies, artemether has proved as efficacious as quinine in the treatment of severe malaria in children, although absorption from the injection site appeared to be impaired in a few very ill, acidotic children.
- Artemisinin can be used orally or rectally.

No important human toxicity of artemisinins has been detected after thousands of treatments. Brainstem damage has been observed in animals given more than 10 times the usual human doses of arte-ether or artemether, but careful examination has failed to reveal any neurotoxicity in clinical practice. Artemisinins reduce gametocyte production. This may in turn reduce the transmission of malaria locally, although probably not to an important degree in hyper- or holoendemic areas. The rapid killing of blood-stage asexual parasites is a property of artemisinin drugs that makes them particularly suitable for combination therapy.

Combination therapy

For many decades in the mid-20th century, chloroquine monotherapy was the standard treatment for malaria. Then *P. falciparum* resistant to chloroquine began to appear and to spread, followed by parasites resistant to many other first-line antimalarial drugs. We now, belatedly, recognize the importance of combining drugs with different modes of action, as is usual practice in the treatment of tuberculosis, leprosy, cancers and HIV infection, in order to delay the development of resistance.

Various combinations of antimalarial drugs are now being evaluated for their capacity to cure patients and to prevent the emergence of drug-resistant parasites. Artemisinin drugs lower asexual parasitaemia rapidly, leaving a much diminished parasite biomass; a second drug can then kill the remaining parasites, with a much reduced chance that a resistant mutation will occur and break through the treatment.

Combination therapies, many of which include an artemisinin drug, are currently being widely evaluated. It seems likely that combination therapy will become a standard first-line therapy for uncomplicated malaria in many countries. It remains to be seen whether the use of combination therapies, especially those containing an artemisinin drug, will reduce the advance of drug-resistant *P. falciparum*. There is emerging evidence that the addition of artemisinin will improve the efficacy of an existing antimalarial treatment that is failing because of increasing resistance.

Co-artem (artemether–lumefantrine)

This is the first commercially available artemesinin combination therapy. It is currently being evaluated by the World Health Organiza-

tion (WHO) in a number of African sites. Standard adult doses are 4 tablets twice a day for 3 days. Absorption of the lumefantrine component is increased considerably by fat and therefore this drug should ideally be taken with food.

Classifying antimalarial drug resistance

Antimalarial drugs can be judged *in vivo* by their efficacy in patients with malaria. They can also be assessed *in vitro* by measuring their capacity to inhibit the growth of cultured *P. falciparum*. An *in vivo* assessment requires the identification of a number of individuals with malaria (fever, parasitaemia and no other cause of illness), who must be observed to take the correct dose of the drug being assessed. Parasitaemia is monitored on days 0, 1, 2, 3, 7 and 14. Further samples on days 21 and 28 may be included, especially if further exposure to malaria infection can be prevented.

Resistance (R) is divided into three grades:
RI. Parasites disappear from the blood by day 3 but reappear by day 14.
RII. Parasites fall to less than 25% of the day-0 density by day 2 or 3, but do not disappear, and are increased in density by day 7 or 14.
RIII. Parasitaemia never falls below 25% of the day-0 level, and may even increase after treatment.

A newer classification allows for the fact that a patient may be completely well, even though parasitaemic, after treatment. In this classification patients are described as having the following.
• *Early treatment failure (ETF)* — if they develop severe illness with parasitaemia during the first 4 days after treatment; or if they remain febrile and parasitaemic throughout those 4 days.
• *Late treatment failure (LTF)* — if, having improved clinically, both fever and parasitaemia recur by day 14.
• *Adequate clinical response (ACR)* — if there was neither ETF nor LTF, and the patient does not develop febrile parasitaemia by day 14; i.e. for an ACR, the patient may have either fever or parasitaemia by day 14, but not both.

The problem of relapse

Relapse in *P. vivax* or *P. ovale* malaria can usually be prevented by giving a course of primaquine, but some strains of *P. vivax* (e.g. from Papua New Guinea) are resistant to normal doses of the drug. Primaquine is a bitter white powder, a synthetic drug of the 8-aminoquinoline group. Tablets of 26.5 mg primaquine diphosphate contain 15 mg primaquine base (or tablets half this size). The dosage is usually expressed as the weight of the base. It is a weak schizonticide. There is action on hypnozoite forms of *P. vivax* and *P. ovale*, and it destroys gametocytes of all species. Side-effects include gastrointestinal disturbance, especially abdominal cramps. Acute haemolysis may occur in G6PD deficiency. Methaemoglobinaemia (cyanosis) is usually only seen with high doses.

For radical cure of relapsing forms of malaria, 30 mg base/day is given for 10–14 days for *P. vivax* and 15 mg base/day for 10–14 days for *P. ovale*. The dose may have to be doubled in some *P. vivax* strains. A single weekly dose of 45 mg base for 6 weeks is better tolerated by G6PD-deficient patients than is daily dosing with the lower dose. For clearing gametocytes of *P. falciparum*, 15 mg/day is given for 5 days. This use of primaquine does not benefit the individual: it has been advocated in order to reduce the transmission of *P. falciparum*. In highly endemic areas this is a waste of time, because most transmission occurs from individuals who are not even known to carry the parasite.

Chemoprophylaxis

Chemoprophylaxis of malaria involves the regular administration of drugs to prevent clinical symptoms. Drugs taken for this purpose act in two ways: as *schizonticides*, so that when the parasites enter the red cells they are destroyed; and *causal prophylactics*, which prevent the development of the PE schizonts in the liver, and may also have blood schizonticidal effects.

Neither the *in vitro* nor the *in vivo* method of assessing drug resistance gives any indication of causal prophylactic efficacy, which can only be assessed by large controlled trials using willing travellers or volunteer populations. In considering malaria prophylaxis, one is always concerned with balancing risk and benefit. It takes a long time for the real incidence of the toxic effects of a new drug to emerge, and just as long to find out how rapidly resistance develops to it. Currently, chemoprophylaxis is routinely advised only for non-immune travellers visiting endemic areas.

Something akin to chemoprophylaxis is recommended for pregnant women in endemic areas. It has been shown in some endemic communities that two or three doses of a drug such as SP, given during the second half of pregnancy (irrespective of maternal illness or parasitaemia), reduces placental malaria and low birth weight in first-born babies. This routine administration of antimalarials occasionally in pregnancy is more accurately termed intermittent presumptive therapy (IPT) rather than chemoprophylaxis. A similar approach to malaria in infancy is under evaluation.

Proguanil (Paludrine, Chlorguanide)

This is a synthetic biguanide. It is a bitter white powder available as hydrochloride. Tablets are 100 mg proguanil hydrochloride, as well as a paediatric 25 mg preparation. It is a slowly acting schizonticide and a causal prophylactic. When a mosquito takes up gametocytes from a patient receiving proguanil, their development in the mosquito is inhibited, so the mosquito fails to become infective.

Proguanil is the safest of all antimalarials: no deaths have ever been recorded from overdose (up to 14.5 g). Occasionally, it can cause heartburn or epigastric pain, but this is minimized by taking the drug after food. Mouth ulcers are an unpleasant side-effect in some people. Gross overdose may cause haematuria. It is safe in pregnancy in a normal dosage, but a folic acid supplement should be given.

It is used as a prophylactic only. The adult dose is 200 mg/day. It is well-tolerated by children, who can take 25 mg/day from infancy, 50 mg/day from age 2, 75 mg/day from age 4 and 100 mg/day from age 6 years. It has been most widely used in combination with chloroquine as malaria prophylaxis for travellers to parts of the world with limited chloroquine resistance.

Chloroquine (for prophylaxis)

The normal adult dosage is 300 mg base/week. There is virtually no danger in taking this dosage of chloroquine for short periods (say, less than 3 years) but, as chloroquine binds firmly to melanin (including the pigment of the retina), there is a theoretical possibility that long-term dosage at this level may lead to retinal damage. This has been well-documented when chloroquine has been given in high dosage for the treatment of rheumatoid arthritis and related diseases. Total dosage (because of binding) may be more important in the genesis of retinotoxic effects than the mean daily dosage. We therefore do not recommend chloroquine for people who require chemoprophylaxis for more than 6 years on 300 mg base/week continuously. Chloroquine is most commonly used in combination with proguanil. However, increasing chloroquine resistance means that this combination is now no longer effective in most parts of the world and it is currently only recommended for parts of South Asia and South America.

Mefloquine (for prophylaxis)

As a result of the spread of chloroquine resistance around the world, mefloquine (alone) is now the prophylactic drug of choice for many areas. Initial anxieties about drug accumulation have diminished, and it is now acceptable to recommend an adult dose of 250 mg/week for periods of a year or more. Contraindications must be borne in mind.

Doxycycline

This long-acting tetracycline is an effective prophylactic against malaria in a dose of 100 mg/day. It is useful in areas where there is resistance to both chloroquine and mefloquine. It should not be used in pregnancy or lactation, nor in young children. An occasional toxic effect is a rash caused by photosensitization.

Atovaquone–proguanil

This recently licensed drug combination is effective against chloroquine-resistant *P. falciparum*. It is a causal prophylactic, preventing development of parasites in the liver and therefore only needs to be taken for a week after leaving the malarious area. The standard dose is one tablet daily. Serious adverse effects are rare.

Other drugs

Sulfadoxine–pyrimethamine (Fansidar) is not used for prophylaxis because of the risk of Stevens–Johnson syndrome. Amodiaquine should be avoided because of a risk of marrow aplasia. Pyrimethamine–dapsone (Maloprim) has been used for many years and is effective in areas of chloroquine resistance, but has largely been superseded by other drugs and is no longer being manufactured. It has been associated with agranulocytosis.

No prophylactic regimen described can be completely depended on to suppress *P. falciparum* malaria, especially in non-immune people. Patients should be warned of this and advised to have an alternative drug available for treatment in the case of failure. Similarly, a non-immune individual developing fever after return from an endemic area, even if he or she faithfully took prophylaxis, may have *P. falciparum* malaria. Nevertheless, if a patient develops malaria more than 4 weeks after leaving a malarious area, it is still most likely that it will be with one of the three non-*falciparum* species, all of which commonly have a long incubation period.

Recommendations for antimalarial prophylaxis in specific geographical locations are constantly changing. Authoritative up-to-date sources such as the WHO or UK guidelines are recommended (see Further reading).

Epidemiology

The epidemiology of malaria has been most studied in the case of *P. falciparum*. The two most important factors are:

1 intensity of transmission (the number of infective bites per year); and
2 the immune response of the host.

Measuring malaria in a community
Traditional methods
It has been customary in the past to characterize the epidemiological situation in a community by describing its malariometric indices. These are established by surveys that, by examining all age groups of the population, determine for each group:
1 *parasite rate* — the proportion of blood films that are positive; and
2 *spleen rate* — the proportion of the group with a palpable spleen.

Morbidity and mortality
It is now recognized that parasite and spleen rates are measures of malaria infection, reflecting the intensity of transmission, but they are not measures of the clinical impact of malaria on the community. It is the morbidity and mortality attributable to malaria that are important as the basis for designing a malaria control programme, and these indicators are equally important in monitoring the effectiveness of control.

Ways of estimating malaria-attributable mortality include hospital studies and 'verbal autopsies'. The latter technique makes use of tested questionnaires to inquire of mothers about the nature of the final illness in any children dying within a specified period before the survey. Unfortunately, the verbal autopsy technique cannot distinguish reliably between malaria and pneumonia or meningitis, and deaths caused by severe malarial anaemia may not be identified. Therefore only an approximate measure of malarial mortality can be obtained.

Stable malaria
Transmission occurs for at least 6 months in the year and is intense. Malarial infection is acquired repeatedly. Children suffer repeated attacks of malaria from the age of a few months onwards. Very young children are partly protected by passive immunity acquired by transplacental pas-

sage of protective maternal IgG. This may modify the severity of the first few attacks, so allowing them to develop some active immunity while still partly protected. Children reaching the age of 5 or 6 years have substantial immunity, but the price of this immunity is that some children will die of malaria before immunity develops. The proportion who do so is likely to depend on many factors, including the intensity of transmission, the availability of drugs and the prevalence of parasite drug resistance. Data on the actual death toll in different populations are still rarely available.

When immunity has been established, older patients may still suffer attacks of malaria, but these take the form only of mild or moderate flu-like episodes lasting a few days. Severe and complicated disease rarely occurs. Nevertheless, in areas of stable malaria, malaria illness episodes in adults may be sufficient to cause absenteeism from work and thus to have an impact on the economy. There is little variation in the incidence of malaria from year to year (hence the word 'stable'), but there may still be pronounced seasonal fluctuations in new cases seen in children. In such an area, there is often a marked rise in the number of children seen with cerebral malaria about 2 weeks after the rains begin.

Unstable malaria
This situation is the antithesis of stable malaria. There are wide changes in transmission, not only throughout each year but also from year to year. This results in the tendency for epidemics to occur (hence the term 'unstable'). The transmission season is typically short and the mosquito population fluctuates widely. Infection is usually so infrequent that no member of the population has the opportunity to develop a significant level of immunity. For this reason, when transmission does suddenly increase (usually because of freak environmental conditions leading to an explosion in the mosquito population), people of all ages are equally susceptible to infection. This results in serious disease or even death, regardless of age. The health of the working community may be disastrously

affected, and the economic effects of an epidemic can be dire.

Global malaria eradication

Attempts at global eradication of malaria have failed and local eradication has only succeeded in a few areas, mostly islands. The main causes of failure have been the following.

1 *Operational*—not all houses were sprayed. There are many causes for this, including lack of cooperation, poor mapping, accelerated destruction of thatched roofs (DDT kills caterpillars and their predators; caterpillars soon reappear but predators do not) and resentment of intrusion into privacy.

2 *Technical*—resistance of mosquitoes to insecticide; behavioural resistance in which mosquitoes fly straight out of the house after feeding, and so do not rest on the sprayed surface; resistance of the parasite to antimalarial drugs.

3 *Political* (not a cause recognized by the WHO)—failure of countries to cooperate; civil war and severe political unrest; lack of a suitable infrastructure on which to build the control programme; political and administrative incompetence.

4 Ill-advised and unpopular pilot schemes.

5 Failure to convince the people of the need for the programme.

Malaria control at present

Global eradication as an objective has been abandoned as unrealistic. Instead, the objective now is the control of disease and mortality that result from malaria. This objective requires:

1 the provision of diagnostic and treatment services as close as possible to where the people live;

2 simple, affordable and safe therapy for uncomplicated disease;

3 prompt recognition and treatment of severe disease, with systems of referral to hospital centres when necessary;

4 education of the population about the features and dangers of malaria;

5 drug prophylaxis or intermittent presumptive therapy for selected subgroups;

6 antivector measures and water clearance where achievable (and not elsewhere); and

7 the use of permethrin-impregnated bed-nets or curtains.

Several randomized controlled trials in endemic areas have demonstrated that widespread use of insecticide-treated nets in a community can reduce child mortality and malaria-related morbidity. How well this works in the context of a routine health service programme (in the absence of a research team) is less certain: some early observations have been discouraging, others have shown significant effectiveness of nets in a programme making use of social marketing to promote net usage.

Vaccines may in the near future be added to the list of effective interventions. Several candidate vaccines have been evaluated or are undergoing trials, but none has yet been found to be both efficacious and practicable.

Individual precautions

Because the anopheline vectors of malaria are night biters, a high degree of protection is given by:

1 covering up the exposed skin in the evenings (of limited value because most infective biting occurs between 10 p.m. and 2 a.m.);

2 the use of insect repellents such as dimethyl phthalate, dibutyl phthalate or diethyl toluamide; and

3 the use of an efficient mosquito net over the bed, preferably impregnated with a synthetic pyrethroid such as permethrin or deltamethrin.

Further reading

Breman JG, ed. The intolerable burden of malaria: a new look at the numbers. *Am J Trop Med Hyg* 2001; **64** (Suppl). [A supplement reviewing the world's malaria problem, with articles on disease burden in childhood and in pregnancy, drug resistance, economic impact and possible long-term sequelae of malaria.]

Gilles HM, Warrell DA, eds. *Bruce-Chwatt's Essential Malariology*. London: Edward Arnold, 2001. [A convenient and erudite volume covering malaria from the point of view of numerous disciplines including clinical medicine, pharmacology, therapeutics, epidemiology, parasitology, immunology, entomology and public health.]

http://www.hpa.org.uk/cdph/issues/CDPHvol4/No2/malaria_guidelinesp.pdf This source has also been published. Bradley DJ, Bannister B. Guidelines for malaria prevention in travellers from the United Kingdom for 2001. *Commun Dis Public Health* 2001; **4**: 84–101. [UK guidelines on prophylaxis.]

World Health Organization. Severe falciparum malaria. *Trans R Soc Trop Med Hyg* 2000; **94** (Suppl): 1–90. [A comprehensive review of clinical features, pathogenesis and management of severe falciparum malaria, with an extensive list of references.]

World Health Organization. *Management of Severe Malaria*. WHO booklet, 2000. [A slim attractive book giving essentials of diagnosis and treatment. Useful for reference—fits easily into a coat pocket—and helpful as text for training seminars or for under- and postgraduate teaching.]

CHAPTER 10

Visceral Leishmaniasis

Epidemiology, 73
Parasite and life cycle, 73
Clinical features of visceral
    leishmaniasis, 74
Differential diagnosis of
    splenomegaly, 74

Viscerotropic
    leishmaniasis, 75
Visceral leishmaniasis and HIV
    coinfection, 75
Investigations, 75

Management, 76
Post-kala-azar dermal
    leishmaniasis, 78
Prevention, 78
Further reading, 79

About 30 species of obligate intracellular proto-zoal parasites of the genus *Leishmania* are re-sponsible for a variety of diseases in humans, collectively known as leishmaniasis. These dis-eases are further classified as visceral, cuta-neous or mucosal according to their principal clinical presentations. A 'leishmaniac' is a per-son who is obsessed with these parasites.

Epidemiology

Visceral leishmaniasis (VL), also known as kala-azar (Hindi for 'black sickness'), is mainly caused by three species belonging to the *Leishmania donovani* complex, each with a characteristic regional distribution:

L. infantum — Mediterranean, Middle East, Central Asia, China;

L. donovani — India, East Africa; and

L. chagasi — South and Central America.

VL may also be caused by *L. tropica* in the Old World and *L. amazonensis* in the New World.

Over 200 million people reside in endemic areas of more than 60 affected countries. There are approximately 600 000 cases reported an-nually, 90% of which occur in Bangladesh, India, Nepal, Sudan and north-eastern Brazil.

Humans are the only known reservoir of *L. donovani* in Bangladesh, India and Nepal. Humans, and possibly rodents, are thought to be the reservoir in East Africa. Wild and domes-tic canines are important reservoirs of *L. infan-tum* and *L. chagasi*. Infection occurs following the bite of an infected female sandfly (*Phlebotomus* spp. in the Old World, *Lutzomyia* spp. in the New World). Transmission may also occur via blood transfusions, infected needles or syringes, and congenitally.

In some regions, such as north-eastern Brazil, expanding urbanization is associated with steadily increasing transmission of leishma-niasis. Globally, HIV is also contributing to the clinical re-emergence of leishmaniasis. Pathogen proliferation and disease progression are mutually enhanced in *Leishmania* and HIV coinfection. In regions of coendemicity, there is likely to be a dramatic shift from subclinical to clinical VL, coupled with an increased rate of transmission of VL in the general population.

Parasite and life cycle

Leishmania amastigotes are spherical or oval bodies measuring 2–4 μm containing two dis-tinct pieces of nuclear chromatin. The larger piece is called the nucleus, the smaller piece the kinetoplast. The sandfly becomes infected by taking up the amastigotes with its blood meal, the amastigotes being in the blood or skin of the infecting host. The amastigotes are liberated in

73

the stomach of the sandfly and begin to multiply by simple fission, eventually forming flagellated metacyclic promastigotes, which are infectious to the new host. This process takes 1–2 weeks, depending on the species. The motile promastigotes migrate from the gut to the proboscis of the sandfly and are injected during feeding. Sandfly saliva inhibits the L-arginine-dependent nitric oxide killing mechanism of macrophages. The promastigotes are ingested unharmed by macrophages and metamorphose into amastigotes. These are distributed in the reticuloendothelial system, where they lodge and multiply by binary fission.

Clinical features of visceral leishmaniasis

Clinical presentation ranges from asymptomatic or subclinical infection to acute, subacute and chronic presentations. The ratio of clinical : subclinical infections is in the range of 1 : 30–100. The incubation period is usually between 2 and 6 months, but ranges from 10 days to more than 10 years. Males are about three times more commonly affected than females.

Onset is usually insidious, with low-grade fever, progressive splenomegaly, hepatomegaly, lymphadenopathy (particularly in Africa), anaemia, anorexia, wasting and increased pigmentation in persons with dark skin (particularly in India).

The liver and spleen are usually firm, regular and non-tender on palpation. The spleen may reach enormous proportions, commonly extending beyond the midline and sometimes into the right iliac fossa. Patients may complain of a dragging discomfort or, less commonly, acute pain as a result of splenic infarcts. Other relatively common features include epistaxis and cough.

Patients with chronic VL may have visited traditional healers and have been subjected to medicinal cuts that subsequently become infected. Infection and ulceration of other superficial wounds is also common at the time of presentation. Intercurrent infections, such as pneumonia, bacillary or amoebic dysentery, and tuberculosis, are particularly important in patients with long-standing disease and may be the main reason for the patient seeking medical care. Other complications of VL include malnutrition, malabsorption, bleeding, nephritis and uveitis.

Some patients present acutely, with an abrupt onset of high swinging fever and other symptoms resembling malaria. Specific cutaneous lesions are uncommon at the time of original presentation in VL. Rarely, a patient may be aware of a painless papule at the site of the infective bite. Post-kala-azar dermal leishmaniasis (PKDL) may occur following treatment and is discussed below.

Case fatality rates associated with VL range from 0–50% of treated cases to 85–90% of untreated cases.

Differential diagnosis of splenomegaly

Massive spleen
• Malaria or hyper-reactive malaria splenomegaly
• Portal hypertension, e.g. caused by schistosomiasis, cirrhosis, etc.
• Lymphoma, leukaemia, myelodysplasia
• Haemoglobinopathies and hereditary haemolytic anaemias
• Splenic hydatid cyst
• Still's disease
• Glycogen storage and other metabolic diseases
• Amyloidosis

Moderate spleen
• Any of the above
• Bacterial endocarditis
• Brucellosis
• Cytomegalovirus
• HIV infection
• Infectious mononucleosis
• Leptospirosis
• Lyme disease
• Relapsing fever

- Syphilis
- Toxoplasmosis
- Trypanosomiasis
- Tuberculosis
- Typhoid
- Typhus

Viscerotropic leishmaniasis

Splenomegaly and fever caused by *L. tropica* infection was noted in American troops involved in the Gulf War in the 1990s. None of these patients developed massive splenomegaly or other features typical of classical VL.

Visceral leishmaniasis and HIV coinfection

Leishmania and HIV coinfection has been recognized since 1986. Most cases are thought to be caused either by reactivation of latent infection or associated with intravenous drug use (IVDU). IVDU-associated *L. infantum* infections are the most common *Leishmania* and HIV coinfections in southern Europe, where *Leishmania* and HIV coinfection accounts for 25–70% of all adult cases of VL and 1.5–9% of patients with AIDS develop VL. Visceralization of *Leishmania* spp. normally associated with cutaneous or mucocutaneous disease is being increasingly reported among HIV-infected patients throughout the world.

Presentation of VL in HIV-positive patients is often atypical and may be a chance finding. Ninety per cent of cases have CD4 counts $< 200 \times 10^6$/L. Atypical clinical features include dysphagia and cutaneous or mucocutaneous lesions. Nodular or ulcerative lesions may affect the tongue, oesophagus, stomach, rectum, larynx or lungs. The course varies from asymptomatic infection to rapidly progressive and fatal disease. Symptoms may be milder and more atypical as the CD4 count falls.

European studies indicate that 30% of patients die during or within 1 month of treatment. The mean survival is 12 months and only 16% survive for more than 3 years. Relapses commonly occur every 3–6 months. The introduction of highly active anti-retroviral therapy (HAART) appears to have reduced the clinical incidence of VL in HIV coinfected patients. However, HAART does not prevent or reduce the relapse rate. Various regimens of antileishmanial drugs, such as pentamidine, liposomal amphotericin B and miltefosine, have been proposed to reduce relapses in HIV coinfected patients.

Investigations

Circumstantial evidence
Full blood count typically reveals anaemia, leucopenia and thrombocytopenia. This is partly explained by hypersplenism. However, a number of other factors may also be important including possible autoimmune mechanisms and bone marrow depression.

Diagnostic work-up should include a coagulation screen or, if unavailable, at least an estimation of the bleeding and clotting times, particularly if a splenic aspirate is planned. Bilirubin and transaminases are (usually mildly) elevated in about 20% of patients, and the alkaline phosphatase is raised in about 40%. Serum albumin is low and globulins are raised, especially IgG. Albuminuria is common but urinalysis is otherwise normal in uncomplicated disease.

The formol gel test (FGT) is a simple test that is sometimes used to provide circumstantial evidence of VL. The FGT is not specific for VL and, when positive, indicates hyperglobulinaemia, whatever the cause. Add 1 drop of concentrated formalin solution (40% formaldehyde) to 1 mL of serum in a test tube. Shake to mix thoroughly. After 20 min at room temperature, the serum becomes a firm opaque jelly (like a cooked egg white) if the test is positive.

Serological evidence
Indirect fluorescent antibody test (IFAT) and enzyme-linked immunoabsorbent assay (ELISA) have sensitivities and specificities above 95%. Although these tests may not be readily available in

remote areas, it is possible to collect blood spots on filter paper, which can be sent elsewhere for testing.

The direct agglutination test (DAT) is also highly sensitive and specific, and easy to perform in the field. However, problems with stability of the antigen have led to unreliable results in some settings. A promising new version of the DAT is currently under development that uses freeze-dried antigen.

The K39 test is a commercially available immunochromatographic strip that uses recombinant leishmanial antigen. This has been shown to be 100% sensitive and 98% specific in India. However, when used to test clinically suspected VL in field conditions in Sudan and Nepal, K39 showed a lack of specificity. Serology may remain positive for years following successful treatment. Serology is unreliable in immunocompromised patients and is positive in only about 50% of patients with *Leishmania* and HIV coinfection.

Parasitological evidence

The gold standard for diagnosis of VL is identification of amastigotes of *L. donovani* spp. These are most readily found in splenic aspirates (>95% positive), bone marrow (>85%), buffy coat (>70%) and lymph node (>65%). In *Leishmania* and HIV coinfected patients amastigotes may be detectable in peripheral blood in 50%, bone marrow in 94% and in skin lesions and other affected tissues. Amastigotes are identified using a Giemsa or other Romanowsky stain. Aspirates may also be cultured on Novy, Mac-Neal and Nicolle's (NNN) medium.

Polymerase chain reaction (PCR) is capable of detecting infection with a single parasite and has been used for the diagnosis of VL in children in Italy and for monitoring relapse in HIV coinfected patients.

Urine antigen tests may prove useful in diagnosis and monitoring response to treatment. The detection of polypeptide fractions of K39 and K26 *Leishmania* antigen in urine of patients with visceral leishmaniasis has been shown to be 96% sensitive and 100% specific; these antigens were not detectable after 3 weeks of treatment, suggesting a good prognostic value.

Performing a splenic aspirate

Splenic aspirate is a straightforward procedure, relatively painless and safe, provided that one excludes patients with a bleeding tendency and those in whom portal hypertension or a splenic hydatid cyst are considered likely in the differential diagnosis. The patient should be comfortable and lying flat with the abdomen exposed. Select an area in the middle of the long axis of the spleen. Clean the skin with an alcohol swab or other antiseptic. Using a 21 gauge needle attached to a 5-mL syringe, insert the needle subcutaneously in line with the long axis of the spleen, draw back the plunger of the syringe to the 1 mL mark to create a negative pressure and swiftly insert the needle into the body of the spleen to a depth of about 2–3 cm at an angle of about 45° to the skin and withdraw immediately while maintaining negative pressure in the syringe. The entire procedure should take only a few seconds.

Having withdrawn the needle, there may be little or nothing visible in the syringe. This is not a problem. Disconnect the needle from the syringe and draw up 2 mL of air. Reconnect the needle and carefully squirt the contents of the needle onto one or more microscope slides and make a smear. The tiny amount of tissue that appears on the slide should be sufficient for diagnostic purposes.

It is probably safest to perform splenic aspirates in a setting where the patient can remain lying down and be monitored for a few hours, and where facilities for transfusion are available if required. However, with experience, outpatient aspirates can be successfully carried out.

Management

Prior to embarking on specific chemotherapy with potentially toxic drugs, it is important to identify and treat intercurrent infections. Attention should also be given to improving the patient's nutritional status.

Pentavalent antimonials (SbV)

Sodium stibogluconate (Pentostam) and meglumine antimonate (Glucantime) are the drugs most commonly used as first-line treatment. These drugs are relatively expensive; however, an effective and cheaper generic version of sodium stibogluconate is now produced in India. The usual dose is 20 mg SbV/kg/day by slow intravenous infusion (the manufacturers of Pentostam recommend a minimum of 5 min) or intramuscularly for 20–40 days, depending on the geographical region. Side-effects include arthralgia, nausea, abdominal pain and pancreatitis. Cardiotoxicity tends to occur with high-dose regimens, particularly with prolonged use, and includes ST segment inversion, prolongation of the QTc interval and fatal arrhythmias. Toxicity, particularly pancreatitis, is increased in HIV-positive patients.

Amphotericin B

Amphotericin B is currently regarded as second-line treatment and is used when antimonials are not appropriate; e.g. in regions with high levels of resistance to SbV, such as Bihar, India. It is usually administered by slow intravenous infusion in 5% dextrose over 4–6 h, commencing at 0.1 mg/kg/day and gradually increasing to 1 mg/kg/day until a *total* dose of 20 mg/kg has been given. Studies in India have shown that there was no difference in infusion-related side-effects if treatment was commenced at 1 mg/kg/day. Side-effects include anaphylaxis, fever, chills, bone pain and thrombophlebitis. Hypokalaemia, renal impairment and anaemia may also occur.

Liposomal amphotericin B (AmBisome)

Amphotericin toxicity is reduced and efficacy enhanced by lyophilization, thereby enhancing distribution in macrophages and reticuloendothelial tissues. The most commonly recommended regimen for immunocompetent patients is 3 mg/kg/day on days 1–5, 14 and 21. Liposomal amphotericin B is very expensive; however, a recent study in India showed that administration of a single infusion (5 mg/kg) or five daily infusions of 1 mg/kg cured 92% of patients. If proved effective in larger trials,

low-dose regimens could make the drug more affordable. In immunocompromised patients with HIV, the dose of liposomal amphotericin B is 4 mg/kg/day on days 1–5 followed by 4 mg/kg/day on days 10, 17, 24, 31 and 38. The relapse rate is high, suggesting that maintenance treatment may be required.

Pentamidine

Pentamidine 4 mg/kg deep i.m. on alternate days for 5–25 weeks may be useful as second-line treatment in regions other than India where there is now significant resistance. Toxicity is common, including sudden hypotension following injection, acute hypoglycaemia, renal impairment and arrhythmias. Long-term irreversible insulin-dependent diabetes occurs in more than 10% of patients treated. Pentamidine may have a role in preventing relapses in patients coinfected with HIV.

Miltefosine

Miltefosine is one of the most promising drugs to appear in recent times for the treatment of leishmaniasis. Originally developed as an oral antineoplastic agent, miltefosine is the first highly effective oral treatment for VL. Studies in India using 100 mg/day (or 2.5 mg/kg/day in children) for 4 weeks achieved 95% cure rates. Side-effects include gastrointestinal upset but this is rarely severe. However, miltefosine is abortifacient and teratogenic and may also reduce male fertility. Miltefosine has a long half-life (2–3 weeks) and a narrow therapeutic index, thus increasing the opportunity for the development of resistance. Combination treatment is now being investigated as a means of mitigating against the development of resistance. Daily miltefosine may also have a role in preventing relapses in *Leishmania* and HIV coinfected patients.

Aminosidine

Aminosidine (paromomycin) at doses in the range 15–20 mg/kg/day i.m. may be used alone for 21 days or synergistically with SbV or pentamidine, thereby allowing a shorter duration of treatment when used in combination therapy with these agents.

Post-kala-azar dermal leishmaniasis

PKDL occurs in about 10% of patients in India, 2–10 years after treatment for VL. Elsewhere, PKDL is uncommon, although an unusually high incidence (> 50%) has recently been reported in southern Sudan, occurring at a mean of 56 days (range 0–18 days) following treatment. Initially, macules and papules appear around the mouth, which gradually spread over the face and sometimes more widely over the trunk and limbs. In time, nodules may develop resembling lepromatous leprosy. The papules and nodules in PKDL are usually packed with amastigotes and patients with this condition, which may persist for more than 20 years, may act as an important reservoir of infection. Prolonged treatment with SbV or other appropriate drug may be required to eliminate infection, although PKDL in Africa often resolves spontaneously without specific treatment.

Prevention

The stated aims of the WHO programme for the surveillance and control of leishmaniasis are, 'to reduce the disease as quickly as possible to such a level that each country can integrate control surveillance activities at both technical and economic levels, into their overall health development activities'. To achieve this goal, WHO has set the following objectives:
• to provide early diagnosis and prompt treatment;
• to control the sandfly population through residual insecticide spraying of houses and through the use of insecticide-impregnated bed nets;
• to provide health education and produce training materials;
• to detect and contain epidemics in the early stages; and
• to provide early diagnosis and effective management for Leishmania and HIV coinfections.
    Achieving these objectives poses many prob-lems. The methods that can be used to control VL depend on the epidemiological situation.

Eliminating or treating the reservoir host

Most success has been achieved where the domestic dog is the main reservoir, efforts being directed to catching and destroying infected dogs. Dogs with the infection look sick, lose their hair and have an enlarged spleen. France has been very active in controlling by this method. However, infected foxes invariably look healthy and are more difficult to control. Where humans are the main reservoir, active case finding and treatment may be considered, but is prohibitively expensive and likely to fail given that the majority of infections are subclinical. Detection and treatment of asymptomatic individuals carries additional cost and raises issues of individual risk–benefit given the toxicity of available drugs.

    There is no vaccine currently available for VL, although the future is brighter. Recent phase II trials in monkeys using alum-precipitated autoclaved Leishmania major with bacille Calmette–Guérin (BCG) have demonstrated protection against L. donovani.

Eliminating or avoiding the vector

Sandflies breed in dark moist habitats, such as cracks in masonry, piles of rubble, caves and in any dark protected sites such as holes in termite mounds or in outside latrines. Sandflies have a short flight range, seldom being found more than 200 m from their breeding place. They do not fly very high and are unlikely to reach people sleeping on the first floor of a building. They normally bite between dusk and dawn, and are small enough to penetrate the mesh of standard mosquito nets. Insecticide-impregnated nets may be more effective, but are relatively costly and do not prevent exposure in many epidemiological settings. Personal use of insect repellents or insecticide-impregnated clothing is generally not an option for people residing in endemic areas.

    In India, successful control was achieved during the period when widespread insecticide spraying of houses for malaria control was in use. The resurgence of infection on an epidemic scale has occurred in several areas many years after the spraying programme was abandoned.

In epidemics and localized outbreaks, residual insecticide spraying of houses and the immediate area around the house is the most effective immediate control measure. Sandflies usually succumb to dichlorodiphenyl-trichloroethane (DDT) and their hopping flight pattern renders them particularly vulnerable. DDT is cheaper than other residual insecticides, but is losing popularity because of concern about the long-term environmental impact.

Further reading

Davies CR, Kaye P, Croft SL, Sundar S. Leishmaniasis: new approaches to disease control. *Br Med J* 2003; **326**: 377–82. [Updated review of key aspects of control, and annotated educational resource list.]

Herwaldt BL. Leishmaniasis. *Lancet* 1999; **354**: 1191–9. [This review includes useful information on the clinical presentation, diagnosis and management of visceral, cutaneous and mucosal leishmaniasis.]

Guerin PJ, Olliaro P, Sundar S, *et al.* Visceral leishmaniasis: current status of control, diagnosis, and treatment, and a proposed research and development agenda. *Lancet Infect Dis* 2002; **2**: 494–501. [This article reviews the current situation and perspectives for diagnosis, treatment and control of visceral leishmaniasis, and lists some priorities for research and development.]

CHAPTER 11

Cutaneous Leishmaniasis

Clinical features, 80   Management, 82   Further reading, 83
Investigations, 81   Prevention, 82

Cutaneous leishmaniasis is among the most important causes of chronic ulcerating skin lesions in the world. The organisms in the host tissues are mainly found in reticuloendothelial cells in the skin, where, as amastigotes, they multiply by simple fission. Microscopically, they cannot be distinguished from *Leishmania donovani*. The parasites, life cycles and vectors are similar to those described for species causing visceral leishmaniasis (VL).

The clinical spectrum of disease may be classified as follows:
• cutaneous leishmaniasis (CL);
• diffuse cutaneous leishmaniasis (DCL);
• leishmaniasis recidivans (LR); and
• mucocutaneous leishmaniasis (MCL), also known as mucosal leishmaniasis (ML).

It is common to attach the terms Old World or New World depending on the region in which the infection is acquired. The following species of *Leishmania* are commonly implicated:
• Old World: *L. tropica, L. major* and *L. aethiopica*; also *L. infantum* and *L. donovani*.
• New World: *L. mexicana* species complex (especially *L. mexicana, L. amazonensis* and *L. venezuelensis*) and *Viannia* subgenus (most notably *L. [V.] braziliensis, L. [V.] panamensis, L. [V.] guyanensis* and *L. [V.] peruviana*); also *L. major*-like organisms and *L. chagasi*.

Ninety per cent of all cases of CL occur in Afghanistan, Brazil, Iran, Peru, Saudi Arabia and Syria, with 1–1.5 million new cases reported annually worldwide. Ninety per cent of all cases of ML occur in Bolivia, Brazil and Peru.

Clinical features

Cutaneous leishmaniasis

There is a variable incubation period, usually several weeks. Infections may be subclinical or clinical. Usually a papule develops at the site of infection, becomes a nodule and subsequently forms an ulcer with a central depression and raised indurated border. This may enlarge to a diameter of several centimetres and persist for months or years before eventually healing, leaving an atrophic scar. Some lesions do not ulcerate but persist as nodules or plaques (Figs 11.1 & 11.2, see colour plate facing p. 146).

Some patients have more than one primary lesion or may develop satellite lesions. A sporotrichoid-like nodular lymphangitis may occur (common with *L. [V.] panamensis* and *L. [V.] guyanensis*) in which there is thickening of the lymphatic channels draining the primary lesion with nodules at intervals along the path. Regional adenopathy may occur and is sometimes bubonic in nature with *L. (V.) braziliensis*. Lesion pruritus or pain, and secondary bacterial infection may also occur.

Diffuse cutaneous leishmaniasis

The following species are usually involved:
• Old World: *L. aethiopica* — Ethiopia, Kenya; and
• New World: *L. mexicana, L. amazonensis, L. venezuelensis*.

DCL closely resembles lepromatous leprosy. A single lesion gives rise to multiple diffuse soft fleshy nodules or plaques containing enormous numbers of amastigotes. Ulceration is unusual, probably because of deficient cell-mediated immunity. There may be extensive depigmentation in the areas of affected skin, increasing the resemblance to leprosy.

Leishmaniasis recidivans

This form is most commonly seen in Iran and Iraq and is also known as lupoid leishmaniasis. LR resembles lupus vulgaris and usually affects the face, sometimes invading mucous membranes. Lesions wax and wane, persisting for 20–40 years, with scarring as they heal. The combination of scarring and signs of active inflammation is characteristic of the condition.

Mucosal leishmaniasis

Also known as espundia, ML is a dreaded complication of New World CL. Most cases are caused by the *Viannia* subgenus, particularly *L. (V.) braziliensis*, *L. (V.) panamensis* and *L. (V.) guyanensis*. The onset is usually a few years after resolution of the original cutaneous lesion, but may occur while the primary lesion is still present or decades later. Haematogenous and lymphatic dispersal results in spread of amastigotes from the skin to the naso-oropharyngeal mucosa. Patients may initially complain of symptoms of chronic nasal congestion. The first perceptible lesion is often a nodule adjacent to the nostril. Granulomatous destructive lesions with chronic ulceration follow. After many years, the nasal septum, other nasal cartilaginous structures and palate may be destroyed, leaving a grotesque cavity in the centre of the face. The risk of mucosal disease following a primary cutaneous lesion is probably less than 5%. Rarely, destructive lesions may occur in the urinogenital region.

The differential diagnosis of ML includes:
- paracoccidiodomycosis;
- histoplasmosis;
- syphilis;
- tertiary yaws;
- leprosy;
- rhinoscleroma;
- midline granuloma;

- sarcoidosis; and
- neoplasms.

Investigations

Cutaneous and diffuse cutaneous leishmaniasis

Parasitological diagnosis is usually made by biopsy of the edge of the ulcer or other lesion. The specimen obtained can be divided in portions for:

1 an impression smear (touch preparation) on a microscope slide that is then fixed with methanol and stained with Giemsa;

2 histopathology (less sensitive than impression smear);

3 culture on Novy, MacNeal and Nicolle's (NNN) medium; and

4 polymerase chain reaction (PCR).

Other techniques that are sometimes used include needle aspirates and dermal scrapings.

Leishmaniasis recidivans and mucosal leishmaniasis

Organisms are usually scanty in affected tissue, therefore PCR and culture are preferred for diagnosis. Serology is generally unhelpful but is more likely to be positive in ML than in CL. The main use of serodiagnostic methods is when CL is suspected clinically but direct diagnostic methods have failed. In endemic areas a high proportion of the population may be seropositive. False-positive results may occur with lepromatous leprosy (LL), so the distinction between DCL and LL cannot be made reliably by serology. Fortunately, bacilli are always easy to find in LL.

Leishmanin test (Montenegro test)

This is a skin test using a killed promastigote suspension as antigen (area and species specific), injected intradermally and read at 48 h, like the tuberculin test. In endemic areas a high proportion of the population will be leishmanin test-positive and may also have healed scars. The test is positive in most cases of established CL and ML. It may be negative in some cases of CL caused by *L. aethiopica* and is usually negative in DCL. A strongly positive test may be useful in diagnosing LR cases, because the routine histology from these lesions is often indistinguishable from lupus vulgaris (cutaneous tuberculosis). The test is negative in active VL but may become positive after successful treatment.

Management

Before commencing treatment, the following issues should be considered:
- the number, size, evolution and persistence of lesions;
- the location of lesion(s) (e.g. on the face);
- whether the patient is at risk of ML; and
- other features (e.g. the presence of nodular lymphangitis).

Treatment of cutaneous leishmaniasis

Cosmetically unimportant lesions caused by non-destructive and non-metastasizing species usually heal spontaneously and therefore may not require active treatment. The following oral agents can be used for treating relatively benign cosmetically unimportant lesions.

Ketoconazole —modest activity against *L. mexicana* and *L. (V.) panamensis*; usefulness against *L. major* infection is unclear.

Itraconazole —better tolerated than ketoconazole but may be less effective, at least against the *Viannia* subgenus.

Fluconazole (L. major).

Miltefosine —currently being investigated for treatment of New World CL.

Various local therapies are sometimes used, including:
- heat treatment or cryotherapy;
- topical amphotericin B (*L. major*);
- intralesional antimony therapy; and
- paromomycin ointment (available in Israel).

Treatment with paromomycin may result initially in increased ulceration, so it is best avoided for ulcers on the face.

Parenteral treatment

Pentavalent antimony therapy (SbV) (i.v. or i.m.) is probably still the best option if optimal effectiveness is important. Short-course pentamidine has been shown to be effective in Colombia where disease is predominantly caused by the *Viannia* subgenus. Old World DCL is treated with a combination of SbV and aminosidine. Response may be poor in some parts of Ethiopia and pentamidine is used as an alternative; however, about 10% of patients are left with diabetes mellitus following treatment. New World DCL is treated with SbV. LR may be treated with parenteral or intralesional SbV, or may respond to heat treatment.

ML treatment is of greater importance. Adequate systemic treatment of cutaneous lesions is assumed (but not proven) to decrease the already low risk of mucosal disease. ML is harder to treat than cutaneous lesions and becomes increasingly so as it progresses. Currently, the best treatment options are SbV (achieves cure rates of about 75% for mild disease and 10–63% for more advanced disease) or conventional amphotericin B. Concomitant corticosteroids are indicated if respiratory compromise develops.

Prevention

In general, the principles of prevention and control are the same as for VL. In the Middle East it is customary for mothers to expose cosmetically unimportant areas of their infants to sandfly bites or to deliberately inoculate them with infected material to render the child immune to that species. The development of effective vaccines is proving difficult. A vaccine using live attenuated *L. major* promastigotes has been produced which appears to be effective although its use carries a small risk of precipitating an aggressive lesion or the development of LR. There has also been some interesting work recently on the development of vaccines against sandfly saliva.

Further reading

Herwaldt BL. Leishmaniasis. *Lancet* 1999; **354**: 1191–9. [This review includes useful information on the clinical presentation, diagnosis and management of visceral, cutaneous and mucosal leishmaniasis.]

CHAPTER 12

Tuberculosis

Microbiology, 84
Epidemiology, 84
Clinical features, 88
Investigations, 90

Management, 90
Prevention and public health
  aspects, 94

Future developments, 96
Further reading, 97

On 24 March 1882, Robert Koch demonstrated *Mycobacterium tuberculosis* to be the cause of the disease tuberculosis (TB). Since then, advances in human understanding of the disease have been major catalysts in the development of modern medicine. In 1993 the World Health Organization (WHO) declared tuberculosis an international emergency—an unprecedented step—and international interest and funding for tuberculosis control has increased since that time. A decade on and into the 21st century, the full genome of *M. tuberculosis* is known, but the microbe still causes more human deaths each year than any other infectious agent (Fig. 12.1). Most of these deaths occur among poor adults and adolescents in the developing countries of the tropics.

Microbiology

Of the mycobacteria, *M. tuberculosis*, *M. bovis* and *M. africanum* are now known to be genetically very similar, have the highest pathogenicity, and are together referred to as the *M. tuberculosis* (MTB) complex. They are anaerobic, non-spore-forming, non-motile bacilli with a large cell wall content of high-molecular-weight lipids. Growth is slow, the generation time being 15–20 h, as compared to well under 1 h for most common bacterial pathogens. Visible growth of yellow colonies in culture, usually on egg-based solid Löwenstein–Jensen medium, takes between 4 and 12 weeks.

Bacilli of the MTB complex are referred to as tubercle bacilli, acid-fast bacilli (AFB) or acid- and alcohol-fast bacilli (AAFB) on the basis of the ability of their lipid-rich cell walls to retain the red carbol-fuchsin stain in the presence of acid and alcohol during the Ziehl–Neelsen staining process. Under oil-immersion light microscopy the stained bacilli appear as slightly bent rods, 2–4 μm long and 0.2–0.5 μm wide. Distinguishing between the three species is impossible by microscopy and difficult by culture. Furthermore, clinical presentations of disease caused by the bacilli are similar. Therefore it is not possible to be precise about the relative contributions of the three species to the sum total of human tuberculosis disease. However, *M. tuberculosis* (MTB) is globally the most prevalent, and widely recognized to cause most of the global burden of disease, particularly in the tropics. This chapter therefore focuses on MTB and not the other less pathogenic mycobacteria often referred to as mycobacteria other than tuberculosis (MOTT). Examples of MOTT include the *M. avium–intracellulare* complex, *M. marinum* and *M. gordonae*.

Epidemiology

Magnitude of the problem

It is estimated that one-third of the global

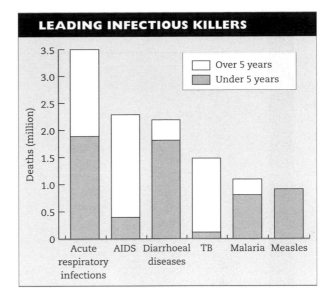

Fig. 12.1 Leading infectious killers: six high-burden diseases cause 90% of total disease deaths. HIV-positive people who have died with TB have been included among AIDS deaths. Source: World Health Organization/CDS (1999).

human population is infected with MTB and that between 7 and 8 million new TB cases, and between 2 and 3 million TB deaths occur per year. MTB is thought to cause one-quarter of avoidable adult deaths in developing countries. The geographical distribution of TB is shown in Fig. 12.2.

Transmission

Humans are the only reservoir of MTB infection and transmission is only possible from individuals with disease. It occurs by droplet nuclei; infectious particles of respiratory secretions aerosolized by coughing, sneezing or talking, which are sufficiently small (around 10 µm, drying to less than 5 µm diameter while airborne) to remain suspended in the air for long periods and reach the terminal air spaces if inhaled.

Infection and immunity

Once MTB droplet nuclei reach alveolar level within the lungs, the bacilli are taken up by phagocytosis into air-space macrophages. Within these cells they are processed into phagosomes which fail to acidify. In this way the bacilli avoid intracellular killing and may survive and multiply for long periods of time. Infected macrophages may therefore carry viable bacilli

in the lymphatics to regional lymph nodes or in the bloodstream to any part of the body.

Both humoral and cell-mediated immune responses are mounted against MTB and are correlated with the development of detectable delayed-type hypersensitivity (DTH) reactions. Rarely, these are manifested clinically in the form of erythema nodosum or phlyctenular conjunctivitis. More usually, DTH to MTB is detected by intradermal injection of mycobacterial purified protein derivatives (PPD)—the basis of the Mantoux, tine and Heaf tests. The extent of local skin erythema, induration and blistering (in vigorous responses) are measured 48 h after injection in order to grade responses. Although individuals clearly vary in their immune capacity to contain or eliminate MTB, it must be emphasized that immune responses to MTB—however generated—are generally not protective against further infection. A common misconception is that PPD skin responsiveness is correlated with the effectiveness of immunity to MTB. It remains unclear which components of cell-mediated and humoral responses to MTB are most important in conferring protective immunity and those components responsible for DTH responses are not necessarily the most useful.

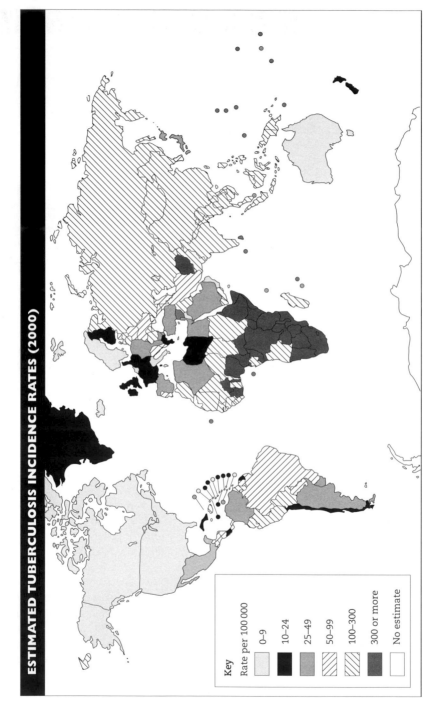

Fig. 12.2 Estimated tuberculosis incidence rates (2000). *Source: Global Tuberculosis Control: WHO Report 2002, World Health Organization, Geneva.*

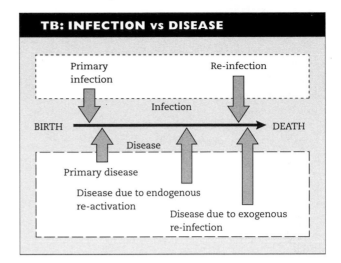

Fig. 12.3 The processes of TB infection and progression to disease.

Progression to disease

In the usual course of events, somewhere between 5 and 10% of people will develop active TB after MTB infection. About 3% develop disease within the first year with the remainder developing disease with ever-diminishing frequency thereafter. More than 90% of MTB infections therefore do not result in disease within a normal human lifespan.

The clinical manifestation of TB among those who develop active disease depends on two things: the state of the immune system and the location of the bulk of the MTB multiplication. In those cases where disease occurs soon after primary infection, the bacilli multiply and spread in the context of a naïve immune system. Primary forms of disease therefore occur at common thoracic sites of initial multiplication; hence pleurisy extending from an alveolar focus and cavitation in hilar lymph nodes. They also tend to disseminate to multiple sites including the central nervous system; hence tuberculous meningitis and 'miliary' tuberculosis. In disseminated disease, mini-granulomas (tubercles) develop around small numbers of bacilli which are widely distributed within tissues.

In those cases where disease occurs a long time after primary infection, either as reactivation of latent infection or as a result of reinfection with a new strain of MTB, the bacilli multiply in the context of a sensitized immune system. The associated DTH responses tend to lead to tissue destruction at the site of multiplication; hence cavitating caseous lesions, in which large numbers of multiplying bacilli are contained by an encircling rim of giant cells and granulomas—the hallmark of tuberculous pathology. These 'postprimary' lesions are most commonly in the apices of the lungs, the theory being that this location provides the most conducive combination of ventilation and perfusion for long-term latency. They may also occur at any site to which bacilli were seeded during initial multiplication around the time of primary infection.

Given sufficient time, postprimary-type disease in the lungs is likely to result in communication between the cavitating pathology and an airway. MTB bacilli can then be aerosolized in droplet nuclei and expelled into the atmosphere when the affected individual coughs, sneezes or talks. Patients with cavitating lung disease are therefore the main sources of new MTB infections. The processes of infection and progression to disease are illustrated in Fig. 12.3.

Risk factors for infection and disease

Risk factors for MTB infection fall into two broad categories:

1 those which put people in an atmosphere where MTB-containing droplet nuclei accumulate in the atmosphere; and

2 those decreasing the ability of alveolar macrophages to incapacitate MTB once taken up.

The first category includes prolonged contact with a person or people with pulmonary TB (especially cavitating disease) and the environmental features associated with poverty. Overcrowded and poorly ventilated living and working conditions are clearly ideal for MTB transmission. As MTB is susceptible to killing on exposure to ultraviolet light, dark and humid conditions such as may be found in mines and prisons also favour transmission. The second category includes anything capable of compromising alveolar macrophage killing of MTB, such as corticosteroid therapy and HIV infection.

Risk factors for disease have in common their ability to impair cell-mediated immunity, particularly those functions dependent on T cells. Examples include HIV infection, malnutrition (particularly vitamin D deficiency) and corticosteroid therapy.

Effect of HIV on the epidemiology of TB

The superimposition of HIV infection in people with pre-existing MTB infection increases the risk of developing tuberculosis from 5–10% over a lifetime to around 15% per year. In addition to this increased risk of reactivation disease, HIV-infected people are at increased risk of acquiring new MTB infections which may also progress to disease. This has meant that in those parts of the world where the prevalence of MTB infection and HIV infection overlap geographically there has been an explosive increase in the number of TB cases which has increased the annual risk of infection for both HIV-infected and HIV-uninfected people. Both HIV infection and MTB infection tend to affect particularly adolescents and adults in the middle decades of life;

their most economically productive years. In many developing countries in the tropics, particularly in sub-Saharan Africa, these two devastating infections overlap both geographically and socially and the resultant impact on livelihoods has been appalling. TB not only arises in conditions of poverty, it is itself a poverty-generating illness.

Clinical features

Pulmonary vs. extrapulmonary disease

Pulmonary features predominate in around 85% of all TB disease presentations. Although in most instances pulmonary disease will be the only obvious pathology, it may be associated with tuberculous pathology in other organ systems. Parenchymal lung disease may extend and include pericardial disease or regional lymph node cavitation. Conversely, both pericardial tuberculosis and tuberculous lymphadenitis may occur in the absence of any concurrent pulmonary pathology and would then be classified as extrapulmonary disease manifestations. Counterintuitively, two forms of intrathoracic TB pathology may be classified as extrapulmonary tuberculosis when they occur in the absence of concurrent parenchymal lung disease: mediastinal lymphadenopathy and pleurisy. This serves to emphasize that any clinical presentation in which pulmonary parenchymal disease is present is classified as pulmonary, and only patients with pulmonary disease, not extrapulmonary disease, are capable of transmitting MTB to others.

Among extrapulmonary presentations, lymphadenitis (Fig. 12.4, see colour plate facing p. 146) and pleurisy are the most common, each accounting for approximately 25% of the total. Genitourinary TB is next at around 15%, followed by miliary and bone TB at around 10%. Meningeal and peritoneal TB each account for less than 5% of extrapulmonary TB disease.

Systemic symptoms

The vast majority of TB presentations, whether

pulmonary or extrapulmonary, are insidious in onset. Varying combinations of the chronic constitutional symptoms of fevers, night sweats, weight loss and malaise (perhaps secondary to an associated anaemia of chronic disease) are common but neither universal nor specific indicators of TB.

Symptoms of pulmonary TB

Beyond the systemic manifestations, the symptoms and signs of TB depend on the site of the major pathology. As pulmonary disease, and not extrapulmonary disease, is both the most common form of the disease and the priority target for public health intervention, it is the main focus for further clinical description here.

Persistent coughing of insidious onset is easily the most common symptom indicating pulmonary TB. As pulmonary pathology advances, the cough becomes more productive of mucopurulent sputum and chest pains may occur with severe coughing. However, it is important to remember that some patients with early pulmonary disease may not produce much sputum. The sputum may be streaked with blood in about 10% of cases, and this usually indicates that the cavitating pathology has led to local damage of small blood vessels. Frank or catastrophic haemoptysis can occur if the larger blood vessels become involved, but this is rare, occurring in fewer than 1% of cases.

Signs of pulmonary TB

Patients with TB may be wasted. Other than this, the signs are dependent on the site and extent of the underlying pathology. Much is often made of chest signs such as 'amphoric breathing' and consolidation. Certainly, the lung damage can be extensive and often includes signs of volume loss, including tracheal shift. The truth is that most patients with pulmonary disease have very few chest signs and, apart from detecting massive pleural effusions that need draining, the slavish pursuit of chest signs is of little use in guiding clinical management. Patients with advanced pulmonary disease in the tropics may have finger clubbing, a sign that is otherwise not often associated with TB in Europe and North America.

Clinical features of selected forms of extrapulmonary TB

Apart from tuberculous lymphadenitis, which usually presents as a unilateral chain of matted lymph nodes that may occasionally ulcerate and discharge, extrapulmonary TB is notoriously difficult to diagnose. This is because non-specific systemic manifestations predominate in the early stages and these forms of disease are not amenable to any investigations that come close to the immediacy and specificity of sputum smear microscopy for AFB. As pathology advances, more useful signs such as meningism, bone damage, serous effusions and fistulae may become apparent.

Effect of HIV on clinical presentations of TB

It is important to remember that there are exceptions to the simplified division of tuberculous disease into the 'primary' and 'postprimary' forms presented above. Postprimary disease may manifest itself as disseminated disease such as miliary TB if the immune system is very compromised by an additional factor such as HIV infection. Postprimary disease arising early in HIV infection, before significant immunocompromise is established, is likely to present with cavitating pathology that is indistinguishable from disease arising in HIV-uninfected individuals. However, in the later stages of HIV infection, as underlying immunocompromise becomes more severe, postprimary TB disease becomes more likely to present in a disseminated or non-cavitating form resembling primary disease. This is why extrapulmonary 'primary-like' presentations of TB, such as pleural effusions, lymphadenitis, TB meningitis, miliary TB (Fig. 12.5, see colour plate facing p. 146) and non-cavitating pulmonary TB, are more common among HIV-infected patients.

Differential diagnosis

Pulmonary tuberculosis in the tropics has a wide differential diagnosis. Some of these, such as pulmonary paragonimiasis, nocardiosis, actinomycosis, coccidioidomycosis and melioidosis, are defined by their geographical distribution. Others, such as *Pneumocystis carinii* pneumonia (PCP) and pulmonary Kaposi's sarcoma (KS), occur in the context of HIV infection. The remainder, including bacterial pneumonias, lung abscess, atypical mycobacteria other than MTB, bronchial carcinoma, bronchiectasis, sarcoidosis, Wegener's granulomatosis and cryptogenic fibrosing alveolitis, are more universal.

The differential diagnosis of the most common extrapulmonary forms of TB, lymphadenitis and pleurisy, is mainly from neoplastic processes such as KS, lymphoma and metastatic bronchial carcinoma.

Investigations

Isolation of MTB by culture from clinical specimens is the gold standard for the definitive diagnosis of tuberculosis. However, because of the slow generation time, mycobacterial culture takes between 2 weeks (modern liquid-based culture techniques) and 12 weeks (more universal, solid-based culture techniques). This is clearly too long to be useful in guiding clinical decision-making. In addition, the laboratory infrastructure required to sustain quality-assured culture of mycobacteria is frequently unavailable in the poorer parts of the tropics.

Sputum smear microscopy for tubercle bacilli is therefore absolutely central to the diagnosis of tuberculosis. Approximately half of all culture-proven cases of pulmonary tuberculosis produce more than the threshold 10 000 organisms per ml of sputum required for detection by microscopy. These smear-positive cases tend to have more cavitating lung disease and are more infectious than smear-negative cases. While smear microscopy is a specific test for pulmonary tuberculosis, it lacks sensitivity, particularly for early disease that has not yet cavitated. Ziehl–Neelsen staining and light microscopy are the most universal techniques used for sputum smear microscopy. Where fluorescence microscopes are available, auramine-phenol staining is easier on the microscopist's eyes and assists with faster reading of specimens.

The main problem in the diagnosis of tuberculosis lies with patients who have clinical features suggestive of pulmonary TB but whose sputum smears are negative for AFB. Unless there are strong clinical indicators of an alternative diagnosis, the decision on whether or not to treat for tuberculosis lies with chest radiography. Unfortunately, chest X-rays of smear-negative TB cases are notoriously difficult to interpret as the features that are most specific to TB (such as cavitation) are frequently absent. The radiological features of pulmonary TB are also particularly difficult in HIV-infected individuals where an increased proportion of culture-proven cases will have a variety of atypical radiographical manifestations including lower lobe consolidation and patchy infiltrates. Films may even be normal.

PPD skin test positivity is used as a marker of MTB infection and high-grade PPD responses are correlated with the presence of active disease. However, false-positives may occur with exposure to non-pathogenic environmental mycobacteria or BCG vaccination. Similarly, false-negatives are a problem when immune responses are blunted, e.g. by HIV, measles, drugs or severe malnutrition. PPD skin testing is mainly used in the diagnosis of TB in children where most disease is of primary type and only rarely smear-positive.

A variety of laboratory tests indicating chronic inflammation, such as raised erythrocyte sedimentation rate and C-reactive protein, or anaemia of chronic disease may help in difficult cases but only have a limited role in the investigation of suspected TB cases in the tropics.

Management

Principles of TB chemotherapy
Tuberculosis treatment aims to:
• cure the patient of TB;
• prevent death from active TB or its late effects;

- prevent relapse of TB;
- decrease transmission of TB to others.

These aims can be achieved while preventing the selection of resistant bacilli in infectious patients through the careful use of modern chemotherapy.

Antibiotic chemotherapy for TB has been built up over the last 40 years around six first-line drugs on the basis of several randomized controlled trials and cohort studies. Four of the drugs are bactericidal (streptomycin, isoniazid, rifampicin and pyrazinamide) while the remaining two are bacteriostatic (ethambutol and thi-

Table 12.1 First-line antituberculosis drugs, standard abbreviations, dosages and adverse effects.

acetazone) and each is referred to by a single capital letter in standard descriptions of different regimens (Table 12.1, which also details main side-effects and doses). These drugs act preferentially on slightly different populations of organisms (Table 12.2).

During the initial, intensive phase of chemotherapy a minimum of three drugs should be administered concurrently to reduce the more rapidly dividing bacillary load. A minimum of two drugs can be used in the continuation phase, aimed at sterilizing lesions containing fewer bacilli with slower generation times.

All modern drug regimens contain rifampicin, isoniazid and pyrazinamide and as yet there is no convincing evidence that regimens shorter than 6 months' duration will reliably cure tuber-

ANTITUBERCULOSIS DRUGS

Essential anti-TB drug (abbreviation)	Mode of action	Main adverse effect(s)	Recommended dose (mg/kg)		
				Intermittent	
			Daily	3×/week	2×/week[*]
Isoniazid (H)	Bactericidal	Peripheral neuropathy Hepatitis	5 (4–6)	10 (8–12)	15 (13–17)
Rifampicin (R)	Bactericidal	Hepatitis Influenza-like syndrome and thrombocytopenia†	10 (8–12)	10 (8–12)	10 (8–12)
Pyrazinamide (Z)	Bactericidal	Arthralgia Hyperuricaemia leading to gout	25 (20–30)	35 (30–40)	50 (40–60)
Streptomycin (S)	Bactericidal	VIII cranial nerve damage—vestibular dysfunction nephrotoxicity	15 (12–18)	15 (12–18)	15 (12–18)
Ethambutol (E)	Bacteriostatic	Optic neuritis	15 (15–20)	30 (25–35)	45 (40–50)
Thioacetazone (T)	Bacteriostatic	Exfoliative dermatitis	2.5	Not applicable	

[*] WHO does not generally recommend twice weekly regimens. If a patient receiving a twice weekly regimen misses a dose of tablets, this missed dose represents a larger fraction of the total number of treatment doses than if the patient were receiving a thrice weekly or daily regimen so there is an increased risk of treatment failure.
† More common with intermittent dosage.

MYCOBACTERIUM TUBERCULOSIS POPULATIONS

	In cavities	In closed caseous lesions	In macrophages
Relative number of organisms/mL	10^7–10^8	10^2–10^4	10^2–10^4
Multiplication	Active/rapid	Slow/intermittent	Slow
Medium	Neutral/alkaline	Neutral	Acid
Most useful drugs	S R H	R H	Z R H

Abbreviations: H, isoniazid; R, rifampicin; S, streptomycin; Z, pyrazinamide.

Table 12.2 Different populations of *Mycobacterium tuberculosis* and their susceptibility to different drugs.

culosis. Monotherapy for tuberculosis disease should never be given as it will lead to the development of antibiotic resistance.

All of the first-line drugs can be given orally except for streptomycin which requires intramuscular injection. Because of the requirement for needles and syringes streptomycin is no longer recommended in the first-line treatment of new TB cases in areas of HIV seroprevalence. Thioacetazone is associated with a high incidence of severe mucocutaneous drug reactions in HIV-infected patients and is therefore only rarely used in modern regimens.

Deciding which treatment regimen to use

It is important to follow national or regional guidelines for chemotherapy. Category I regimens are used for new patients who are unlikely to harbour resistant organisms. Category II regimens are reserved for patients who have previously received some form of antituberculous chemotherapy and are therefore more likely to harbour mycobacteria that have become resistant to one or more of the first-line drugs. It is therefore important to take a careful history about previous treatment before starting a patient on TB chemotherapy. Some examples of WHO-approved Category I and II regimens are shown in Table 12.3. Category III

regimens have been recommended for use in less severe, smear-negative forms of disease, but for simplicity there has been a move towards the use of Category I regimens for all new cases, regardless of smear status.

A 'trial of therapy' as a way of confirming a TB diagnosis is not recommended, unless the situation is life-threatening. In some instances, non-TB infections may respond to the broad-spectrum antibiotic effect of drugs such as streptomycin and rifampicin. There is also an increased risk of chaotic ingestion of drugs and the consequent development of drug resistance.

Monitoring treatment

Patients with smear-positive pulmonary TB should be monitored by sputum smear examination: once at the end of the intensive phase; once during the continuation phase; and once at the end of therapy. It is unnecessary and wasteful of resources to monitor using chest radiography. For patients with smear-negative pulmonary TB and extrapulmonary TB, clinical monitoring is the usual if somewhat unsatisfactory way of assessing response to treatment. Routine monitoring by mycobacterial culture of sputum is not feasible in developing countries in the tropics.

At the end of the intensive phase, most patients will have negative sputum smears. Such patients can then start the continuation phase of treatment. If sputum smears remain positive at this stage despite careful adherence to treatment it may indicate that the patient had a par-

ANTITUBERCULOSIS REGIMENS

TB treatment category	TB patients	Alternative TB treatment regimens	
		Intensive phase	Continuation phase
I	New smear-positive PTB or New smear-negative PTB with extensive parenchymal involvement; new cases of severe forms of extra-pulmonary TB	*2EHRZ (SRHZ) 2EHRZ (SRHZ) 2EHRZ (SRHZ)	4HR $4H_3R_3$ 6HE
II	Sputum smear-positive relapse; treatment failure; treatment after interruption	*2SHRZE/1HRZE 2SHRZE/1HRZE	5HRE $5H_3R_3E_3$
III	New smear-negative PTB (other than in Category I); new less severe forms of extrapulmonary TB	*2HRZ 2HRZ 2HRZ	4HR $4H_3R_3$ 6HE
IV	Chronic case (still sputum-smear positive after supervised retreatment)	Not applicable Refer to WHO guidelines on management of multidrug-resistant TB	

Note: Standard code for TB treatment regimens.
There is a standard code for TB treatment regimens. Each first-line anti-TB drug has an abbreviation (shown in Table 12.1). A regimen consists of two phases. The number before a phase is the duration of that phase in months. A number in subscript (e.g. $_3$) after a letter is the number of doses of that drug per week. If there is no number in subscript after a letter, then treatment with that drug is daily. An alternative drug (or drugs) appears as a letter (or letters) in brackets.
*Preferred option.

Table 12.3 WHO-approved antituberculosis regimens.

ticularly heavy initial bacillary load. Rarely, this is an indication of drug-resistant TB which will not respond to Category I treatment. Whatever the reason, the initial phase should be prolonged for a third month. The patient then starts the continuation phase. If smears remain positive after a month of the continuation phase, this constitutes treatment failure and the patient should be restarted on a full course of a Category II regimen.

At the end of the treatment course treatment outcomes are recorded according to one of six categories as shown in Table 12.4. This allows for systematic cohort analysis and reporting of cure rates—an important part of TB control (see below).

The question of isolation

Routinely admitting smear-positive TB patients in order to 'isolate' them from the community is not an absolute requirement. In most cases, any onward community transmission of MTB from a smear-positive case will have occurred by the time the diagnosis is established and modern chemotherapy will render such patients non-infectious by the end of the second week of treatment in more than 95% of cases provided the initial isolate is fully drug sensitive. Protecting others from infection is, on the whole, best achieved by careful chemotherapy rather than physical isolation. Admitting patients to hospital should only be necessary where the patient is severely ill and needs full hospital care and it must be remembered that a TB case is more likely to come into contact with individuals who are vulnerable to MTB infection (such as those

Cured
Patient who is smear-negative at, or 1 month prior to, the completion of treatment and on at least one previous occasion

Treatment completed
Patient who has completed treatment but without smear microscopy proof of cure

Died
Patient who died during treatment, regardless of cause

Failure
Smear-positive patient who remained or became smear-positive again 5 months or later after commencing treatment

Defaulted
Patient whose treatment was interrupted for 2 months or more

Transferred out
Patient who has been transferred to another reporting unit and for whom the treatment outcome is not known

Table 12.4 Recording standardized treatment outcomes in smear-positive pulmonary TB.

infected with HIV) in hospital than in the general community. If hospital admission is necessary then this should be to a dedicated well-ventilated TB ward located away from other inpatients. Individual isolation rooms for TB patients are mostly unavailable and unnecessary in countries with high TB incidence.

Adjunctive corticosteroid therapy
Many advocate the concurrent use of high-dose corticosteroids with TB treatment for large pleural and pericardial tuberculous effusions. There are some trials indicating that this is helpful for rapid relief of symptoms and for reduction in complication rates but systematic reviews suggest that the evidence is not strong and is certainly not available for HIV-infected patients with TB. The potential disadvantages of corticosteroid therapy, including pharmaco–kinetic interactions with TB drugs and reactiva-

tion of other latent infections (such as *Strongyloides*), should be weighed carefully against potential benefits.

Prevention and public health aspects

The WHO-recommended DOTS strategy is the internationally recognized approach to TB control. Bacillus Calmette–Guérin (BCG) vaccination and isoniazid preventive therapy are mentioned briefly for completeness.

BCG vaccination
The BCG vaccine is the world's most frequently administered vaccine and it has been available since the 1920s. It is a live attenuated vaccine which is given intradermally. Unfortunately, there is little evidence that it provides long-lasting protection against the development of pulmonary TB. This appears to be particularly true from the trials conducted in the tropics. None the less, BCG is still included in the Expanded Programme of Immunization mainly because there is some evidence that it protects against disseminated forms of TB in children. It has also been shown to be protective against leprosy.

Isoniazid preventive therapy
The rationale behind preventive therapy is to eradicate latent infection in PPD-positive people before it develops into active disease. Several placebo-controlled trials in HIV-negative people infected with MTB have shown that daily isoniazid given for 6–12 months substantially reduces the subsequent risk of tuberculosis disease. However, preventive therapy has not been recognized as a cost-effective universal approach to TB control but instead has been focused on individuals at increased risk of developing active disease. Such individuals are usually identified by skin testing and are either contacts of known smear-positive index cases or people who are occupationally exposed to infection (such as nurses and doctors).

In recent years a series of randomized controlled trials have indicated that isoniazid preventive therapy also reduces the risk of subsequent tuberculosis in HIV-infected individuals with latent MTB infection. Significant hurdles in operationalizing this as a TB control measure have yet to be overcome.

DOTS strategy for TB control

The objectives of TB control are to reduce mortality, morbidity and disease transmission and to prevent the development of drug resistance. The strategy recommended to meet these objectives is to provide standardized short-course chemotherapy under direct observation at least during the initial phase of treatment to, at least, all identified smear-positive TB cases (the sources of infection). The success of this strategy depends on the implementation of a five-point package.

1 *Direct* smear microscopy for case detection among symptomatic patients self-reporting to health services.

2 *Observation* of therapy for administration of standardized short-course chemotherapy to ensure adherence.

3 *Treatment* monitoring through a standardized recording and reporting system allowing continuous assessment of treatment results.

4 *Short course* chemotherapy through a system of regular drug supply of all essential antituberculosis drugs, which should be free to patients at the point of delivery.

5 Government to ensure a sustained approach to policy and funding.

Tuberculosis control activities should aim to meet two internationally agreed targets: a cure rate of 85% and a case detection rate of 70%. Greatest emphasis has been placed on the cure rate target because achieving this will lead to less acquired drug resistance, which makes future treatment of TB easier and more affordable. The standardized outcome reporting categories described above make it possible to conduct quarterly cohort analyses of treatment outcomes and hence to report cure rates. In recent years, there has been considerable debate over the extent to which direct observation of therapy (DOT) is required to ensure patient adherence to therapy. In its purest form DOT (distinct from the five-point DOTS policy package) dictates that a health care worker should hold a given patient's TB treatment so that the patient can swallow every dose under the watchful eye of the health care worker—at least for the intensive phase. Many now accept that this system does not always make it easy for patients to adhere to every prescribed dose of treatment. If, for example, patients live far from the health care worker, they will inevitably incur considerable direct and opportunity costs in the daily travel required. Many innovative approaches to DOT have been piloted, including DOT by grocery store keepers in rural areas, workplace DOT and DOT by respected family members. The guiding principle is to make it as easy as possible for patients to stick with therapy for the full course.

Although TB transmission is reduced every time a smear-positive case is cured, overall reductions in prevalence and incidence depend on progress towards the 70% case detection target and this has proved more problematic than the cure rate target. Estimating the expected number of infectious cases in a given population conventionally relies on extrapolation from an estimate of the annual risk of infection (ARI). This is usually derived from PPD skin test surveys, but the reliability of these is diminished where HIV prevalence and/or BCG coverage is high. In addition, the relationship between the ARI and the expected number of cases where HIV prevalence is high is not known. A rough guide, in the absence of HIV, is to expect 50 infectious (smear-positive) cases each year from every 100 000 head of population when the ARI is 1%. In most developing countries in the tropics the ARI is likely to be 2% (or more).

If reliable estimates of expected numbers of infectious cases are required, then full-scale community-based prevalence surveys are required. These are expensive and time-consuming and most countries have not had the resources to undertake them. Overall, most countries do not have reliable data against which to measure their progress towards the case detection target.

The process of case detection needs close

attention, especially if it is to work for those who need it most, i.e. the poor. Smear microscopy is central in the DOTS strategy: no patients should start short-course chemotherapy unless they have had their sputum checked. The recommended approach is termed 'passive' case-finding; patients who are prompted by their symptoms to present to health facilities are then encouraged to submit a total of three sputum specimens. The process of sputum submission takes the patient a minimum of 24 h and three visits to the health facility followed by another visit to collect the results and, usually, one further visit to start treatment. Once again these repeat visits may be impossible for particularly disadvantaged patients such as the poor and women in some traditional cultures. It is important therefore to make adequately staffed quality-assured microscopy facilities as accessible as possible in populations with high TB incidence. In some instances this means establishing extra laboratory services (but no more than one such facility for 100 000 people, as the volume of work is then not sufficient for staff to retain skills). In other instances this means ensuring excellent logistics for transporting specimens and results between microscopy centres and collection sites.

There is increasing interest in improving case detection including moves to re-examine the potential role of 'active' case-finding for certain high-risk populations. Mass screening of populations by radiography has not been thought cost-effective relative to opportunist 'passive' case-finding as described above but it, or related approaches (sometimes called intensified case-finding), may be appropriate in particular situations.

Future developments

New diagnostics

There is clearly a need for a new diagnostic tool that is as robust and specific as smear microscopy, less dependent on laboratory skill and infrastructure, more sensitive and more immediate in read-out. To date there are no obvious candidates for widespread use in developing countries.

New interventions

No new drugs look set to replace the current gold standard 6-month short-course chemotherapy and there is no obvious replacement for the BCG vaccine. Nevertheless, the recent introduction of three- and four-drug fixed-dose combinations to supplement the existing two-dose combinations of rifampicin and isoniazid should make it easier for patients to adhere and harder for drug resistance to emerge. Trials of M. vaccae as an adjunct to chemotherapy have been disappointing.

Multidrug-resistant TB

Wherever TB chemotherapy is delivered in a chaotic manner, drug resistance develops and some organisms accumulate resistance to successive drugs. When resistance to both isoniazid and rifampicin are detected in vitro, the isolate is categorized as multidrug resistant (MDR). Patients with disease caused by such organisms are unlikely to be cured by first-line regimens in Categories I, II or III. Second-line regimens are being piloted through 'DOTS-Plus' pilot schemes around the world, but the core message is to work hard not to allow drug resistance to develop in the first place.

HIV and TB

One of the challenges facing developing countries in the tropics is the management of TB in populations with high HIV prevalence. HIV-infected patients who develop TB are, on the whole, more difficult to diagnose in a timely fashion. Although their TB seems curable using short-course chemotherapy, they are at increased risk of dying during and after TB therapy. This mortality appears to be multifactorial and includes late initiation of treatment. This may be partly a result of the difficulties of diagnosis but stigma and other factors also contribute. Intercurrent HIV-related complications such as bacterial and fungal infections also contribute to mortality. One randomized controlled trial in West Africa has suggested that administration of cotrimoxazole prophylaxis concurrently with short-course chemotherapy can help reduce mortality in HIV-infected patients with TB. This may be one medical intervention that could go some way to improving TB care for HIV-infected patients while concurrent use of anti-TB drugs and antiretroviral therapy remains so complicated. Either way, it must be emphasized that the most significant gains for HIV-infected patients with TB in the tropics are likely to be found through strengthening core services resulting in timely diagnosis of and careful clinical care for TB patients.

Further reading

Cochrane Library 2002, Issue 4, ISSN 1464–780X. [Updated every 6 months, this is available via www.nelh.nhs.uk/cochrane.asp from computer sites registered within the UK. This includes systemic reviews on DOT, use of isoniazid preventative therapy and use of adjunctive steroid therapy.]

Enarson DA, Rieder HL, Arnadottir T, Trébucq A. *Management of Tuberculosis: A Guide for Low Income Countries*, 5th edn. 2000. [Available from International Union Against Tuberculosis and Lung Disease, 68 Boulevard Saint-Michel, 75006 Paris, France. See www.iuatld.org. Examples of model reporting forms are included, as well as a technical guide to sputum smear microscopy.]

WHO Stop TB Department. *Treatment of Tuberculosis: Guidelines for National TB Programmes*, 3rd edn. 2003. WHO/CDS/TB/2003.313. [Available from the World Health Organization, Via Appia, Geneva, Switzerland and downloadable from www.who.int/gtb a website that includes details of drug therapy as well as a full description of the DOTS strategy.]

CHAPTER 13

HIV Infection and Disease in the Tropics

The viruses, 98
Differences between HIV-1 and
   HIV-2, 98
Testing for HIV, 99
Epidemiology, 100

Transmission, 100
Control strategies, 102
Mechanisms of disease, 103
Staging HIV disease, 104
Natural history, 104

Clinical problems, 105
Prophylaxis and antiretroviral
   therapy, 109
Further reading, 110

Over 90% of new human immunodeficiency virus (HIV) infections now occur in developing countries. Wherever HIV is prevalent, hospitals and clinics are faced with a large and escalating burden of disease. HIV and consequent acquired immune deficiency syndrome (AIDS) is now the leading cause of death in sub-Saharan Africa and the fourth biggest killer worldwide, making HIV a major tropical disease.

Specific guidelines, current seroprevalence data, epidemiological reviews, summaries, reports and updates are published regularly by the Global Programme on AIDS (GPA) which is part of the World Health Organization (WHO). These are available (usually free of charge) from the GPA Document Centre, WHO, 1211 Geneva 27, Switzerland. Also the WHO and UNAIDS have comprehensive websites (see Further reading).

Most countries have national AIDS programmes which provide help and management protocols relevant to local conditions and resources.

The viruses

There are two distinct types, HIV-1 and HIV-2, as well as several closely allied species. These two viruses originated in the chimpanzee and sooty mangabey, respectively. It is likely that HIV-1 and HIV-2 infected humans as a result of these primates being hunted for food. This probably happened many times before the social and environmental conditions were present to spark off the HIV epidemic. Theories linking the introduction of HIV into humans via polio vaccination campaigns in Africa in the 1950s have been disproved.

Differences between HIV-1 and HIV-2

There are clear differences between HIV-1 and HIV-2 in genomic structure and in the antibody response to infection. Although not always easy, the two infections can usually be separated serologically. HIV-1 is rapidly spreading round the world and is universally distributed, whereas HIV-2 is much less common and largely restricted to West Africa.

The two viruses are transmitted in the same way but HIV-2 seems less transmissible. Where HIV-1 and HIV-2 coexist, HIV-1 infection is rapidly overtaking HIV-2 in prevalence. Dual infections can occur and there is no evidence that one infection protects against the other. Both viruses cause the same immune defects and are associated with a similar disease. HIV-2 takes several years longer than HIV-1 to cause significant immunosuppression or death, and diseases such as Kaposi's sarcoma do not usually occur in HIV-2 infected individuals. The main importance of HIV-2 in areas where it is prevalent is to ensure that the kits used for blood tests can detect both viruses. Differentiation is unimportant for individual patient management because treatment is not altered.

The rest of the chapter refers to HIV without differentiating between the two types.

Testing for HIV

Whom to test?

It is important to have a local policy on HIV testing that reflects local needs, resources and conditions. As a minimum, all blood donors should be tested each time blood is donated, and the hospital should always have sufficient kits to provide this minimum service.

In the past, health care professionals in developing countries may have taken a nihilistic view towards HIV testing, as there was nothing to offer the individual with a positive result. Now, with wider availability of primary prophylaxis, specific protocols for management of common problems in HIV-infected individuals, vertical transmission prevention programmes and, in some cases, antiretroviral therapy the balance has changed.

The starting point should be voluntary testing and counselling (VTC) facilities. These should be easily accessible and provide a confidential service. They should be staffed by trained counsellors who can help clients decide on the pros and cons of accepting a test. Certain groups can be targeted, such as commercial sex workers and their clients. VTC facilities also act as centres for education on preventing HIV infection. Pregnant women should be encouraged to be tested to allow them to make informed decisions about this and future pregnancies. In many areas nevirapine is available to help prevent mother–child transmission. Testing critically ill inpatients has little value and a test should only be considered if it is going to alter management or provide prognostic information for the individual. It is unnecessary to test all TB patients unless different drug regimens are used, e.g. avoiding thioacetazone, which can

cause serious skin reactions in HIV-infected patients. Confidentiality is difficult to safeguard when HIV-negative patients receive one therapy and HIV-positive patients another.

Clinical AIDS surveillance alone will provide a falsely low estimate of the impact of HIV disease unless supplemented by HIV testing of blood. All monitoring can be carried out on stored samples and so can be undertaken without disruption to services.

What test to use?

Some tests are specific for HIV-1 or HIV-2, whereas others can identify both types. All are highly specific and sensitive if the manufacturer's guidelines are followed correctly; the kits are as accurate as the laboratories using them.

The most widely used tests identify specific anti-HIV antibodies. This can be done by the ELISA method, a variety of rapid and/or simple colorimetric or agglutination tests that do not require a laboratory or electricity, and the Western blot. There is no single test that is suitable for all circumstances. ELISA testing is best suited for regular processing of large numbers of samples so that complete plates (usually 90 samples) can be run. It is unsuitable for laboratories with limited facilities and fridge space, irregular electricity supply and inadequately trained or supervised technical staff. With bulk purchasing, a single ELISA test costs around £0.60 (2002 prices).

Rapid and simple tests, such as particle agglutination or dot immunoassay tests, can be performed in less sophisticated laboratories. Many are available as single kits and so can be efficiently used when small numbers of samples need testing. Some can be stored safely at room temperature. With central purchasing, unit costs can be kept quite low, depending on the kit.

Western blots are expensive (over £20), can be difficult to interpret and standardize, and are merely serology tests elongated on a piece of blotting paper. The WHO no longer recommend Western blotting for confirmation, suggesting instead that if it is necessary then the combination of ELISA with a simple or rapid assay is as reliable and much cheaper.

With antiretroviral therapy likely to become available in developing countries, tests for monitoring will be needed. Viral load assays are prohibitively expensive at around £25 per test, as are CD4 counts; cheaper alternatives are being evaluated.

GPA issues regular updates on HIV test kits for use in developing countries.

Epidemiology

Surveillance

Surveillance is carried out in order to monitor the extent of HIV infection and disease in a given region or community. In poor tropical countries, few resources are available for epidemiological monitoring. At the start of the HIV epidemic in Africa, surveillance was only able to show gross changes and to monitor relatively crudely the arrival and subsequent spread of infection. With experience and institutional strengthening, surveillance is now much more accurate.

Specific at-risk groups include female sex-workers (prostitutes), clients at sexually transmitted disease (STD) clinics, workers such as migrant labourers and long-distance lorry drivers, and intravenous drug users. It is important to monitor these groups as they often drive the epidemic in certain areas. Groups more representative of the general population include pregnant women attending for antenatal care, blood donors, military recruits and newborn infants, although all these groups have inherent biases. Ideally, a population-based sample should be tested at regular intervals to accurately determine prevalence and incidence for a particular region. This provides the most valuable information but population-based testing is difficult and expensive to organize. Also it may not be acceptable to the local population as it involves testing people in their homes and collecting a lot of demographic information that individuals may find intrusive. It is important as a minimum to record age and sex in all surveys.

Disease surveillance is usually carried out in hospital and often concentrates on counting cases of AIDS as defined by the WHO in the provisional clinical case definition for Africa. HIV seroprevalence can be measured in specific groups such as hospital admissions, adults with active TB or pneumonia or cadavers in the hospital or district mortuary.

Seroprevalence

With rapid spread of infection (and delays in reporting and analysis) current figures quickly become out of date. By the end of 2001, WHO estimated 40 million people were living with HIV/AIDS worldwide. During the same period around 5 million new infections occurred including 800 000 children. About 3 million people died of HIV/AIDS in 2001. Seventy per cent of people living with HIV/AIDS reside in sub-Saharan Africa. Sub-Saharan Africa is the hardest hit by HIV/AIDS, accounting for 68% of all new infections and 77% of all AIDS deaths in 2001. Table 13.1 shows the estimated number of cases by region.

Some of the largest increases in prevalence are occurring in former Eastern block countries, largely driven by injecting drug users. Latin America has seen a drop in mortality, partly because of the introduction of low-cost antiretrovirals in some countries such as Brazil. This may act as a model for other areas, although it is sobering to note that in North America and western Europe, where mortality has dropped considerably since the widespread use of antiretroviral therapy, the incidence of new infections has been stable or is increasing.

Transmission

Sexual transmission

In nearly all tropical countries the most important way HIV is transmitted is by heterosexual sex. The risk of acquiring HIV after sexual intercourse is hard to quantify. There are numerous factors such as the age and sex (young women seem more susceptible, perhaps because of an immature genital tract) of the individual as well as the factors mentioned below. However, it is

ESTIMATED POPULATION WITH HIV/AIDS

World region	Estimated number of people with HIV/AIDS in millions (percentage)
Sub-Saharan Africa	28.1 (70%)
South and South East Asia	6.1 (15%)
Latin America	1.4 (3.5%)
East Asia and Pacific	1 (2.5%)
Eastern Europe and Central Asia	1 (2.5%)
North America	0.94 (2.3%)
Western Europe	0.56 (1.4%)
North Africa and Middle East	0.44 (1%)
Caribbean	0.42 (1%)
Australia and New Zealand	0.015 (0.04%)

Table 13.1 Estimated number of people with HIV/AIDS at end of 2001 by region.

thought that the risk is between 0.01 and 0.1 per sexual act.

Risk factors

There are several factors that markedly increase the risk of transmission, the most important being other STDs that cause ulceration, chancroid (Haemophilus ducreyi) in particular, as well as primary syphilis and genital herpes simplex. STDs that cause inflammation and discharge, such as gonorrhoea, Chlamydia and perhaps trichomoniasis also increase the chance of sexual transmission of HIV. In fact, anything that disrupts the vagina's normal flora may cause increased risk of transmission. A recent study using nonoxinol '9', a vaginal spermicide, actually increased the incidence of HIV infection amongst the women using it. Cervical erosion may be a risk factor in women.

For a man, being uncircumcised may be a factor that increases risk; it certainly increases the risk of acquiring other STDs. A recent review (see Further reading) showed that the preva-

lence of HIV infection was much lower in areas where the men practised circumcision.

Vertical transmission

In the developing world, around 20–30% of children born to HIV-infected mothers will themselves become infected. Ten per cent will be infected transplacentally, 5% during delivery and around 10% as a result of breast-feeding. Factors associated with transplacental transmission include later stage of infection of the mother and high viral load. During birth the presence of an STD and prolonged labour will increase transmission. Breast-feeding does lead to HIV transmission and in the West formula feeding is recommended to all HIV-infected mothers. In resource-poor settings, formula feeding is not practical. Even when formula is provided free, adequate facilities for its sterile preparation are rarely available. The benefits of breast milk in reducing infant mortality from pneumonia and diarrhoea probably outweigh the risk of HIV infection. There have been some confusing messages recently over whether to recommend breast-feeding to HIV-infected mothers in developing countries and about what sort of breast-feeding. Some studies have shown that exclusive breast-feeding carries a reduced risk of HIV transmission compared to mixed feeding, which is commonly practised in Africa. Until definitive studies are completed or viable alternatives become available it is best to counsel the mother realistically on what safe options are available; in most of Africa this will mean breast-feeding.

Transmission by infected blood

Two groups are at particular risk: patients receiving blood transfusions and people who inject drugs and share needles and syringes. Improperly sterilized injection equipment in hospitals and other health facilities is another (unquantifiable) risk.

Screening blood donors for HIV has greatly reduced the chance of HIV transmission through transfusion. Errors can occur in HIV testing and some donors may be in the 'window' phase with an acute infection that is not yet serologically recognizable. This can be a problem in areas of high incidence.

There is a small risk to health care workers exposed to HIV-infected blood. In a typical needlestick injury, where the skin is punctured but the inoculum is small, the risk of acquiring HIV is about 0.3%.

Control strategies

Control strategies need to be targeted at specific groups. Condom promotion and distribution is important for preventing new infections both in the community as a whole, as well as in sex workers and their clients who can be a core of HIV transmission in an area. Condoms are proven to be effective in preventing most STDs. They may not always be culturally acceptable and they rely on the male partner's cooperation. Female condoms are also effective but are cumbersome to use and may not be accepted by some partners.

Treatment of STDs has been shown to reduce STD transmission in some studies. There is debate over the likely benefits of mass STD treatment campaigns; many areas are using syndromic management to simplify STD treatment and ensure that the important diseases are treated in a single visit.

Targeting core groups such as commercial sex workers and their clients is important but education for school children is also a high priority. Previously, sex education was not widespread in Africa and girls would not find out about contraception or STDs until they were attending antenatal programmes. Sex education and HIV awareness has to be tailored to the target audience as Western-style campaigns and materials are often inappropriate for developing countries.

Ensuring a safe blood supply is an important minimum standard in HIV prevention.

VTC can act as an entry point to HIV services and as place for disseminating information about HIV prevention and risk reduction for those found to be HIV infected. Women found to be infected can be counselled about future

pregnancies and pregnant women can also be informed of ways to reduce transmission to their unborn child. Through various donation programmes, the provision of nevirapine to reduce mother–child transmission is now becoming more common. This can be a powerful incentive for a pregnant mother to accept an HIV test. Studies have shown that nevirapine given as a single dose to the mother in labour and the child within 72 h of birth can reduce transmission by over 50%. The argument some governments have used for not implementing this strategy is that until antiretrovirals can be provided for the mother one risks creating a large group of orphans.

Antiretroviral therapy can be considered to be a method of preventing HIV transmission. By lowering an individual's viral load their chances of infecting others are reduced. Provision of antiretrovirals is being undertaken by some developing countries and certain institutions, as will be discussed latter.

Other methods of prevention include vaginal microbicides, which have had a chequered past but are being evaluated in large prospective trials, and vaccines. Currently, there are several vaccines in stage 2 clinical trials.

Cost and likely impact

There are few studies looking directly at the economic impact of HIV/AIDS on Africa. The data that are available show that HIV/AIDS is having a marked negative impact on GDP. It has been estimated that national average growth rates have been reduced by 2–4% per year across the continent.

Can antiretrovirals reverse this? Targeting specific economically active groups (e.g. miners in South Africa) may be one strategy. It is hoped that by ensuring the people who generate the country's revenue and exports are kept productive, governments will be able to continue improving education and health services in an attempt to control the epidemic.

More widespread use of antiretrovirals is beyond the budget of most African countries at present. It has been estimated that treating 25% of the HIV-infected population in sub-Saharan

Africa would prevent an overall 3-year decline in life expectancy but would cost £13 billion (see Further reading). This is realistically beyond most countries' health budgets unless the price of antiretrovirals falls substantially and the costs of monitoring and delivering the drugs can be reduced. Several large trials looking at these issues are currently underway, including the use of structured treatment interruptions.

Mechanisms of disease

Pathogenesis of HIV infection

The main cell population infected by HIV is lymphocytes that carry the CD4 antigen on their surface. This is because the CD4 molecule is acting as the receptor to which the virus can initially attach before entering the cell. CD4 lymphocytes are T-helper cells. The key concept in understanding the pathogenesis of HIV infection is the selective loss of function and progressive depletion of T-helper lymphocytes. This is a dynamic process with millions of T cells being made daily and millions being killed daily by HIV. HIV slowly depletes the body's capacity to replace the killed T cells. The speed with which this process occurs can be predicted from the individual's viral load.

T-helper lymphocytes have an important role in the regulation of the cell-mediated immune response and also cooperate with B cells in the production of antibody. Loss of CD4 cells by HIV infection disrupts both cell-mediated and humoral immunity. This is shown by the loss of delayed hypersensitivity to such skin test (recall) antigens as purified protein derivatives (PPD) or tuberculin, Candida and mumps antigen; and by polyclonal B-cell activation with hypergammaglobulinaemia.

Other cell populations can be infected including macrophages, which may be important reservoirs of HIV outside the blood and may carry HIV to different organs, including the CNS. Cytokine secretion by infected macrophages is aberrant and may have a role in chronic fever, wasting and enteropathy. Active replication of HIV is evident in lymph nodes at all stages of infection and B cells may be non-specifically activated.

Different strains of HIV may differ in virulence in the cell types that are preferentially infected. A single HIV infection can generate many different antigenic variants; it is possible that in the course of infection variants will eventually emerge that can escape an increasingly exhausted immune system.

Progressive immunosuppression

With the progressive destruction of one part of the immune system, a distinct form of immunosuppression develops. As with other immune deficiency syndromes, a relatively limited number of organisms are able to exploit the specific immune defect and commonly cause disease in seropositive individuals.

It is unclear why some pathogens are so characteristic of HIV and others not. Different pathogens characteristically occur in the early and later stages of HIV disease. A few conventional pathogens cause clinical disease in the early as well as the later stages of HIV disease; Mycobacterium tuberculosis, Streptococcus pneumoniae, non-typhi Salmonella (NTS) and the varicella-zoster virus are the most important. Disease is caused by acute infection or by reactivation of a dormant focus.

In the early stages of HIV, when immune function is relatively preserved, the main abnormality is a much higher attack rate with clinically typical disease presentation and a normal response to standard treatment. In later stages of HIV and in AIDS itself clinical presentations become atypical and there is a diminished response to standard therapy.

Opportunistic pathogens (relatively avirulent organisms that only usually cause disease in individuals with disrupted immune systems) are only seen in the later stages of HIV disease when much more severe immunosuppression has developed. Immune surveillance is also abnormal and specific cancers can develop.

The opportunistic infections that dominate clinical practice in industrialized countries are Pneumocystis pneumonia, disseminated M. avium, cytomegalovirus retinitis and toxoplasmosis. The malignancies are non-Hodgkin's lymphoma, primary CNS lymphoma and Kaposi's sarcoma. Cervical neoplasia is also more common in HIV-infected women. These unusual infections and malignancies are so characteristic of (late) HIV disease that they can be grouped together and used to define AIDS.

Staging HIV disease

CD4 lymphocytes are found in peripheral blood and the normal count in a seronegative person is about 1000×10^6 CD4 cells/L. The absolute CD4 count is a useful way of staging HIV infection and assessing the immune status of a patient, but as CD4 counts are seldom available in the tropics they will not be discussed further here.

WHO have proposed a clinical staging system for HIV disease. The four categories are:
Stage 1: Asymptomatic or persistent generalized lymphadenopathy (PGL).
Stage 2: Mild disease with minor weight loss, minor mucocutaneous manifestations, one episode of herpes zoster or recurrent upper respiratory tract infection.
Stage 3: Intermediate disease with weight loss of more than 10% of body weight, unexplained chronic diarrhoea or fever, oral *Candida*, leukoplakia, pulmonary tuberculosis or severe bacterial infection.
Stage 4: Severe disease equivalent to AIDS.

This system may be of some use for individual prognosis but is of little help in clinical management.

Natural history

The following account of illness caused by HIV infection highlights those areas where the illness in the tropics differs from that reported in North America and Europe.

Acute seroconversion illness
Initially in acute infection most adults develop a high viraemia and a marked fall in CD4 count; they are highly infectious. During seroconversion, neutralizing antibodies appear and viraemia is greatly reduced. A minority of adults experience an acute seroconversion illness which resembles glandular fever. A wide range of skin rashes are seen and with transient CD4 depletion oral *Candida* and even opportunistic infections can occur. Because of the relatively non-specific features, the seroconversion illness is seldom recognized in tropical countries.

The latent phase
After seroconversion the CD4 count rapidly rises to near normal and the individual feels well. Active viral replication is taking place in the reticuloendothelial system during this time and viral load assays can be used to quantify this. Lymph node architecture is progressively disrupted and nearly 50% of adults develop PGL; this has no prognostic significance.

With time the CD4 count falls and immunosuppression slowly but inevitably progresses. The rate of progression is extremely variable but can be predicted from the viral load. It is during this latent infection that most onward transmission of HIV takes place. Latency can be a very difficult concept to convey in counselling.

Early HIV disease
In the relatively early stage of HIV infection, when CD4 counts are only moderately reduced, many individuals start to experience specific symptoms of HIV or develop a disease typical of early HIV infection.

The HIV symptoms can be weight loss, night sweats, pruritic skin rash, unexplained fever or chronic diarrhoea. Early HIV disease can be: relatively trivial, such as oral *Candida* or oral hairy leukoplakia; painful and disabling but not life-threatening, such as herpes zoster; or life-threatening bacterial or mycobacterial infections. This stage of disease is sometimes referred to as AIDS-related complex (ARC).

The serious early HIV diseases are pneumococcal infection (pneumonia and sinusitis), TB (pulmonary and lymphatic) and NTS infections (often bacteraemic). In general, clinical presentation is straightforward and the response to therapy good.

In most tropical regions early disease predominates, whereas it is of relatively minor importance in industrialized countries. This is because of the much higher exposure in poor, overcrowded tropical communities to respiratory and diarrhoeal pathogens.

Late HIV disease or AIDS

The important AIDS-defining opportunistic infections in tropical countries are cryptosporidiosis and isosporiasis in the bowel and cryptococcosis and toxoplasmosis in the CNS.

Extrapulmonary and disseminated TB (Fig. 13.1, see colour plate facing p. 146), severe bacteraemic pneumococcal disease and disseminated salmonellosis are all common. Other Gram-negative septicaemia, especially *Escherichia coli*, are increasingly being recognized. Mixed infections are also frequent.

Early HIV disease is more common than late disease or AIDS in many tropical communities. Much early HIV disease, such as lobar pneumonia or pulmonary TB, is clinically typical and is only recognized as associated with HIV by serological testing. Studies that just focus on 'clinical AIDS' will fail to identify nearly all early HIV disease.

There are two reasons why early HIV disease predominates. There is intense exposure to TB, pneumococci and salmonellae and few patients are surviving long enough to develop marked immunosuppression and late disease or AIDS. High mortality occurs in the early stages of HIV disease from clinical problems that are easily treated and cured in centres with better facilities and resources. The time from seroconversion to death—usually 10–12 years in developed countries—may be as short as 4–5 years in the tropics. Although there are no reliable data to show there that is a big difference between the time from acquiring HIV to death in developing countries, it is now thought that untreated survival times may be similar.

Clinical problems

This section describes the common and important clinical presentations of HIV disease in the tropics and discusses simple investigations, management and treatment. It is assumed that therapy is limited to a range of cheap broad-spectrum antimicrobials, and that antibiotic sensitivity testing is not routinely available.

Skin problems

Skin problems are common in HIV-positive patients and cause considerable discomfort and morbidity.

Generalized pruritus

Many adults develop a chronic itchy maculopapular rash. In dark skins, many lesions become hyperpigmented and even nodular. Some cases are caused by scabies, when treatment with benzyl benzoate or lindane is effective; it may need to be repeated and should include family members. The patient should be instructed to wash in warm water if possible then apply the benzyl benzoate and repeat the treatment a week later. In those who do not respond, antihistamines such as chlorphenamine are of limited use but may restore normal sleep. A topical corticosteroid sometimes eases itching.

Herpes zoster

Shingles is extremely common and is often the first recognizable HIV-related problem. Where HIV disease is widespread, lay people often recognize the implications of a zoster eruption and realize they are probably HIV-infected. Zoster affecting more than one dermatome is virtually diagnostic of HIV.

Adequate analgesia is important: codeine phosphate 30–60 mg every 6 h is usually effective. Persistant pain may respond to amitriptyline. Daily dressing may be necessary. Aciclovir is too expensive for poor countries.

Skin infections

Minor wounds can develop into deep-seated

and necrotic lesions which may cause bacter-aemia. Poor wound healing or unusual skin in-fections can indicate underlying HIV infection. If antimicrobials are necessary, the best choice is a penicillinase-resistant penicillin such as flu-cloxacillin. Erythromycin or chloramphenicol may also be used.

Fungal nail infections and extensive tinea cor-poris, pedis and cruris are common. These may respond to a topical antifungal cream but often require an oral agent such as triazoles or griseo-fulvin, which may not be available.

Acute cough and fever

Acute respiratory infections are amongst the most important clinical problems that occur in HIV-infected adults in the tropics. Most patients who present with acute cough and fever have community-acquired bacterial pneumonia. This has always been an important disease in the tropics but HIV has dramatically increased the incidence of bacterial pneumonia. *Pneumocystis carinii* pneumonia is uncommon in adults but is now recognized as a problem in HIV-infected infants.

There is usually a short history of cough, spu-tum (purulent, rusty or blood-stained), pleurit-ic pain, fever and rigors, marked toxicity and shortness of breath. Many cases will have had a previous episode. Examination reveals a sick and toxic patient, often lying on the side of the pneumonia. The patient is sweaty, taking rapid shallow breaths and coughing frequently. Some patients may be mildly jaundiced, have meningeal irritation, confusion and be shocked. Most will have lobar consolidation with bronchial breathing.

Nearly all infections are pneumococcal. *Haemophilus influenzae* is sometimes isolated but, in a series from Nairobi, this was usually coinfecting a primary pneumococcal pneumo-nia. Mixed bacterial and mycobacterial infec-tions are relatively common in HIV disease.

Pneumonia can be confirmed by chest X-ray but this is wasteful if physical signs are definite. Gram-staining of sputum is useful. Pus cells con-firm that a good specimen has been obtained; diplococci should be abundant. Treatment is the same as for HIV-negative patients: benzylpeni-cillin 0.6–1.2g parenterally 6 hourly for 5–7 days. An early switch to oral therapy is easier for the patient. If the patient is cyanosed, oxygen (if available) should be given by face mask; if shocked, intravenous fluids are vital.

The prognosis in HIV-related pneumonia is usually excellent. When patients present late with extensive disease mortality is significantly higher. If the patient is not responding there are several possibilities. There may be coinfection with NTS, *Haemophilus* or TB; an empyema is forming (especially with inadequate initial therapy); or the initial diagnosis is wrong and the patient has pulmonary TB. Changing to chloramphenicol or starting antituberculous therapy may be indicated. As *Pneumocystis* is un-common in most areas and many salmonellae are resistant, cotrimoxazole is not recom-mended. If an empyema is identified it must be drained. Remember that a liver abscess can present with fever and cough. It may be identi-fied on ultrasound if available or by noting a raised hemidiaphragm. These will most com-monly be amoebic abscesses and will respond to metronidazole.

Chronic cough with fever

Chronic respiratory problems are common and highly associated with underlying HIV infection. Most patients have pulmonary TB. The main dif-ferential diagnosis is recurrent or partly treated bacterial pneumonia, which is very common in late-stage patients. Pulmonary Kaposi's sar-coma can occur with skin lesions that are usual-ly obvious although there may only be palatal lesions, which carry a very poor prognosis.

Chronic cough and fever should be easy symptoms to identify; many patients will have weight loss, night sweats, weakness and haemoptysis; lower-lobe disease is frequent with widespread crepitations; effusions are much more common in HIV. TB frequently re-curs so some patients will have had adequate previous therapy and represent with sputum-positive disease within a few months.

The diagnosis of TB is more difficult to con-firm by radiology or microscopy when there is underlying HIV infection. Classic upper-lobe cavitary disease is much less common and lower-lobe consolidation is more frequent. Fewer cases are smear-positive for acid-fast bacilli. TB culture is of little help because the result is so delayed. Prior to HIV, patients could be left untreated with smear-negative disease because they were a negligible public health risk for onward transmission and would usually return a few months later with smear-positive disease. This cannot be done with HIV-TB and many patients need to be started on therapy on clinical suspicion alone. National treatment guidelines should be followed. Patients with TB/HIV-related disease respond well to short or standard-course therapy, unless they have end-stage overwhelming disseminated TB.

If bacterial pneumonia is suspected, a therapeutic trial of benzylpenicillin or ampicillin should be started. If a good response is seen then TB can be excluded. If not, TB therapy should then be initiated.

Thioacetazone toxicity frequently occurs in HIV-positive patients. Some 10–20% may develop skin reactions, usually erythema multiforme, and some progress to Stevens–Johnson syndrome. Because of this, national treatment guidelines have been changed in several countries and thioacetazone has been withdrawn, usually replaced by ethambutol. TB cases can be offered screening and those who are HIV-positive can be given ethambutol. If there is a limited supply of ethambutol, it can be kept for patients who have started to react. The drug reaction starts with return of fever, itchy skin, non-specific rash then target lesions, usually within 6 weeks of commencing treatment. Stopping thioacetazone at this point will usually prevent severe Stevens–Johnson syndrome from emerging. The main challenge in therapy is to ensure good compliance. Most HIV cases have relatively few organisms in the sputum and are therefore only slightly infectious. A secondary epidemic of TB from the new HIV-associated cases is therefore unlikely.

High fever without focus

Fever is a frequent symptom in HIV-infected in-

dividuals. While some patients seem to have no obvious cause for the fever, fever usually indicates infection. Careful history and examination can sometimes reveal a focus, especially in the CNS, joints or soft tissue (pyomyositis is relatively common and requires drainage); or a cardiac murmur in an intravenous drug-user suggests endocarditis. Patients often have chronic middle ear disease and sinusitis, which may be the cause of the fever. Patients can have chronic symptoms such as diarrhoea, dry cough or skin lesions and then acutely develop a high swinging fever but no additional focus.

Malaria must always be excluded. Usually a high fever without focus in the tropics in a patient with underlying HIV indicates a high-grade bacterial or mycobacterial infection. Again, remember the possibility of amoebic liver abscess.

In HIV-infected patients this clinical presentation is sometimes referred to as an enteric fever-like illness and is very common. NTS rather than Salmonella typhi are important. Disseminated TB is increasingly recognized but M. avium is rare. Without blood culture, salmonella bacteraemia cannot be reliably diagnosed. In Africa about 10% of all HIV-positive adults presenting to hospital will have NTS bacteraemia and a further 5% may have other Gram-negative sepsis (including E. coli and classical S. typhi). Disseminated TB will only be diagnosed postmortem.

The next problem is how to differentiate the two, and this cannot be done clinically. A positive blood culture will obviously identify Gram-negative sepsis but a negative culture can occur if antibiotic therapy has recently been given, or it may indicate disseminated TB. Many cases of disseminated TB are anergic with minimal pulmonary lesions. Mycobacterium tuberculosis can take weeks to be positive in a blood culture, unlike M. avium.

How can these sick patients be managed without any microbiology support? Experience is that, with awareness of the possibility of NTS bacteraemia, prompt broad-spectrum antibiotic therapy backed up by intravenous fluids can reduce an 80% mortality to about 25%.

Knowing the antimicrobial sensitivity pattern can further reduce mortality to about 15%. In marked contrast, very few HIV-positive patients with *M. tuberculosis* mycobacteraemia and none with *M. avium* in Africa usually survive, even with the prompt use of quadruple therapy. Disseminated mycobacterial disease seems to be an agonal problem in Africa, and may be so in other tropical regions.

This suggests the following strategy. For any patient with an enteric fever-like illness, exclude a silent pulmonary/pleural/pericardial/abdominal/renal TB focus by a chest X-ray, urine microscopy and, if possible, ultrasound. Then concentrate on the treatment of NTS bacteraemia. First-line blind therapy can be ampicillin and gentamicin, or chloramphenicol. A quinolone would be the best choice but may not be available or affordable, but remember that oral ciprofloxacin has equivalent bioavailabilty to many intravenous antibiotics. If there is little or no improvement after 3 or 4 days (like typhoid itself, NTS bacteraemia can take several days to respond), drug resistance may be a problem. Switch to whichever first-line treatment was not used initially.

Chronic diarrhoea

'Slim disease' is what many people equate with African AIDS. It was the first clinical problem specifically associated with HIV infection by Ugandan investigators in 1985, and named by the local patients and their carers, who recognized a major new disease in their community.

Chronic diarrhoea with profound wasting is easy to identify and is highly associated with underlying HIV infection. Widespread metastatic disease, advanced TB (consumption), Addison's disease and untreated insulin-dependent diabetes present with wasting but not usually diarrhoea. While diarrhoea and wasting are the most obvious problems on the wards in Africa, if comprehensive studies of hospital admissions are carried out they account for only 10–20% of the HIV-related workload. In some hospitals the number of 'slim' patients is dropping as carers recognize that hospitals have little to offer.

Home care is increasingly seen as a more appropriate option.

The diarrhoea is usually painless, watery and without blood or mucus. It can be variable or intermittent. Profound weight loss is clinically obvious and may be 20% or more of the premorbid weight. High fever is not typical and should suggest NTS septicaemia or agonal disseminated TB.

Studies from various African centres have shown a variety of stool pathogens: about 15% will have *Cryptosporidium*, 15% enteropathic bacteria, 10% *Isospora* and 10% widespread TB with faecal mycobacteria. Stool viruses may be found in some but microsporidium is probably rare. Amoebae, *Giardia* and helminths are not important. No pathogen is identified in many cases. The hyperinfection syndrome with *Strongyloides stercoralis* is not associated with HIV infection.

Specific treatment is very limited. Some salmonellae, *Shigella* and *Isospora* will respond to high-dose cotrimoxazole, so a trial of therapy may be indicated. Blind treatment with albendazole has also been shown to reduce diarrhoea but is very expensive. There is no effective treatment for cryptosporidium and disseminated TB is unlikely to respond. Codeine phosphate 30 mg four times daily may offer some symptomatic relief. It is now almost universally acknowledged that 'slim' disease is best managed by a home-care programme.

Other specific problems

Many patients have painful oral *Candida*. Antifungal agents such as nystatin are effective where available but expensive; gentian violet mouthwashes may give some relief.

Oesphageal *Candida* can be treated with fluconazole, which is now available free in some African countries via a donation programme.

Kaposi's sarcoma (Fig. 13.2, see colour plate facing p. 146)

Only 4% of the first 5000 reported AIDS cases in Uganda had Kaposi's sarcoma. Although the cancer is endemic in East Africa, the high rates noted in homosexual men in the USA have not been seen, although Kaposi's sarcoma is now the most common cancer in many African countries. Kaposi's sarcoma is caused by human herpesvirus 8, but in most cases immunosupression such as HIV also needs to be present for Kaposi's sarcoma to occur.

It is usually easy to diagnose clinically. Unmistakable purple or violet raised plaques and sometimes nodular lesions can occur anywhere in the skin. The roof of the mouth is a typical site. The lesions are painless and do not itch. Lymphadenopathy is common and may occur in the absence of skin plaques. Lesions can disseminate, particularly to the lungs. Biopsy is not necessary unless the diagnosis is uncertain.

Some patients with few lesions progress slowly and will die from other causes. Other patients progress rapidly over several months. Widespread pulmonary disease is an ominous sign. The drugs used for chemotherapy are expensive, and palliative only. In many cases Kaposi's sarcoma responds to antiretroviral therapy.

CNS disease

Cryptococcal disease varies in incidence across Africa. It commonly presents as a headache with little neck stiffness, there may be behavioural change easily mistaken for psychiatric illness. Where available it can be diagnosed by India ink staining of CSF although this will miss over 30% of cases. Amphotericin B is rarely available, although fluconazole may be available via donation programmes. Headache should be treated with adequate analgesia, and may also respond to repeated lumbar puncture.

Toxoplasmosis may present with a generalized encephalitis, but a relatively mild headache and an altered mental state are most common. Other signs include hemiparesis, ataxia, cranial nerve lesions, generalized incoordination, seizures and confusion. Fever is variable. Diagnosis is difficult with limited resources.

One autopsy study from West Africa found evidence of cerebral toxoplasmosis in 15% of HIV-infected cadavers and was considered a prime cause of death in 10%. However, in Kenya the experience is that clinically obvious encephalitis is rare (less than 5%).

Paediatric disease

Little work has been done on paediatric HIV disease in the tropics. It seems likely that there will be a strong association with pathogenic bacteria, particularly *Streptococcus pneumoniae*, *Haemophilus influenzae*, NTS and *M. tuberculosis*: they are already recognized as important problems in children with HIV infection in industrialized countries. One would expect higher rates of otitis media, primary TB, acute lower respiratory disease, acute and chronic diarrhoeal disease and perhaps pyogenic meningitis.

Many children with late HIV disease present with failure to thrive and malnutrition. Wasting rather than oedema is seen and the child often has an associated maculopapular or nodular rash with active excoriation (consider scabies) and generalized lymphadenopathy. The HIV-positive child will not respond to nutritional rehabilitation. *Pneumocystis carinii* pneumonia is now recognized as a relatively common cause of disease in African children. Many centres with limited resources now promote early discharge of such children because the prognosis is so poor. The mother will need counselling, even if not HIV-tested.

Prophylaxis and antiretroviral therapy

Antiviral therapy

Antiviral therapy was thought to be too expensive and too impractical to deliver in resource-poor settings until recently. There is now a realization that morally, ethically and economically it will have to be provided to the areas that are experiencing the brunt of the HIV epidemic. Drug prices are falling and in some cases companies are introducing dual pricing for developing countries.

Some countries such as Thailand and Brazil have started to manufacture their own generic antiretovirals in contravention of trade agreements. In Brazil, which is classed as a middle income country, this has led to a fall in HIV-related morbidity. Even in more resource-limited countries, antiretrovirals are being considered for certain groups of people. In Botswana, mining companies are supplying drugs to their employees and families. This makes economic sense as it is cheaper to keep a skilled employee healthy rather than have to keep training replacements. In Botswana it is proposed to supply antiretrovirals to the population at large for the same reasons. Again it has been calculated that this makes financial sense, e.g. currently teacher training colleges are admitting two people for every required teacher, as one will die of HIV during training and antiretroviral use should prevent this. Similar economic arguments are being used in Southern Africa by an increasing number of private companies and may be soon by government departments, such as the health sector where loss of nurses and doctors to HIV is taking its toll on service provision. Outside of the very high prevalence areas of Southern Africa the economic argument may be less strong.

What sort of regimen to provide and how to monitor individuals is to be the subject of a number of trials. Intermittent therapy seems attractive as it can be easier to deliver and is cheaper, but problems of resistance need to be addressed.

Prophylaxis for health care staff following accidental needle stick exposure is available in many places and is effective at reducing transmission. Following an injury, the wound should be washed and encouraged to bleed. The source patient should be tested for HIV but if the test is not immediately available there should be no delay in giving prophylaxis. In areas where antiretroviral use is not widespread, a 4-week course of zidovudine plus lamivudine is standard therapy.

TB chemoprophylaxis

A 6-month course of isoniazid has been shown to be effective chemoprophylaxis against TB. The main problems with this intervention are first proving that the patient does not have active disease and, secondly, the logistics of identifying individuals and delivering the treatment to them.

In many settings prophylaxis is only provided to children who are still being breast-fed by mothers who have sputum-positive disease.

Preventing pneumococcal disease

A recent study of 23-valent pneumococcal polysaccharide vaccine in Uganda showed a detrimental effect on survival in those receiving the vaccine. This study emphasizes the importance of testing interventions in the target population and not basing guidelines for Africa on studies conducted in the West.

Other infections

Following two studies in Abidjan, Côte d'Ivoire, cotrimoxazole prophylaxis is now recommended for people living with HIV/AIDS in Africa. Cotrimoxazole has activity against several bacteria as well as *Isospora belli, Pneumocystis carinni* and *Toxoplasma gondii.* Although the Abidjan studies showed an improvement in survival and decreased morbidity in patients given cotrimoxazole, there is debate as to whether these findings are applicable to the whole of Africa, given different resistance levels to the drug across countries. There are no NTS vaccines licensed for human use. Fluconazole available through donation programmes can be used as secondary prophylaxis for patients who have had cryptococcal disease.

Further reading

Ansary MA, Hira SK, Bayley AC, Chintn C, Nyaywa SL. *A Colour Atlas of AIDS in the Tropics.* London: Wolfe Medical Publications, 1989.

Bailey RC, Plummer FA, Moses S. Male circumcision and HIV prevention: current knowledge and future research directions. *Lancet Infect Dis* 2001; **1(4)**: 223-31. [A comprehensive review of the possible role of circumcision in HIV prevention.]

Dixon S, McDonald S, Roberts J. The impact of HIV and AIDS on Africa's economic development. *Br Med J* 2002; **324**: 232-4.

French N, Nakiyingi J, Carpenter LM et al. 23-valent pneumococcal polysaccharide vaccine in HIV-1 infected Ugandan adults: a double-blind randomized and placebo controlled trial. *Lancet* 2000; **355**: 2106-111. [A study showing that pneumococcal vaccine was associated with a worse outcome in HIV-infected patients in Africa.]

Gilks CF, Otieno LS, Brindle RJ et al. The presentation and outcome of HIV-related disease in Nairobi. Q J Med 1992; **267**: 25–32. [A description of the clinical course of HIV-infected patients in Africa.]

Hopper E. The River: A Journey Back to the Source of HIV and AIDS. Boston, MA: Little, Brown, 1999. [An account of the early cases of AIDS and some insight into how the epidemic spread.]

Mofenson LM, McIntyre JA. Advances and research directions in the prevention of mother to child HIV-1 transmission. Lancet 2000; **355**: 2237–44. [A summary of mother–child prevention studies.]

Sharp P, Bailes E, Chaundhuri EE et al. The origins of acquired immune deficiency syndrome virus: where and when? Phil Trans R Soc Lond B 2001; **356**: 867–76. [A description of the probable origins of HIV.]

Taegtmeyer M, Chebet K. Overcoming challenges to the implementation of antiretroviral therapy in Kenya. Lancet Infect Dis 2002; **2**: 51–3.

Wood E, Braitstein P, Montaner J et al. Extent to which low level use of antiretroviral treatment could curb the AIDS epidemic in Sub-Saharan Africa. Lancet 2000; **355**: 2095–100. [An economic argument for the use of antiretrovirals.]

www.unaids.org [A comprehensive website with general information, technical reports and annually updated statistics of the global HIV/AIDS pandemic.]

CHAPTER 14

Filariasis and Onchocerciasis

Onchocerciasis, 112   Filariasis, 116   Further reading, 119

Onchocerciasis

Onchocerciasis or 'river blindness' is caused by the filarial worm *Onchocerca volvulus*. It is a major cause of blindness in tropical Africa and of skin disease throughout its distribution in Africa, Yemen and Central and South America (Fig. 14.1).

Life cycle

Adult female *Onchocerca volvulus* worms are threadlike, about 40 cm long, and live in human subcutaneous tissue, sometimes coiled within fibrous nodules, where they produce living microfilariae. The microfilariae are about 350 μm long and migrate through the skin and often into the eyes. Microfilariae only develop further if they are taken up by biting blackflies of the genus *Simulium*. These flies can transmit infection after a week or more and deposit infective larvae as they bite another person. Adult *O. volvulus* live for 12 years on average but sometimes up to 17 years and microfilariae live for about 1 year.

Epidemiology

Because *Simulium* flies breed in rapidly flowing freshwater and bite nearby, people with onchocerciasis are concentrated around suitable rivers and among those, such as farmers, whose occupation brings them into contact with the fly. The *Simulium damnosum* complex contains the chief vectors in Africa and although they concentrate around rivers some flies may be blown for great distances and reinvade other river systems that have been cleared of vectors in the past. Flies in some savannah regions transmit onchocercal strains very likely to cause blindness. *Simulium naevei* flies, which attach their eggs to crabs but do not fly far, were an important cause of onchocerciasis in Kenya in the past but this focus has been eradicated with insecticides. *Simulium ochraceum* and *S. metallicum* are vectors in the Americas and frequently breed in smaller rivers within coffee plantations.

Clinical features

The pathology of onchocerciasis is almost entirely caused by immunological reactions to dying microfilariae. Adult worms cause few symptoms. Some people do not react to microfilariae and remain asymptomatic carriers for long periods. The incubation period is usually about 15–18 months.

Skin (Figs 14.2–14.4, see colour plate facing p. 146)

Intense itching is the chief symptom and may precede any signs. Early signs are as follows.
• Papular rashes situated asymmetrically in relation to the sites of adult worms. Each papule represents the tombstone of a microfilaria.
• Papules are erythematous in light-skinned people but blackish in those with dark skins.
• Sometimes secondary bacterial infection and excoriations.
   Later signs include the following.
• Thickening, oedema, blackening and lichenification.
• Atrophy with loss of elastic fibres causing premature ageing (presbydermia).

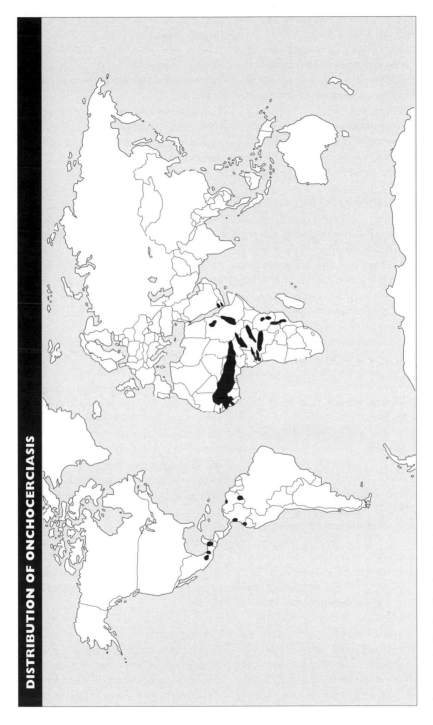

Fig. 14.1 Distribution of onchocerciasis.

• Spotty loss of pigment, especially over the front of the shins (leopard skin).
• In the Yemen 'sowda' is the name given to a form of disease with marked blackening of the skin, enlarged lymph nodes but no demonstrable microfilariae because of intense immunological responses.
• Regional lymph nodes are often enlarged.
• The combination of atrophy and large nodes can result in a 'hanging groin' of loose skin folds, sometimes complicated by femoral hernia.

Subcutaneous nodules (Fig. 14.5, see colour plate facing p. 146)

Adult onchocercal worms migrating subcutaneously are often arrested over bony prominences and enclosed in fibrous tissue. Nodules may measure several centimetres in diameter and enclose a number of live or dead coiled female worms. Nodules are firm and at first lie free but may later become deeply attached.
• In Africa, nodules are found chiefly around the pelvis or the knees. Examine particularly over the pelvic brim, the sacrum and femoral greater trochanters.
• In the Americas, nodules are more often found over the head.
• Ultrasound examination may demonstrate impalpable nodules.

Eye disease

Itching, redness and excess lachrymation are symptoms of early disease. Late disease leads to all degrees of loss of vision and blindness.

Anterior eye disease
• Punctate keratitis is caused by reactions to the death of microfilariae in the cornea. Each may produce a 'snow flake' opacity. Many of these opacities clear.
• Pannus forms as blood vessels invade the cornea from the sides and below. The pannus may cover the pupil—sclerosing keratitis—and cause blindness. This is particularly likely among heavily infected individuals in the African savannah.
• Iritis leads to a loss of the pigment frill and to synechiae that cause a deformed, often pear-shaped pupil. Secondary cataracts occasionally result.

Posterior eye disease
• Encountered in both savannah and forest regions.
• May present as widespread chorioretinitis with pigmentary changes.
• Often accompanied by optic atrophy, which is sometimes the only finding.
• All degrees of visual loss may result but a common presentation is with 'tunnel vision'; the loss of peripheral vision is very disabling.

Eye disease is particularly likely when adult worms are near the eyes and head nodules predict this risk. Because disease progress is relatively slow, most people are blinded in middle life but with heavy infections blindness can occur in people in their twenties.

General health

Skin disease may reduce a person's chance of marriage and resources spent on medication may cause significant economic loss. Very heavy infections can cause growth retardation and perhaps dwarfism. An association with convulsions has been postulated from East Africa but the evidence remains inconclusive. Although onchocerciasis does not kill directly, blind people in village communities have a shortened lifespan and onchocerciasis has led to abandonment of fertile riverside ground in some areas.

Diagnosis

Skin onchocerciasis is frequently misdiagnosed as scabies; the papular eruptions of HIV infection should also be considered. History of exposure in a known focus of onchocerciasis is helpful.

Finding microfilariae

Snips of skin should be taken without blood contamination. Take snips from the vicinity of subcutaneous nodules or in Africa from the lateral aspects of the calves, thighs, the hip region or the iliac crests. In the Americas, snips from the shoulder tip or outer canthus of the eye may be more valuable because microfilariae tend to concentrate in the upper part of the body. Up to six snips may be needed to be reasonably sure that infection cannot be detected.

Techniques

1 Clean skin with alcohol and allow to dry; lift up a small piece of skin on a sharp sterile needle. Slice off a piece 1–2 mm^2 with a sterile scalpel or razor blade. The piece should be deep enough to show white dermis and capillaries should ooze blood into the site. Place the skin piece in 0.2 mL saline in the well of a microtitre plate. Fluid from the plate may be examined after half an hour to look for active microfilariae with the low power of the microscope. The plate should then be allowed to incubate for 24 h at room temperature and be re-examined for emergent microfilariae which will by then usually be immobile.

2 This technique is essentially the same but more elegant as it uses a Walsar corneoscleral punch to obtain the snip. However, these punches are expensive and difficult to keep sharp and sterilize.

Count the numbers of microfilariae in each skin snip. You may need to stain some to differentiate microfilariae of onchocerca from other skin or blood microfilariae. Polymerase chain reaction (PCR) techniques sometimes demonstrate microfilariae in skin snips from which no microfilariae have emerged.

3 Use a slit lamp to look for active microfilariae in the anterior chamber of the eye after the patient has put his or her head between his or her knees for 5 min in order to bring microfilariae into view.

Other evidence

• There is usually a moderate eosinophilia in early onchocerciasis.
• Serological methods remain poorly standardized and subject to cross-reaction with other helminth infections. ELISA tests to detect IgG4 antibodies reacting to specific recombinant antigens appear reasonably specific and may detect infection before skin snips become positive.
• Subcutaneous nodules can be removed surgically to demonstrate adult worms, or aspirated with a needle to look for microfilariae.

• PCR tests or immunoblots for parasite antigen in blood are available in research settings.
• The Mazzotti reaction is sometimes useful to detect lightly infected patients but is potentially dangerous. It consists of giving diethylcarbamazine 50 mg orally and recording the development of an itching papular skin reaction within 24 h; this may be accompanied by fever, limb oedema and even hypotension or worsening of eye damage. Only use this test for patients with negative skin snips and normal eyes.

Treatment

Ivermectin, a macrocyclic lactone drug originally introduced for veterinary purposes, is the drug now in common use. The drug kills microfilariae but has little effect on adult worms except that it reduces embryogenesis. (Is this the first oral contraceptive for worms?) Fortunately, the relatively slow rate of microfilariae killing does not commonly produce severe Mazzotti reactions. However, a temporary increase in itching, papular eruptions and sometimes oedematous limb swelling, headache and fever occur and may need treatment with paracetamol or even a brief course of prednisolone.

Give ivermectin orally at a dose of 150 µg/kg body weight. Avoid use in pregnancy and breastfeeding although teratogenesis has not been demonstrated. Carefully avoid the use of ivermectin in patients who also have heavy *Loa loa* infections as fatal cases of encephalitis have occurred. Because adult worms are not killed, repeated doses of ivermectin at 6 monthly or yearly intervals are needed for several years. Doxycycline has been shown recently to kill the endosymbiotic *Wolbachia* organisms in filarial species. A dose of 200 mg doxycycline daily for 4–6 weeks can maintain freedom from microfilariae for up to 2 years.

Control

• A major control effort in West Africa by WHO and the World Bank achieved interruption of transmission in a large area at considerable cost by aerial spraying of rivers with insecticides.
• Current efforts depend upon the widespread distribution by community participation of iver-

mectin donated freely by the drug firm Merck. There is fair, but not conclusive, evidence for the effectiveness of this programme in reducing blindness and skin disease. Whether the programme can be sufficiently rigorous and sustained to interrupt transmission for long enough to allow onchocerciasis to die out (15–20 years) remains highly controversial.

• There remains a need for a safe and effective macrofilaricidal drug.

Filariasis

Lymphatic filariasis

Filariasis caused by *Wuchereria bancrofti*, and in some areas of Asia by *Brugia malayi* or *B. timori*, affects about 120 million people in the tropics (Fig. 14.6). While many of them have no symptoms, a significant minority develop elephantiasis or other disabling symptoms. WHO has targeted the elimination of filariasis by the year 2020.

Bancroftian filariasis
Life cycle
Adult *W. bancrofti* are threadlike worms living in the lymphatics of the groin and scrotum or sometimes those of the arm. Males measure about 4 cm long and females about 10 cm. Their sheathed microfilariae migrate to the bloodstream where they are usually present only during the night and disappear into the pulmonary capillaries by day (nocturnal periodic). In some Polynesian islands microfilariae are found in greater numbers in daytime blood (diurnal subperiodic). The periodicity of microfilariae in the blood relates to the habits of the local vector mosquitoes.

Many different mosquitoes including *Culex*, *Anopheles*, *Aedes* and *Mansonia* species may act as vectors but the chief vector in towns is *Culex quinquefasciatus* which breeds in drains and polluted water and bites at night. Filariform larvae migrate to the mouthparts of the mosquito after 10 days or more and are injected into a new human host during feeding. There is no multiplication of larvae in the mosquito. Adult worms may live for more than 10 years and microfilariae for a year.

Clinical effects
The effects of infection with *W. bancrofti* depend critically upon the person's immunological response. Many people do not react and remain carriers of microfilariae for years without clinical symptoms, although even these may have dilated lymphatics. This lack of reaction may be caused by blocking antibodies of IgG4 subclass and is more likely in children of carrier mothers. Those who respond very vigorously develop tropical pulmonary eosinophilia (see Chapter 30); others with an intermediate level of response develop some of the symptoms described below.

Early disease
• The incubation period is usually 8–16 months but variable.

• Recurrent episodes of fever (over years) associated with:

1 Lymphangitis, typically radiating distally from an enlarged tender lymph node and not spreading upwards from an infected lesion. Areas chiefly affected include:
   (a) leg (thigh and groin);
   (b) arm (axilla and upper); and
   (c) breast (women) or spermatic cord (funiculitis, in men).

2 Transient oedema of limb.

3 Episodes of epididymo-orchitis or of peritonitis may occur.

Lymphatic obstruction
This occurs after years of recurrent episodes but recent evidence emphasizes that much obstructive disease results from bacterial secondary infection of damaged lymphatics.

• Hydrocoele in males is often the most common presenting feature.

• Persistent lymphadenopathy and in some areas splenomegaly.

• Elephantiasis.
   (a) The initial stage is pitting oedema which gradually becomes permanent.
   (b) The skin thickens and firms with subcutaneous fibrosis.

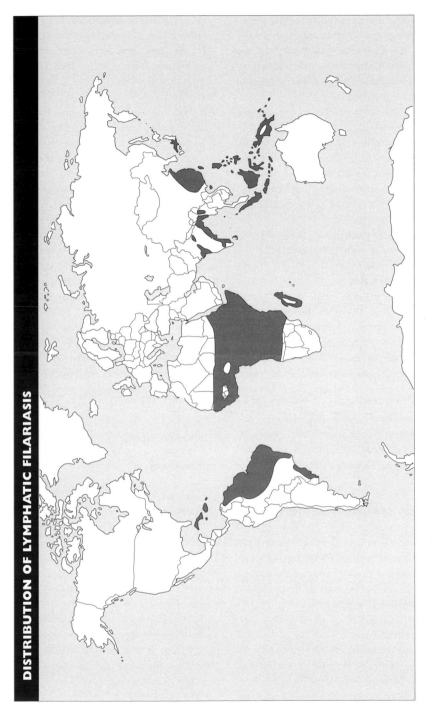

Fig. 14.6 Distribution of lymphatic filariasis.

(c) Nodular verrucose skin changes follow.

(d) The legs and scrotum are commonly involved, but in the Pacific sub-periodic area breast and arm elephantiasis are frequent.

• Varicose lymphatics may show on the skin or rupture:

(a) onto the scrotum causing wet, secondarily infected skin;

(b) into the renal collecting system, causing chyluria with initially bloodstained and later milky urine and sometimes wasting from loss of fats (the chyluria is often intermittent);

(c) into the abdominal cavity causing pain and chylous ascites; and

(d) into a large joint causing filarial arthritis, e.g. a painful warm knee with effusion.

Brugian filariasis

Brugia malayi is a cause of filariasis in certain rural areas of Asia. There are two main forms: the nocturnal periodic form in swampy areas from India to Korea and Japan; and the nocturnal subperiodic form in damp forests of South East Asia. The subperiodic form has an animal reservoir in monkeys, cats and pangolins. *Mansonia* species of mosquito are the major vectors although anophelines may also be involved.

Clinically, brugian filariasis is less severe than bancroftian disease. Elephantiasis is usually limited to below the knees and scrotal involvement is less common.

Diagnosis of lymphatic filariasis

Elephantiasis may be caused by anything that causes chronic lymphatic blockage and must not be equated with filariasis. If elephantiasis is present in many community members the cause is likely to be either filariasis or podoconiosis caused by aluminium silicates.

1 Marked eosinophilia is usual in the early stages of infection only.

2 Look for microfilariae in the blood at the appropriate time. For nocturnally periodic *W. bancrofti* this is between 10 p.m. and 2 a.m. Methods with increasing sensitivity depending upon the volume of blood studied are:

(a) Examination of a thick blood film stained with Giemsa and made from 20 µL blood.

(b) Examination of anticoagulated blood in a white blood cell counting chamber.

(c) Filtration of about 5 mL blood through a membrane filter which is then stained and examined.

Microfilariae of *W. bancrofti* and *B. malayi* are sheathed but can be distinguished from each other by the arrangement of their nuclei.

3 Remember that most symptomatic patients do not have microfilariae in their peripheral blood although there are many microfilariae carriers in the towns or villages where they live.

4 Tests for antigens of *W. bancrofti* in blood based on monoclonal antibodies are very useful because they are not dependant upon the time of taking blood and are positive in symptomatic patients as well as in microfilariae carriers. Commercial card tests such as the immunochromatographic (ICT) filariasis test are available.

5 Antibody tests including complement fixation tests (CFT) tests using *Dirofilaria* antigens are less specific as they cross-react with other helminths and may not indicate current infection. Very strongly positive antibody tests are helpful in diagnosing tropical pulmonary eosinophilia.

6 Ultrasound examination of the upper thighs and groins can often detect characteristic movements of adult worms.

Treatment

Chemotherapy

1 Diethylcarbamazine citrate (DEC) has been used for many years and is effective in both killing microfilariae and damaging adult worms. Repeated doses of DEC over months or years are most effective against adult worms but the standard treatment is with 6 mg/kg DEC in divided doses orally daily for 12 days. As reactions to dying worms are common, especially in brugian disease, it is wise to start with a dose of 50–100 mg and work up to the full dose over a few days. Reactions consist of fever, head and muscle pains, nausea and later painful lymphadenopathy or lymphangitis and, rarely, abscess formation; sometimes scrotal nodules form around dying worms.

2 Albendazole 400 mg twice daily for 3 weeks has also been shown to have a macrofilaricidal effect.

3 Ivermectin kills microfilariae but not adult worms and is most useful for community control of microfilariae carriers in a dose of 400 μg/kg, when it is now often combined with a single dose of albendazole 400 mg.

Management of obstructive disease

1 Hydrocoeles and scrotal elephantiasis are best managed surgically.

2 Elephantiasis of the legs can sometimes be improved with very careful skin washing and hygiene with antibiotic treatment of any secondary infection. Antibiotic prophylaxis with phenoxymethylpenicillin is useful if there are recurrent streptococcal infections. Elevation of the limb with massage and compression bandaging to reduce oedema is useful but is often not tolerated in the humid tropics. Surgical treatment of limb elephantiasis is not straightforward, leaves scars and is often unavailable to poor people.

3 Chyluria usually subsides with bed rest, but sometimes requires surgical disconnection of lymphatics to the urinary tract.

Control

1 Control of mosquito breeding is helpful, particularly in towns.

2 Insecticide-impregnated mosquito nets protect against night-biting mosquitoes and are important in areas of nocturnally periodic transmission.

3 Elimination of microfilariae carriage by chemotherapy is feasible and effective in the long term. The addition of DEC (0.33%) to table salt has proven very useful where the salt supply can be controlled. Mass treatment with single doses of DEC + albendazole or ivermectin + albendazole is also very effective but depends upon good cooperation by the local population over several years. Cooperation may be limited by side-effects from dying worms; these are worse in areas of brugian filariasis. The expulsion of intestinal parasites by these drugs is clear to local people and aids their cooperation. Mass treatment is often most effective when given monthly but 6 monthly or annual treatment is usually more practicable.

Further reading

Dreyer G, Medeiros Z, Netto MJ et al. Acute attacks in the extremities of persons living in an endemic area for bancroftian filariasis: differentiation of two syndromes. *Trans R Soc Trop Med Hyg* 1999; **89:** 517–21. [A useful article which points out that many attacks are caused by bacterial infection.]

Hoerauf A, Büttner DW, Adjei O, Pearlman E. Onchocerciasis. *Br Med J* 2003; **326:** 207–10. [Endosymbiotic *Wolbachia* bacteria in worms cause pathology and can be therapeutic targets.]

Ismail MM, Jayakody RL, Weil GJ et al. Long-term efficacy of single-dose combinations of albendazole, ivermectin and diethylcarbamazine for the treatment of bancroftian filariasis. *Trans R Soc Trop Med Hyg* 2001; **95:** 332–5. [This paper investigates the best treatment for microfilariae carriers.]

Vincent JA, Lustigman S, Zhang S, Weil GJ. A comparison of newer tests for the diagnosis of onchocerciasis. *Ann Trop Med Parasitol* 2000; **94:** 253–8. [A useful look at newer diagnostic methods.]

WHO. *Onchocerciasis and its Control: Third Report of an Expert Committee.* Technical Report Series no. 852, Geneva: WHO, 1995.

CHAPTER 15

African Trypanosomiasis

Parasites, 120
Life cycle, 120
Disease, 120
Clinical picture, 122

Diagnosis, 123
Treatment, 124
Epidemiology, 125

Sleeping sickness control and
surveillance, 126
Further reading, 126

African trypanosomiasis is caused by species of *Trypanosoma brucei*. There are three morphologically identical parasite species:
1 *T. brucei brucei*, confined to domestic and wild animals.
2 *T. brucei gambiense*, causing gambiense sleeping sickness in West and Central Africa.
3 *T. brucei rhodesiense*, causing rhodesiense sleeping sickness in East and Southern Africa.

Transmission is by the bite of tsetse flies (members of the genus *Glossina*) and the flies are only found in Africa. In general, the infected areas are found south of the Sahara and north of the Zambezi (Fig. 15.1).

Parasites

The parasites are flattened and fusiform in shape, like slender pointed leaves, 12–35 μm long and 1.5–3.5 μm broad. They are actively motile, using a thin fin-like extension from the main body, the undulating membrane, to propel themselves. The form of the parasite found in humans is the trypomastigote, in which the kinetoplast is posterior to the nucleus, and from which the flagellum arises. The flagellum runs along the free edge of the undulating membrane and usually projects in front of it, sometimes extending as far again as the creature's body.

Life cycle

This is the same for both species. Trypomastigotes from the infected host are taken up by the tsetse fly during a blood meal. In the stomach of the fly the parasites multiply by simple fission, penetrate the gut wall and migrate to the salivary glands. There the morphology changes, the kinetoplast coming to lie just in front of the nucleus, and the creatures are now called epimastigotes (crithidia). The infective trypomastigote (the metacyclic trypanosome) is found in the saliva about 20 days after the original infecting blood meal, and the fly remains infective throughout its normal lifespan of several months.

Disease

Local effects
Metacyclic trypanosomes injected during tsetse feeding multiply in the extracellular space and lymphatics before becoming disseminated by the bloodstream. This local multiplication may cause a marked inflammatory reaction — the trypanosomal chancre.

The trypanosomal chancre appears 3 or more days after the bite and typically increases in size for 2 or 3 weeks, at the end of which time it begins to regress. The presence of a chancre is

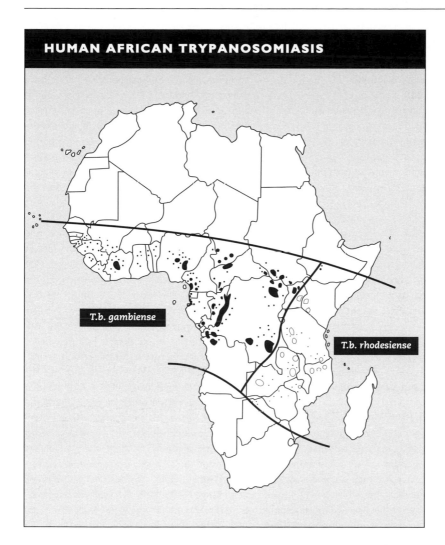

HUMAN AFRICAN TRYPANOSOMIASIS

T.b. gambiense

T.b. rhodesiense

Fig. 15.1 Distribution of human African trypanosomiasis.

much more common in *T.b.rhodesiense* infection than in *T. b. gambiense* infection. Local lymphadenopathy may be found in the region of a bite.

Systemic effects
Multiplication of trypanosomes in the lymphatics leads to parasitaemia 5–12 days after the bite (haemolymphatic or early stage). Waves of parasitaemia are associated with fever. Parasites may

then enter the central nervous system (CNS) via the choroid plexus or by transcytosis across endothelial cells to cause a lymphocytic meningoencephalitis (late stage).

In general, *T.b.gambiense* is better adapted to the human host than is *T. b. rhodesiense*. *Trypanosoma brucei gambiense* is therefore relatively well tolerated; the illness it causes tends to be subacute or chronic and parasitaemia may even be asymptomatic. In contrast, *T.b.rhodesiense* is normally a zoonotic infection which is transmitted to humans 'accidentally'. It causes pronounced systemic effects; parasitaemia usually causes severe incapacity and the course of the illness is relatively rapid.

Immune response and pathogenesis

The main response to trypanosomal infection is antibody production, particularly IgM. Antibody production initially controls parasitaemia, but antigenic variation in parasite surface antigens means that immune control is incomplete and this leads to successive waves of parasitaemia, which may explain the fluctuating nature of the illness. In the brain and other organs (e.g. heart or serous membranes), perivascular infiltration with lymphocytes, plasma cells and macrophages and characteristic morular cells occurs. Microglial and astrocyte proliferation may be associated with neuronal destruction and demyelination in the brain.

Clinical picture

Trypanosoma brucei gambiense
Early stage

Fever, headache and joint pains are the main early symptoms, sometimes accompanied by fleeting areas of cutaneous oedema. A small proportion of patients with parasitaemia are asymptomatic. Lymph glands become enlarged, the most prominent glands often being in the posterior triangle of the neck (Winterbottom's sign). Odd skin rashes sometimes occur (visible only in relatively unpigmented skins), usually taking the form of areas of circinate erythema. There may be generalized pruritus, and characteristic thickening of the facial tissues giving a sad or strangely expressionless appearance. The spleen enlarges to a moderate size in many cases.

This early stage usually lasts many months, sometimes even over 2 years. Occasionally, patients with *T. b. gambiense* develop a rapidly progressive toxaemic disease that is fatal before the CNS is involved. However, most deaths occur after CNS invasion unless the patient develops an intercurrent infection.

Late stage symptoms and signs

Symptoms and signs of disturbed cerebral function predominate. Behavioural changes are common: a patient whose personal habits were previously fastidious becomes careless about appearance; his or her speech becomes coarse and temper unpredictable and he or she may behave in a socially unacceptable way. Psychiatric manifestations of agitation or delusions may become severe enough to mimic mania or schizophrenia. Sleep becomes disordered in that the patient sleeps badly at night but falls asleep during the day.

In the early evolution of this change, the patient can be readily awoken and responds by conversing fairly normally. As time goes by, sleeping periods may become longer until the patient is sleeping most of the time, and may even fall asleep while eating. At this stage, speech and motor functions in general are usually severely disturbed. Weight loss may occur because of inadequate nutrition unless the family makes strenuous efforts to help with feeding.

Focal CNS signs may develop, but there is usually more diffuse evidence of CNS disease, especially relating to extrapyramidal and cerebellar functions: widespread tremors involving the limbs, tongue and head, spasticity (mainly of the lower limbs), ataxia and sometimes choreiform movements. Convulsions are relatively uncommon. Kérandel's sign (delayed hyperaesthesia) may occur in which, following firm pressure on the tissues overlying a bone, there is a definite delay before the patient shows any sign of pain. In advanced cases, the tendon reflexes are often grossly exaggerated and the plantar responses may be extensor. Death usually occurs within a few months of CNS involvement becoming manifest, but may be delayed for up to a year.

Trypanosoma brucei rhodesiense
Symptoms and signs

The parasite usually produces a more acute and virulent infection than does *T. b. gambiense*, with fever and systemic symptoms prominent. Serous effusions, especially into the pleural and pericardial spaces, are common and myocarditis occurs. In the early stages, *T. b. rhodesiense* may cause hepatocellular jaundice and mild anaemia, and severe anaemia may soon develop. Both liver and spleen may be slightly enlarged, and lymph gland enlargement (seldom so prominent as in *T. b. gambiense*) is most common in the inguinal, axilliary and epitrochlear glands. *T. b. rhodesiense* may be fatal within a few weeks of the onset, often as a result of death from myocarditis before the CNS is involved.

The picture in the late stage of *T. b. rhodesiense* infection is much like that of *T. b. gambiense* but occurs early in the course of the disease and is more rapidly progressive. Clinical features of CNS involvement are similar to *T. b. gambiense*, but death is more rapid and neurological features are more pronounced than behavioural changes.

Diagnosis

Early stage disease

The diagnosis is usually made by demonstration of parasites. A number of methods may be used.
1 Examination of stained or unstained thick blood films. Films can be stained with a Romanowsky stain as for malaria or be examined wet (simply place a coverslip over a drop of blood) when the disturbance of the red cells produced by the movement of the trypanosomes can be detected using a dry 40× objective (Fig. 15.2, see colour plate facing p. 146).
2 Concentration methods are used to detect scanty parasitaemia, best among which are microscopy of the buffy coat following centrifugation using the microhaematocrit (MHCT) and the recently developed, more sensitive quantitative buffy coat (QBC) technique as well as the minianion exchange column technique (MAEC). Examination of blood films is most useful for *T. b. rhodesiense* infection. The organisms may also be isolated by inoculation into special culture media or into animals.
3 Gland puncture. This is of most use in *T. b. gambiense* where posterior cervical lymphadenopathy is common and gland aspirates may be positive when there is no peripheral

parasitaemia. A needle is inserted into an enlarged node held between thumb and finger. The flow of gland juice up the needle can be improved by massaging the gland while the needle is *in situ*. The juice is then expressed on to a slide, using a syringe containing air, and examined immediately.
4 Bone marrow aspiration. This is useful in the early stages when other methods are negative.
5 The chancre. Trypanosomes can be recovered by aspiration from the chancre or from the regional glands draining the chancre if they are enlarged, before the blood is positive.

In *T. b. gambiense*, trypanosomes may be difficult to find in the blood, especially in late infections. The longer the duration of infection, the more difficult it is to find trypanosomes.

Late stage disease (CNS involvement)

This may be diagnosed clinically on the basis of neurological signs but cerebrospinal fluid (CSF) examination should be performed in all patients following one or two doses of suramin or pentamidine to clear parasitaemia (and hence reduce the risk of parasites being introduced into the CSF from the blood). The deposit from 5–10 mL of centrifuged fluid should be examined as soon as possible for motile trypanosomes or made into a smear, dried, fixed and stained with a Romanowsky stain. Late stage disease is diagnosed by the presence of trypanosomes in the CSF cell or by a raised CSF cell count (> 5/mm^3) or increased protein level.

Immunological diagnostic methods

A number of serological methods are available. The card agglutination test for trypanosomes (CATT) is technically simple to carry out, and gives good results in most areas of *T. b. gambiense* but is of no value in *T. b. rhodesiense*. It is a valuable test for screening populations as the results are obtained within 30 min. The disadvantages at present are the problems of sensitivity and specificity of the antigen, as trypanosomes share antigens with several other protozoa and bacteria. More recently, the card indirect agglutination test for trypanosomes (CIATT), which detects circulating antigens, has been used for the diagnosis of both *T. b. gambiense* and *T. b. rhodesiense*. This technique may also be useful to follow responses after treatment.

Routine laboratory findings

A mild normochromic anaemia is common. The white blood count (WBC) is usually normal, but the erythrocyte sedimentation rate (ESR) is usually raised above 50 mm/h and sometimes to over 100 mm/h. Serum and CSF IgM levels are usually very high.

Treatment

A number of different drugs have activity against trypanosomes. Most are relatively toxic and there is an urgent need for the development of new drugs for the treatment of sleeping sickness. Not all drugs penetrate the CSF and different drugs are therefore used for the treatment of early and late stage disease. It is important to treat coexisting infections and anaemia prior to using specific treatment: many advocate routine antihelminth and antimalarial therapy.

Early *Trypanosoma brucei gambiense* and *Trypanosoma brucei rhodesiense* infections

Suramin is the drug of choice for treating first stage *T. b. rhodesiense* infection. It is also effective in *T. b. gambiense* but pentamidine is now most commonly used to treat early *T. b. gambiense* infection. Neither of these drugs penetrate the CSF.

Suramin is administered intravenously. Following a test dose of 5 mg/kg, 20 mg/kg (max 1 g) should be given on days 3, 10, 17 and 24. Suramin is usually well tolerated, but fever, nausea and proteinuria may occur. Infrequent idiosyncratic anaphylactic reactions also occur.

Pentamidine can be given intramuscularly or intravenously; intravenous administration avoids painful local tissue reactions. Normal doses are 4 mg/kg/day for 7–10 days. The major reaction is syncope and hypotension; hypoglycaemia may also occur.

Late stage disease

Melarsoprol can be used for the treatment of late stage *T. b. gambiense* and *T. b. rhodesiense* infections. Eflornithine is a newer agent that is only effective in *T. b. gambiense*.

Melarsoprol is a trivalent arsenic compound that is given intravenously and is active against blood, tissue and CNS trypanosomes. There are many different treatment schedules but melarsoprol is normally administered as three or four series of three injections separated by 7 days at doses that range from 1.2 to 3.6 mg/kg (see Table 15.1 for commonly used schedule). Recent studies suggest that shorter courses may be as effective for *T. b. gambiense*. Melarsoprol therapy is normally preceded by 1–2 doses of suramin to clear blood, lymph and tissue trypanosomes.

Melarsoprol is a toxic drug. The major side-effect is a serious encephalopathy (reactive arsenical encephalopathy), which occurs with a frequency of 2–10% and a case fatality rate of up to 50%. The danger of severe toxic effects is minimized by improving the patient's general condition, as already described; prophylactic

TREATMENT SCHEDULE

Day	Drug	Volume (ml)	Volume (mg/kg)
1	Suramin	2.5	5
3		5.0	10
5		10.0	20
7	Melarsoprol	0.5	0.36
8		1.0	0.72
9		1.5	1.1
16	Melarsoprol	2.0	1.4
17		2.5	1.8
18		3.0	2.2
25	Melarsoprol	3.0	2.2
26		4.0	2.9
27		5.0	3.6
34	Melarsoprol	5.0	3.6
35		5.0	3.6
36		5.0	3.6

Table 15.1 Treatment schedule for an adult with late-stage *T. b. rhodesiense*.

corticosteroids reduces the risk of an encephalopathy in *T. b. gambiense*. Other side-effects include peripheral neuropathy.

Eflornithine is used for the treatment of late stage *T. b. gambiense* infection at a dosage of 400 mg/kg/day in divided doses for 14 days. It is relatively expensive and difficult to administer but less toxic than melarsoprol. The common side-effects (gastrointestinal symptoms and anaemia) do not usually require treatment to be stopped.

Monitoring cure

Patient symptoms should resolve after treatment. Despite the severity of the symptoms in advanced late cases, the degree of functional recovery after successful chemotherapy is remarkable. It may take 6 months or more for the CSF cell counts to fall below 5/μL and for normal protein concentrations to occur. Failure of these parameters to reach normal may be the first indication that treatment has been unsuccessful. Full cure cannot be assumed unless a 2-year follow-up has been completed. If treatment of patients with CNS involvement has been delayed, a variable degree of neurological defect will persist. This most commonly takes the form of intellectual impairment.

Relapse

Treatment of relapse can sometimes be difficult. Relapse in *T. b. gambiense* following treatment with suramin or pentamidine is often treated with melarsoprol; eflornithine can also be used. Relapse in *T. b. rhodesiense* is usually treated with a second course of melarsoprol. Nifurtimox is a drug used for the treatment of Chagas' disease that is being evaluated for its role in the treatment of relapse.

Epidemiology

Approximately 40 000 cases of sleeping sickness are notified to the World Health Organization (WHO) each year although it is estimated that between 300 000 and 500 000 individuals are infected. A lack of resources and civil conflict in many of the heavily affected areas has led to an increase in cases as previously successful control programmes have broken down.

Trypanosoma brucei gambiense infection

Humans are much the most important reservoir of infection, although the pig and other animals are naturally infected in some parts of West Africa. Infection is spread from human to human by the bite of riverine tsetse flies (*palpalis* group), which breed along the banks of rivers and lakes. Infection tends to occur where human activities bring humans into contact with the fly, such as at river crossings and sites used for the collection of water, and when fishermen come into contact with flies on the river or lake shores. Village-sized and larger outbreaks occur, sometimes amounting to epidemics. The spread of epidemics tends to be linear, following the distribution of flies along the course of rivers, or affecting islands in lakes.

Trypanosoma brucei rhodesiense infection

Trypanosoma brucei rhodesiense is usually a zoonotic infection in members of the antelope family, especially the bushbuck. It can be maintained as a zoonosis in the animal population in the absence of human cases, and is transmitted by tsetse species dwelling in savannah and woodland habitats (*morsitans* group). It is a particular hazard to those who spend long periods in enzootic areas in pursuit of their livelihood, such as hunters and honey-gatherers. A more recent group at risk is the growing band of tourists intent on seeing African game in its natural habitat.

Although *T. b. rhodesiense* cases tend to be sporadic, epidemics do occur, especially in East Africa around Lake Victoria. In these epidemics tsetse populations build up adjacent to human populations containing active cases. The usual *morsitans* group vector is replaced with a *palpalis* group tsetse. Domestic cattle are infected, develop a chronic parasitaemia and act as a reservoir host. The current epidemic in south-east Uganda is caused by peridomestic breeding

Glossina fuscipes in thickets of the exotic plant *Lantana camora*.

Sleeping sickness control and surveillance

There are two major components of sleeping sickness control:
1 detection and treatment of cases; and
2 vector control.

In *T. b. rhodesiense* areas, patients who present with symptoms of early parasitaemia (passive surveillance) can be treated at local rural centres: in epidemics, rapid deployment of active surveillance using blood film screening and the establishment of effective local treatment centres is important. In *T. b. gambiense* areas, limited clinical symptoms in the early stages require active surveillance. Individuals can be screened using gland aspiration or rapid antigen tests (e.g. CATT).

Vector control is best achieved using insecticide-impregnated traps and targets. Sterile insect release methods may also be useful in reducing vector populations. Residual insecticide application to *Glossina* resting sites, insecticide spraying and the clearing of riverine habitat have been used in the past but resource and environmental considerations means that these can no longer be considered. In epidemic situations, treatment of the cattle reservoir by cattle trypanocides may also be a strategy for prevention of human sleeping sickness.

In all endemic areas the disease should be made notifiable to a central trypanosomiasis control unit. Specialized staff can then be sent promptly to the area and steps taken (such as active case finding and treatment) to prevent the development of an epidemic. This strategy has proved very effective in the past in Ghana, Nigeria and Uganda.

Further reading

Mulligan HW, ed. *The African Trypanosomiases*. London: George Allen and Unwin, 1970. [An old but still very useful classical clinical description.]

World Health Organization. *Control and Surveillance of African Trypanosomiasis*. Technical Report Series no. 881. Geneva: World Health Organization, 1998. [Current advice on treatment and surveillance.]

CHAPTER 16

South American Trypanosomiasis (Chagas' Disease)

Parasite, life cycle and
    pathogenesis, 127
Clinical features, 127

Diagnosis, 128
Treatment, 128

Epidemiology and control, 128
Further reading, 128

Parasite, life cycle and pathogenesis

South American trypanosomiasis occurs in humans and a large number of wild and domestic animals and is widespread in Central and South America. It is caused by *Trypanosoma cruzi*, which differs from trypomastigotes of the *T. brucei* group in having a large kinetoplast. Trypanosomes in the blood of the mammalian host are taken up by triatomine bugs (reduviid, assassin bug, kissing bugs), which bite at night. All stages feed on blood but only adult bugs can fly. Organisms multiply in the hindgut of the bug as epimastigotes and develop into metacyclic trypanosomes which are excreted in the faeces of the bug during feeding. Infection is acquired by rubbing faeces of the bug into a wound or conjunctiva; infection can also be acquired by transfusion or congenital infection.

In the host, trypomastigotes multiply at the site of the bite, enter the bloodstream and enter a variety of tissue cells, particularly neuroglia and muscle cells. Parasites develop as intracellular amastigotes and form pseudocysts: rupture of these pseudocysts causes inflammation, tissue damage and further dissemination. Most pathological effects are chronic, probably related to a combination of tissue damage, neuronal loss and an autoimmune response.

Clinical features

Acute Chagas' disease

This is most common in children but may occur at any age: only one-third of individuals are symptomatic. Penetration and local multiplication of the parasite at the site of entry may cause an area of cutaneous oedema (chagoma) or orbital oedema (Romaña's sign) if entry is via the conjunctiva. A febrile reaction may develop 1–2 weeks later with the development of lymphadenopathy, hepatomegaly and splenomegaly. Rarely, death may occur at this stage as a result of cardiac damage or meningoencephalitis, especially in children. If symptomatic, the acute phase lasts for 1–3 months and resolves spontaneously.

In untreated patients, asymptomatic low-level parasitaemia may continue for many years: (indeterminate phase); 15–40% of patients will develop chronic Chagas' disease.

Chronic Chagas' disease

Chronic disease normally occurs 10–20 years after initial infection. Classical manifestations are as follow.

1 Cardiac disease. Biventricular cardiomyopathy or cardiac rhythm disturbance (often heart block).

2 Mega-oesophagus or megacolon as a result

of destruction of the intramural parasympathetic nerve plexus. This presents as aspiration pneumonia or intractable constipation and abdominal distension.

3 Similar mega disorders of other hollow muscular viscera such as small bowel and ureter may occur resulting from nerve damage.

Immunocompromise

HIV infection or the use of immunosuppressive drugs may lead to the reactivation of latent infection causing severe myocarditis or neurological problems.

Diagnosis

Parasitological techniques

1 *Microscopy*. In the acute phase, parasites can usually be easily found on thick or thin films; centrifugation techniques increase the sensitivity.

2 *Culture*. Parasites can be cultured, but specific media and expertise are required.

3 *Xenodiagnosis*. Low-level parasitaemias can be detected by allowing uninfected bugs to feed on patients. Three to four weeks later, the bugs are dissected to look for gut infection.

4 *Biopsy*. Amastigotes may be demonstrated in pathological specimens.

Serological techniques

IgM and life-long IgG responses may be detected by a number of techniques, including complement fixation test and ELISA. Cross-reactivity with other parasitic diseases and autoimmune disorders leads to poor specificity.

Treatment

Acute stage

Nifurtimox and benzidazole suppress parasitaemia, shorten the course of the acute illness and prevent acute neurological and myocardial complications. However, elimination of parasites and prevention of chronic disease only occurs in 50–70% of patients.

Indeterminate and chronic phase

Although treatment has traditionally been thought to have little effect in this phase, there is emerging evidence that benzidazole may be of benefit in clearing parasitaemia in some patients in the intermediate phase. Chronic phase treatment does not appear to be of benefit: cardiac complications require symptomatic treatment and insertion of pacemakers is often necessary.

Epidemiology and control

Trypanosoma cruzi occur in a large number of mammalian species, but the most common wild hosts are rodents or small marsupials. Many species of triatomine bugs simply maintain infection amongst wild animals, but some species have become adapted to living in human dwellings, leading to human infection when infection is transmitted from domestic animals. A single adobe dwelling can harbour thousands of bugs and up to 50% of bugs may be infected. Chagas' disease can also be transmitted by transfusion and congenital infection occurs in up to 10% of seropositive women.

Control of the disease can be achieved by the use of seroprevalence surveys to determine areas at risk and spraying of pyrethroid insecticides. Improvement in the standard of housing is also important. Elimination of cracks in mud walls or replacement of natural material roofing with iron sheets reduces available habitats for the bugs. In a number of South American countries, such activities have reduced the incidence of Chagas' disease by between 60 and 99% over the past 20 years.

Further reading

Prata A. Clinical and epidemiological aspects of Chagas' disease. *Lancet Infect Dis* 2001; **1**: 92–100. [Good general review of South American trypanosomiasis.]

CHAPTER 17

Schistosomiasis

Parasitology, 129
Epidemiology, 131
Clinicopathological
   features, 132

Investigation, 136
Management, 137
Prevention and public health
   aspects, 139

Future developments, 139
Further reading, 139

Schistosomiasis is often known as 'bilharzia' or 'bilharziasis' after Theodor Bilharz who first described the parasite in humans and found the adult flukes in a human postmortem in Egypt in 1851. In ancient Egyptian papyri the typical symptom of chronic *Schistosoma haematobium* infection, haematuria, is described and the disease named; its hieroglyph being a dripping penis. Schistosome eggs have been recovered from both Chinese and Egyptian mummies, showing that the infection was present in both of these early civilizations. Today schistosomes remain distributed throughout the tropics and flourish wherever freshwater bodies, both natural and man-made, create habitats for the appropriate snail vectors.

Three main species of schistosome affect humans with different geographical distributions (Fig. 17.1).

1 *Schistosoma haematobium* causes urinary schistosomiasis. It is scattered throughout Africa, parts of Arabia, the Near East, Madagascar and Mauritius.

2 *Schistosoma mansoni* is mainly found in Africa and Madagascar. It was exported by the slave trade to parts of South America, the Caribbean and Arabia, where permissive snail vectors were present.

3 *Schistosoma japonicum* is found in China, the Philippines and Sulawesi. There is a small focus in the Mekong river on the east border of Thailand. *Schistosoma mansoni* and *S. japonicum* cause dis-

ease of the bowel and liver. (*Schistosoma intercalatum* is a minor species confined to West Africa. It inhabits the veins of the lower bowel and produces terminal-spined eggs.)

Parasitology

The adult flukes causing human schistosomiasis are worm-like creatures 1–2 cm long which inhabit parts of the venous system of humans. The male worm resembles a rolled leaf in having a groove on his ventral surface in which the longer, more slender female is held *in copulo*. Both sexes are actively motile. The worms sometimes live for 30 years, but their normal lifespan is probably 3–5 years.

Life cycle

Fertilized adult females lay eggs in the terminal venules of the preferred host tissues. Their bodies obstruct the vessel and so impede the escape of eggs into the circulation. Most of the eggs penetrate the vessel wall and enter the tissues. Movements of the walls of the hollow viscus involved (as well as other factors) propel the eggs towards the lumen, from which they escape to the outside world — in the urine in the case of *S. haematobium*, and in the stools in the case of *S. mansoni* and *S. japonicum*.

The shapes of the eggs of each species are distinctive, and each contains a ciliated miracidium.

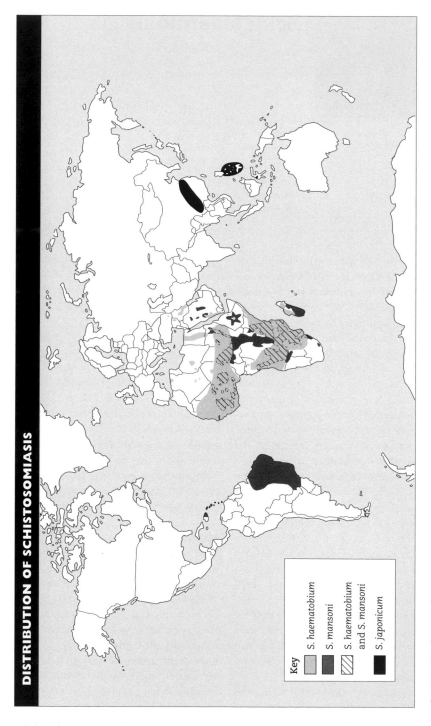

Fig. 17.1 Distribution of schistosomiasis.

This hatches out in freshwater, and swims in search of a suitable snail intermediate host. Many species of snail host are known, but in general:

1 *Schistosoma haematobium* requires an aquatic sinistral turretted snail of the genus *Bulinus*;

2 *Schistosoma mansoni* requires a flat aquatic 'ramshorn' snail, most commonly of the genus *Biomphalaria*; and

3 *Schistosoma japonicum* requires a small amphibious operculate turretted snail, usually of the genus *Oncomelania*.

The miracidium penetrates the body of the snail and begins a complicated asexual replicative cycle that results, a few weeks later, in the release of minute fork-tailed cercariae into the water. As cercariae are about 200–500 μm long, they are just visible to the naked eye. They emerge from the sporocyst inside the snail in response to light. A snail may shed cercariae for many weeks. The cercaria is infective to the definitive host. If it finds no suitable host within 24–48 h, it dies. If it contacts human skin, however, the cercaria penetrates, sheds its tail and body, and enters the circulation with the new name—a schistosomule. The schistosomule reaches the liver through the lungs, by passive intravascular migration. Once in the liver, it begins to feed and grow, and in 1–3 months develops into a mature fluke in an intrahepatic portal vein. The mature males and females couple, and then migrate to their final habitats. It is easy to understand how *S. mansoni* and *S. japonicum* find the way to their homes in the lower mesenteric veins, as they have only to travel straight down the portal vein. However, it is a mystery how *S. haematobium* reaches the vesical plexus.

The time elapsing between cercarial penetration and the passage of eggs is the prepatent period. It can be as short as 4 weeks with *S. mansoni*, usually 12 or more weeks with *S. haematobium* and somewhere in between with *S. japonicum*. The prepatent period is sometimes very prolonged in light infections; perhaps if there are few worms in the liver, the sexes have difficulty finding each other.

Epidemiology

Magnitude of the problem

Some three hundred million people are infected with schistosomes throughout the tropics wherever freshwater bodies (particularly lakes, dams and irrigation systems) support large snail populations near concentrations of human habitation.

Requirements for transmission

1 Contamination of water with viable eggs from a reservoir host.

2 Presence in the water of susceptible snail intermediate hosts.

3 Suitable environmental conditions for development in the snail.

4 Human exposure to water containing cercariae.

Reservoir hosts

In all three schistosomes, humans are the main reservoir. Rodents and baboons may be able to maintain *S. mansoni* infection sometimes. Many animals are susceptible to *S. japonicum* infection, including domestic animals such as the horse and dog.

Epidemiological patterns

In most infected communities, infection is most common and heaviest in children between 10 and 15 years old. Because children have the highest egg output and are more likely to contaminate water, they are usually the most important reservoir of infection. Exposure may be occupational, such as occurs among workers on irrigated farms and fishermen. Transmission is often focal, and neighbouring villages may have greatly differing endemicities because of this.

Exposure and immunity

There is evidence that some degree of resistance to superinfection develops in schistosomiasis. It is certainly incomplete, and may depend for its maintenance on the continued presence of some living schistosomes in the body. It does not seem to be antibody-mediated, and is probably directed against the schistosomule stage. The survival of adult worms in the circulation may be partly related to their ability to incorporate host antigens in their integument (surface).

The log–normal distribution

In a population apparently exposed to a uniform risk of infection, some people will be found to be very heavily infected and others very lightly infected. If egg output is accepted as being related to the number of adult worms present, the distribution of worms in the population is not 'normal'. It will always be found that some of those infected have an infection with perhaps 100 times as many worms as those with the most common level of infection. If a frequency–distribution plot of egg output is carried out, the usual Gaussian curve will be seen to be distorted by having a greatly extended 'tail' to the right of the graph. The curve can be made to resemble a 'normal' curve if, instead of the egg count, the logarithm of the egg count is plotted on the x axis. This sort of distribution is called a log–normal distribution, and applies to the abundance of almost all non-replicative parasites in humans and animals.

There is no generally accepted explanation, but it could be related to the host's first exposure to infection. If the initial challenge was with a large number of parasites, at a time when no immunity existed, a large population could become established in the absence of immune opposition. On the other hand, if the first exposure was to a small number of parasites, the subsequent development of immunity could resist further infective challenges, and the total number of parasites would then remain low.

Infection in children

In hot countries, children naturally play in water. This recreational exposure is sometimes the most important source of infection. At puberty, exposure often diminishes as modesty develops at the same time as sexual awareness. However, infection may still be acquired during activities such as personal bathing, washing clothes and utensils, and in the pursuit of irrigated farming.

This change in behaviour is one factor in the tendency of the infection to diminish after puberty, but it is not the only one. Acquired immunity also appears to reduce the likelihood of a given exposure to cercariae leading to an established infection.

Progression to disease

Most people infected with schistosomiasis die of an unrelated disease. In many parts of the world, although the prevalence is high, the adverse effects of the infection are difficult or impossible to demonstrate. The notion that the infection always causes general debility and malaise, in the absence of more specific effects, is wrong. In the absence of reinfection, the tendency is for most of the worms to die within a few years and for pathology related to the eggs (see below) to resolve. In only a small proportion of cases will progressive pathology develop. These are mainly those with heavy infections and re-exposure to infection over a period of many years. Treatment can certainly modify the natural history of the infection and even advanced cases may show a surprising degree of improvement after chemotherapy.

Clinicopathological features

Effects of cercarial penetration

Cercariae may cause an itchy papular rash ('swimmer's itch' or 'fisherman's itch') as they penetrate the dermis. This is seldom seen in endemic areas. A conspicuous cercarial rash is more often caused by avian or other schistosomes not otherwise pathogenic in humans. It is quite common in northern Europe, North America and South East Asia.

Initial illness:
acute schistosomiasis

An initial febrile illness is sometimes recognized following the first exposure. It does not develop in very light infections, and is seldom recognized in residents of endemic areas. It is mainly a problem in immigrants or visitors encountering a large cercarial challenge for the first time.

The illness comes on 4 or more weeks after infection, and is usually self-limiting. The theory is that as the worms begin to lay eggs, soluble antigen (Ag) leaks out of the eggs and into the circulation. While antibody (Ab) production lags behind antigen release, moderate antigen excess prevails. This favours Ag–Ab complex formation, with the development of generalized immune complex disease. Because the antigen is soluble and distributed by the bloodstream, the effects are more general than local. The immune complex disease hypothesis probably also explains why acute schistosomiasis has been reported more frequently in S. japonicum where egg production per worm pair is heavier than in S. haematobium and S. mansoni.

Features of acute schistosomiasis

The condition is sometimes called Katayama fever, after the prefecture in Japan where it used to be common. Some or all of the following may occur.

- Fever.
- Urticaria.
- Eosinophilia.
- Diarrhoea.
- Hepatomegaly.
- Splenomegaly.
- Cough and wheeze.
- Cachexia.

Perhaps it is seldom recognized in children in endemic areas because immune tolerance develops in utero, because of transplacental passage of antigen. Spontaneous recovery may be related to restoration of Ag–Ab balance as the infection matures and antibody production increases.

Importance of the eggs: those that get away

Eggs that escape from the body enable the life cycle to be completed. Their passage through the bladder in S. haematobium typically causes terminal haematuria, the cardinal symptom of the infection. In heavy infections, irritation of the bladder may cause dysuria. In S. mansoni and S. japonicum corresponding effects may occur in the bowel: diarrhoea and blood. More com-

monly, the presence of a little blood is noticed in an otherwise normal stool. In most infections, no bowel symptoms are noticed. Through these mechanisms of blood loss, schistosomes may contribute to the development of iron-deficiency anaemia in some individuals.

Importance of retained eggs: the main pathology

The serious mischief in humans arises from tissue reaction to retained eggs. This reaction, which follows sensitization to egg antigens, is a circumoval granuloma. It results from combined humoral and cell-mediated attack on the egg, and the granuloma occupies several hundred times the volume of the egg itself. Its characteristics are epithelioid and giant cells, as well as lymphocytes and eosinophils arranged in concentric fashion around the egg. The cellular content diminishes with time, to be replaced by fibroblasts and a collagenous scar. Precipitation of Ag–Ab complex on the egg surface helps activate inflammation.

The duration of the vigorous cellular response to a single egg lasts a few weeks. If egg laying is stopped by chemotherapy, the cellular component of the granuloma usually resolves in 2 or 3 months. Not all granulomas lead to scars (Fig. 17.2). These pathological processes occur, with variations, in all the schistosome infections. They help to explain the specific features of each of the species described next.

Clinical features of Schistosoma haematobium

Obstructive uropathy

Eggs become deposited in the bladder and nearby organs, not singly, but usually in clutches. This is because a female schistosome may occupy the same site for long periods, during which time she lays several hundred eggs a day. The eggs give rise to a granulomatous lesion up to several centimetres in diameter. Most commonly these fleshy lesions form in the bladder mucosa, where they simulate tumours and are called pseudopapillomas. They may be sessile (flat) or pedunculated (on stalks). Smaller deposits of eggs cause lesions a few millimetres in diameter,

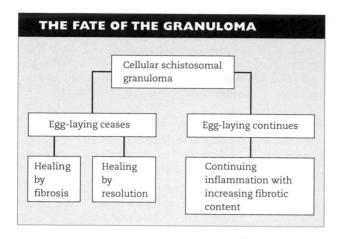

Fig. 17.2 The fate of the granuloma.

resembling tubercles. When granulomas form near the ureteric orifices or in the ureters themselves, the ureters may become obstructed. This is the cause of early obstructive uropathy.

The secondary effects are hydroureter, in which the ureter becomes dilated and elongated, with varying degrees of hydronephrosis. In the most severe cases, kidney drainage may be so impaired as to cause uraemia.

It used to be thought that all the changes of obstructive uropathy were irreversible. It is now known that in the early cellular phase of the granuloma, complete resolution may follow effective chemotherapy. Longitudinal follow-up has shown that, provided reinfection does not occur, spontaneous resolution without significant scarring may also occur.

Calcification of the bladder

This is common in S. haematobium because of calcification of the eggs, not of the bladder itself. A calcified bladder outline on X-ray is fully compatible with normal bladder function. Calcification tends to disappear over the years, provided reinfection ceases or is greatly reduced, presumably because of the continued passage of calcified eggs in the urine. Eggs in the ureters may be calcified in heavy infections, and the calcification may demonstrate the dilatation and tortuosity characteristic of hydroureter.

Genital schistosomiasis

Eggs from S. haematobium may be deposited at various sites throughout the urogenital system. Urethral papillomatous lesions have been reported in men and boys, and some men report changes in ejaculate consistency with or without blood in the semen. Semen microscopy in these cases often reveals eggs, but the effect on sperm counts, function and fertility have not been systematically investigated. Female genital schistosomiasis is increasingly recognized to include inflammatory lesions arising around deposited eggs in the vulva, vagina, cervix and fallopian tubes. The lower lesions can sometimes be mistaken for malignancies while the higher lesions are associated with sterility.

Schistosoma haematobium and the lung: schistosomal cor pulmonale

Eggs escaping from the pelvic veins into the caval circulation reach the lungs. In heavy prolonged infections, granuloma formation may cause obstruction in pulmonary arterioles. Pulmonary hypertension, right ventricular hypertrophy and congestive heart failure may follow. Cyanosis develops from vascular shunting in the lungs.

Clinical features of *Schistosoma mansoni*

Most patients with S. mansoni infections have few or no symptoms. The liver is often enlarged, the spleen only in the presence of portal hypertension or during the initial illness. Severe clinical effects, except those caused by ectopic worms (see below), are only seen in heavy infections.

Pseudopolyposis of the colon

In severe S. mansoni, granulomas in the large gut may develop into papilloma-like outgrowths of the mucosa. They may ulcerate and bleed, and cause symptoms of dysentery. There is no proven causal relationship to colonic carcinoma, and strictures do not form.

Schistosoma mansoni and the liver: schistosomal liver fibrosis

Severe long-standing S. mansoni infections cause a characteristic liver disease, 'Symmers pipestem fibrosis'. Large numbers of eggs escaping from the lower mesenteric veins and reaching the periportal regions cause a granulomatous response that leads to gradual occlusion of the intrahepatic portal veins. Portal hypertension follows, but liver cell function is not disturbed until very late in the pathological evolution. The clinical features are:

- enlargement of liver and spleen; and
- bleeding from oesophageal varices.

Patients tend to survive their bleeds much better than patients with true cirrhosis (e.g. caused by hepatitis B or alcohol), because of the well-preserved hepatocellular function. Also, because the serum albumin level is well maintained, ascites is not typical until the terminal stages of the disease. In late cases, hepatic perfusion may be so impeded that peripheral liver ischaemia occurs. Then, features of true cirrhosis may develop. When portocaval shunts are well established, the eggs of S. mansoni may bypass the liver in large numbers, and so reach the lungs. In some cases, they may be numerous enough to cause schistosomal cor pulmonale.

Clinical features of Schistosoma japonicum

This resembles S. mansoni but, for an equal number of worms, the infection is more severe. The parasite is less well adapted to humans, the circumoval granuloma is very large, and the egg output of each female worm is greater than that in S. mansoni. The initial illness (Katayama fever) may be prolonged and sometimes fatal. Many Chinese workers believe that S. japonicum can cause carcinoma of the colon, but most other experts consider the case is unproven.

Neuroschistosomiasis

Some worm pairs wander from their usual habitats and take up residence elsewhere. The chances of this happening are increased in heavy infections. The most important site for ectopic worms is the central nervous system (CNS), such as in the paravertebral venous plexus or the cerebral cortical veins. All three species may occasionally be found at these sites but this is, by comparison with pathology associated with worms in their usual sites, a rare occurrence.

In the paravertebral plexus, egg laying leads to the development of a granuloma in the constricted space of the spinal canal. The clinical syndrome is spinal cord compression or a cauda equina lesion. If treated promptly, full functional recovery may occur. When it does not, ischaemic injury may be the cause. In the brain, large localized granulomas produce symptoms and signs indistinguishable from a cerebral tumour. Diagnosis usually requires sophisticated brain imaging.

Schistosomiasis pathology at unusual sites

Schistosomiasis cases that do not have overt antemortem pathology have been shown at postmortem to have schistosome eggs in virtually all organs, but without associated inflammation. The relation of eggs found in the brain in this way to symptoms such as epilepsy and neurosis remains speculative. The factors that lead to damaging granuloma formation in some instances but not others are not well understood. Despite this conundrum, pathology associated with schistosomal eggs has been reported on rare occasions in diverse sites including skin, peritoneum and even bone.

Investigation

Direct diagnosis

This is the only approach that can lead to a definitive diagnosis. The adult worms are inaccessible, so the aim is to find living eggs. The miracidium inside the egg dies within 4 weeks.

Schistosoma haematobium

Something about bladder wall activity means that most eggs are voided around midday. Specimens collected between 10 a.m. and 2 p.m. are most likely to contain eggs. For quantitative surveys, it is very important to standardize urine collection times. Exercise has no demonstrable effect on egg output. There is no significant or reliable concentration of eggs in any part of the urinary stream. The methods for finding eggs depend on their high specific gravity (sedimentation) or their size (filtration).

Urine sedimentation
Eggs are sedimented by natural gravity (30 min in a conical glass; the sediment is then aspirated by Pasteur pipette and examined under coverslip using lens power × 10), or by artificial gravity (10 mL urine in 15 mL centrifuge tube, spun for 3 min at 1500 r.p.m., arm radius not critical; the deposit is then examined as before).

Living eggs are translucent, and the miracidium is recognizable. Flame cells can be seen flickering. The viability of the eggs can be checked by adding them to boiled (cool) water in a flask. Emerging miracidia are visible in light shone across the neck. Normal-looking eggs usually hatch. Opaque (calcified) eggs do not, and do not themselves signify active infection (live worms).

Urine filtration
Urine is passed through a filter by vacuum or pressure. An entire 24-h urine collection can be filtered. There are several variants of this method including a miniature membrane version that allows the eggs to be detected unstained. Advantages of the method are that it is sensitive, accurate for counts and a permanent record is available. Disadvantages are cost and time.

Schistosoma mansoni *and* Schistosoma japonicum

Direct smear examination is not sufficiently sensitive (e.g. for an output of 100 000 eggs per day; a stool of 200 g; a smear of 2 mg; the average count is 1 egg per smear). There is a 1 in 3 chance of finding no eggs in a patient who could be harbouring about 1000 *S. mansoni* worm pairs. Instead, more sensitivity is achieved through concentration techniques such as formol-ether, thiomersal, iodine and formol (TIF) glycerol sedimentation (the simplest) or a modified Kato smear.

Biopsy techniques for all schistosomal infections

A small piece of rectal mucosa can be removed by biopsy forceps or curette under direct proctoscopic vision. It is placed on a slide under a coverslip and examined under lens power (× 10). *Schistosoma haematobium* eggs are often trapped in the rectal mucosa, but may be calcified. It can be difficult to identify living eggs. Histology is not used for diagnosis as a routine. Serial sections are often needed as only the central slices of a granuloma will contain parts of the egg.

Indirect diagnosis

As less toxic drugs have become available, the imperative of making a definitive diagnosis before therapy has been reduced. Although all the indirect means of diagnosis suffer more or less from a lack of specificity, treatment is increasing based on this approach.

Immunodiagnostic tests for all species

There are numerous tests for detecting circulating antischistosomal antibody including CFT, IFAT, ELISA and several others. Although the better ones correlate well with the results of direct diagnostic methods, they all suffer from the following disadvantages to a greater or lesser extent.

1 They give no indication of the intensity of infection.

2 They do not distinguish between past and present infection.

3 They are not species specific.

4 Most require high technology and are often 'in-house' in academic institutions rather than commercially available for widespread use.

5 They do not reliably become positive until 3 months after infection.

Immunodiagnostic tests capable of detecting the presence of circulating antigen would be of much greater use to the clinician and epidemiologist. Unfortunately, this field has not progressed to the point where tests are available for routine clinical use in the tropics.

Approaches to diagnosis of different schistosomiasis clinical syndromes

Acute schistosomiasis

In the initial illness, the association of fever and eosinophilia with the other symptoms should raise the question of worms, as should the patient with diarrhoea and eosinophilia, although other worms such as *Strongyloides stercoralis*, *Capillaria philippinensis* and *Trichuris trichiura* can cause the same symptoms. In the differential diagnosis of these, direct diagnosis by examination of the stools is paramount.

Eosinophilia is not always prominent in acute schistosomiasis. In addition, acute schistosomiasis occurs at the onset of initial egg production, so eggs are rarely found in urine or stool and the antibody detection tests are not reliably positive at this stage of infection. For these reasons the diagnostic process includes elimination of other causes of fever such as malaria.

Schistosoma haematobium

In areas endemic for S. *haematobium*, the presence of haematuria (provided menstruating females are excluded) correlates well with the passage of schistosome eggs in the urine. With a dipstick-type test, provided it can detect both free haemoglobin and discrete red cells, the number of false-positives and false-negatives is very low. The false-positives are partly explained by glomerulonephritis and partly by the passage of dead eggs by patients whose worms are dead.

Radiological changes in the urinary tract may be very suggestive. Almost pathognomonic is the ring-like calcification of the bladder (Fig. 17.3, see colour plate facing p. 146), which may also involve the ureters, prostate and seminal vesicles. Multiple, rounded filling defects produced by pseudopapillomas in the bladder are also very typical. Ultrasonography is clearly important in detecting obstructive uropathy. Otherwise, unaccountable pulmonary hypertension in an endemic area should also arouse suspicion of schistosomiasis.

Schistosoma mansoni *and* Schistosoma japonicum

The presence of colonic polypi in an endemic area incriminates schistosome infection as the most likely cause, as does the syndrome of portal hypertension with normal liver function tests. In recent years, ultrasonography of the liver has been used to detect the typical pipe-stem fibrosis and alteration to liver shape and size in order to grade liver pathology.

Neuroschistosomiasis

The most useful clue, in cases with disease caused by ectopic worms or metastatic eggs, is the presence of eosinophilia. Unfortunately, eosinophilia is not invariably present, so immunodiagnostic tests may be particularly helpful.

Management

It is helpful to reach a definitive diagnosis with a direct test that confirms the presence of living worms before starting treatment. In practice, therapy is increasingly accepted on the basis of indirect evidence such as the results of urine dipstick detection of haematuria in an endemic country, or ELISA detection of antibody (where technology allows).

Drug treatment

All the available drugs (with the exception of artemisinin derivatives; see below) act on adult worm pairs only. After effectively eliminating the worms, the speed of resolution of the immunopathology induced by the eggs depends on how established the tissue damage has been.

Praziquantel (Biltricide)

This isoquinoline compound currently eclipses all other chemotherapy for schistosomiasis because of its ease of administration, lack of toxicity and price. It is effective against all human schistosomes. There has been some debate about the development of resistance in areas of intense transmission in West Africa, but the case for resistance remains unproven as yet. It is given in a dosage of 40 mg/kg as a single oral dose, which is sufficient for all species. Some argue that 30 mg/kg for two or three doses may be necessary for S. mansoni or S. japonicum infections. Side-effects include giddiness and minor gastrointestinal disturbances. No serious toxicity has been reported, but unexplained abdominal pain and short-lived bloody diarrhoea are troublesome in heavy S. mansoni infections. The drug should be used with caution during pregnancy and breast-feeding.

Other drug options

Metriphonate (active against S. haematobium only) and oxamniquine (active against S. mansoni only) are occasionally still used in some countries. Recent years have seen a rising interest in the use of the artemisinin derivatives in both treatment for and prophylaxis against schistosomiasis—particularly in China. This group of drugs appears to have effects against schistosomula as well as adult worms.

Management approaches for specific presentations

In the tropics, praziquantel is most frequently used to clear adult worms in patients presenting with symptoms caused by retained eggs in tissues, or as part of mass chemotherapy (see below). The specific presentations peculiar to individuals from non-endemic areas who pick up infections during travel need mention.

Acute schistosomiasis

Praziquantel is often used in the management of this condition but the speed of its effect on symptoms is variable. On the whole, this is a self-limiting illness caused by an excess of egg antigen triggering aggressive immune responses. These will persist for a variable length of time even after the adult worms have been killed, and in severe cases adjunctive corticosteroid therapy is occasionally advocated. It is wise to give a second dose of praziquantel 3 months after the first in order to clear worms that were only maturing during the initial illness.

Asymptomatic infection

Travellers who have one-off significant freshwater exposure (e.g. during water recreational activity such as snorkelling, wind-surfing or scuba-diving) are often screened for schistosomal antibody even when they have no symptoms. It is common therefore for praziquantel treatment to be offered on the basis of a positive antibody test alone. On the whole this is a reasonable approach, but its overall effectiveness and cost-effectiveness in preventing later pathology has not been assessed and there are some pitfalls.

1 Screening before an adequate time (3 months) has elapsed since exposure.

2 Assuming that antibody tests can be used to monitor cure and that titres will fall to negative after treatment.

Neuroschistosomiasis

Praziquantel is used to kill the adult worms, but the offending circumoval granulomas in the nervous tissue will take a while to resolve and there is usually (unfounded) concern that the immunopathology will worsen on treatment. Adjunctive corticosteroid therapy is therefore the norm. In the case of cerebral involvement with epilepsy, it may take many months before anticonvulsant therapy can be withdrawn.

Monitoring treatment

It can be assumed that most light infections will be cured with a single praziquantel dose. However, when a direct diagnosis detecting viable eggs has been made, it is wise to check that egg production has ceased (in the absence of rein-fection) at a 3-month follow-up.

Prevention and public health aspects

The schistosomiasis life cycle can be attacked at various sites. Although some sites have proved more vulnerable than others, combined approaches, where possible, have most impact but are rarely implemented in a sustained fashion.
1 Contamination of water.
2 Intermediate host.
3 Human contact with infection.

Reducing contamination of water
The main methods used are:
1 health education;
2 provision of sanitation;
3 prevention of access to transmission sites; and
4 reduction of egg excretion by the definitive hosts (humans) by drug treatment.
Of all these measures, the one most immediately successful in most circumstances is mass chemotherapy (see below).

Attack on snails
Permanent results are possible if the habitats can be eliminated. It has been achieved in Japan and many parts of China by drainage and landfill. Temporary results are obtained with the application of poisonous chemicals (mollus-cicides such as niclosamide) to snail habitats. If used alone, this method is usually disappoint-ing. The number of infected snails may not be reduced in proportion to the total snail reduction. Disadvantages include cost, the need to reapply chemicals for an indefinite period, and undesirable effects such as killing fish. It is most effective in highly controlled environ-ments, such as irrigated agricultural estates, and when used in combination with chemotherapy. Some enthusiasm has been generated around endod, a naturally occurring plant product which is rich in saponins and has molluscicidal activity.

Reducing contact with infection
The necessity for contact can be reduced by the provision of a safe water supply for washing and drinking through chlorination or filtration which will clear water of cercariae. This will not pre-vent recreational or occupational contact. At-tempts to fence off transmission sites are usually unsuccessful. Health education is important.

Mass chemotherapy
When community prevalence is more than 40%, mass chemotherapy is attractive. There are various refinements depending on the local cir-cumstances. Examples include targeting school-aged children because they have the highest intensity of infection or, in the case of S. haema-tobium, targeting only those with dipstick posi-tive tests for haematuria. Whatever the initial approach, follow-up repeat rounds of treat-ment are needed for sustained effects.

Future developments

New diagnostics
As with all important tropical infections, there is clearly a need for a new diagnostic tool that is as robust and specific as direct microscopy for eggs, less dependent on laboratory skill and in-frastructure, and more sensitive. However, as mass chemotherapy can be conducted without definitive diagnosis, the pressure to develop this tool has diminished and there are no obvious candidates for widespread use in developing countries.

New interventions
No new drugs look set to replace the current gold standard of single dose praziquantel. While global attention is focused on coordinated approaches to HIV/AIDS, TB and malaria, it is proving hard to mobilize broad and sustained public health campaigns against schistosomiasis.

Further reading

War on the Worm. *Parasitology Today* 1998; 14: 379–436. [This is a series of review articles covering the major areas of schistosomiasis, as well as future trends.]

WHO web-site http://www.who.int/health-topics/schisto.htm [Includes useful background and country-specific information, including the downloadable, but not reproducible without WHO permission, Report of the WHO Informal Consultation on schistosomiasis control. Geneva: 2–4 December 1998, WHO/CDS/CPC/SIP/99.2.]

Xiao SH, Booth M, Tanner M. The prophylactic effects of artemether against Schistosoma japonicum infections. Parasitology Today 2000; 16: 122–6. [A review concluding that artemether is a novel and effective drug against S. japonicum schistosomula, and also has schistosomicidal activity against other schistosome species. The successes recorded in trials with this drug in China offer new hope for occupationally exposed people such as flood-control workers, fishermen, and others who have benefited very little from on-going control strategies.]

CHAPTER 18

Leprosy

Epidemiology, 141
Microbiology, 141
Immune response in
  leprosy, 142

Clinical features, 143
Leprosy classification, 143
Diagnosis, 145
Management, 146

Prevention of disability, 148
Control and prevention, 149
Further reading and other
  resources, 150

Leprosy is a chronic granulomatous disease caused by *Mycobacterium leprae*. The principal manifestations of disease are anaesthetic skin lesions and peripheral neuropathy with peripheral nerve thickening. The clinical form of the disease in any individual depends on the degree of cell-mediated immunity (CMI) expressed by that individual towards *M. leprae*. High levels of CMI with elimination of leprosy bacilli produces the tuberculoid form of disease, whereas absence of CMI results in lepromatous leprosy. The medical complications of leprosy result from nerve damage, immunological reactions and bacillary infiltration. Nerve damage accompanying leprosy is a particularly serious complication because this will remain with the patient for the rest of his or her life and causes considerable morbidity. Currently available drug treatments are highly effective in clearing viable bacilli but do not prevent nerve damage. Leprosy has a long history as a deforming disease and leprosy patients the world over are frequently stigmatized and ostracized. Words such as 'leper' should be avoided and using the term Hansen's disease may reduce stigmatization.

Leprosy must be considered in the differential diagnosis of a patient who has lived in the tropics and who presents with any chronic or bizarre acute skin lesion, peripheral neuropathy or acute immune complex rheumatological syndrome.

Epidemiology

The infection is spread from human to human by droplets. There is a long clinical incubation period of 2–5 years for tuberculoid disease and 8–11 years for lepromatous disease. The geographical distribution is patchy, with 70% of cases in India. There is a steady global incidence of 650 000 new cases per year, and in 2000 there were 19 330 new cases detected. Age, sex and household contact are important determinants of disease. In the major leprosy endemic areas the childhood case rate remains at about 17%, indicating ongoing high transmission levels. HIV infection is not a risk factor for disease acquisition or severity.

Microbiology

Mycobacterium leprae is an obligatory intracellular parasite which cannot be cultivated *in vitro*, although it can be grown in the armadillo and in footpads of nude mice. *M. leprae* has a doubling time of 12 days and is a remarkably hardy organism, remaining viable in the environment for up to 5 months. It has a highly resistant cell wall composed of lipids, carbohydrates and proteins. Phenolic glycolipid *M. leprae* is species-specific. Numerous protein antigens have been identified as important immune targets using antibody and T-cell screening. The *M. leprae* genome was completely sequenced in 2001. The organism has lost many genes and survives on only a few biochemical pathways. Only 40 genes are unique to *M. leprae* and analysis of these genes will inform us about the unique biology of this organism.

Immune response in leprosy

The host immune response to *M. leprae* is crucial in determining either disease or immunity and the type of disease. The T cells and macrophages of the cell-mediated immune system have an important role in processing, recognition and response to *M. leprae* antigens. Antibodies to *M. leprae* antigens are produced but these do not appear to have any useful role in protection. Several stages in the immune response are recognized:
• phagocytosis of *M. leprae* by macrophages;
• presentation of *M. leprae* antigens in association with human leucocyte antigen (HLA) class II molecules;
• binding of antigen specific T cells via the a/b T-cell receptor;
• activation of T cells and production of interleukin 2 (IL-2) and T-cell proliferation;
• IL-2 activates CD4, CD8, natural killer (NK) cells and macrophages; and
• interferon-γ (IFN-γ) is produced and activates bactericidal mechanisms within the macrophage.

Granuloma formation results from mycobacterial persistence with continued cytokine release. The leprosy granuloma has a core of macrophages, epithelioid cells and giant cells, with lymphocytes surrounding the core, and is dependent on tumour necrosis factor α (TNF-α) from activated macrophages and T cells.

The immunological and clinical effects vary across a spectrum between two 'poles' of presentations (Fig. 18.1). In tuberculoid disease

Fig. 18.1 The immunological features of the different types of leprosy.

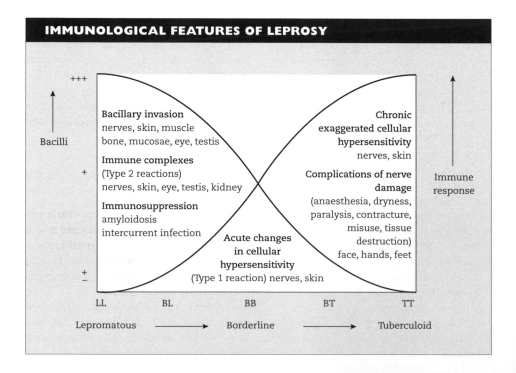

IMMUNOLOGICAL FEATURES OF LEPROSY

+++

Bacilli

+

+
−

Bacillary invasion
nerves, skin, muscle
bone, mucosae, eye, testis

Immune complexes
(Type 2 reactions)
nerves, skin, eye, testis, kidney

Immunosuppression
amyloidosis
intercurrent infection

**Acute changes
in cellular
hypersensitivity**
(Type 1 reaction) nerves, skin

**Chronic
exaggerated cellular
hypersensitivity**
nerves, skin

**Complications of nerve
damage**
(anaesthesia, dryness,
paralysis, contracture,
misuse, tissue
destruction)
face, hands, feet

Immune
response

LL   BL   BB   BT   TT

Lepromatous ⟶ Borderline ⟶ Tuberculoid

(TT), CMI is active and contains infection, so that few bacilli are found in tissues and CD4 cells and their cytokines (IL-2, IFN-γ) predominate (Th1 response). At the other pole of lepromatous leprosy (LL), the CMI response is poor, there are many bacilli in the tissues and responses include both CD4 and CD8 cells. CD4 cells produce IL-4, IL-5, IL-6 and IL-10 (Th2 response) and immunoglobulin G (IgG), IgM and IgA levels are also elevated. The unresponsiveness in LL disease is because of specific T-cell anergy. It may be caused by T-cell non-activation, suppression or clonal deletion and also involves defective macrophage function. Borderline states (borderline tuberculoid, BT; borderline leprosy, BB; borderline lepromatous leprosy, BL) are intermediate between these poles.

There is no current immunological test that can determine whether a person has protective immunity against *M. leprae*.

Clinical features

The cardinal signs of leprosy are skin lesions, anaesthesia and thickened peripheral nerves.

Skin lesions (Figs 18.2–18.5, see colour plate facing p. 146)
The most common skin lesions are macules or plaques; more rarely, papules and nodules are seen. In LL, diffuse infiltration of the skin often occurs. Lesions may be found anywhere, although rarely in the axillae, perineum or hairy scalp. The number of lesions indicates the ability of the CMI to limit the spread of bacilli. Tuberculoid patients have few hypopigmented lesions, while lepromatous patients have numerous sometimes confluent lesions. The few tuberculoid lesions are usually asymmetrical; more numerous lesions are likely to be distributed symmetrically.

Nerve damage (Figs 18.6 & 18.7, see colour plate facing p. 146)
Only the peripheral nervous system is affected. Damage to peripheral nerve trunks is common and leads to motor weakness in the muscles supplied and to regional sensory loss. Sensory and autonomic fibres in skin lesions are also affected. The principal sites of peripheral nerve involvement are ulnar (elbow), median (wrist), radial cutaneous (wrist), common peroneal (knee), posterior tibial and sural nerves (ankle) and the facial nerve (zygomatic arch). All these nerves should be examined for enlargement and tenderness. Nerve function impairment occurs before, during and after treatment. In field cohort studies, 16–56% of newly diagnosed patients had functional nerve impairment.

Leprosy classification

Classification of the stage of disease helps to predict the future likelihood and types of reaction that a patient may experience, and also guides the content and duration of specific chemotherapy. Polar forms of disease are stable but borderline disease is unstable. BT/BL disease is associated with severe large nerve damage caused by Type 1 'reversal' reactions and neuritic reactions, and LL/BL patients suffer erythema nodosum leprosum (ENL) (Type 2) reactions. Multibacillary disease requires longer treatment with more drugs to prevent relapse or the development of drug resistance.

The Ridley–Jopling classification uses clinical and microbiological features of the patient, which mirror the immunological state (Table 18.1).
• Skin lesions
  Number
  Distribution and symmetry
  Definition and clarity
  Anaesthesia
  Loss of sweating and hair growth
• Peripheral nerve involvement
• Mucosal and systemic involvement
• Bacillary load
A simpler World Health Organization (WHO) field classification merely divides patients into those with few lesions (paucibacillary) or those with more (multibacillary) if skin smears are not available (see Management section).

CHARACTERISTICS OF POLAR LEPROSY

	Tuberculoid (TT)	Lepromatous (LL)
Skin and nerves		
Number and distribution	One or a few sites, asymmetrical	Widely disseminated
Skin lesions		
Definition:		
Clarity of margin	Good	Poor
Elevation of margin	Common	Never
Colour:		
Dark skin	Marked hypopigmentation	Slight hypopigmentation
Light skin	Coppery or red	Slight erythema
Surface	Dry, scaly	Smooth, shiny
Central healing	Common	None
Sweat and hair growth	Impaired early	Impaired late
Loss of sensation	Early and marked	Late
Nerve enlargement and damage	Early and marked	Late
Bacilli (bacterial index)	Absent (0)	Many (5 or 6+)
Natural outcome	Healing	Progression

Table 18.1 Main clinical characteristics of polar leprosy.

Tuberculoid leprosy

Infection is localized and asymmetrical. The skin lesions are few, hypopigmented and have sharp borders. Anaesthesia is usually present in the lesion and is often accompanied by loss of sweating, indicating local autonomic nerve damage. The cutaneous nerve on the proximal side of the lesion is frequently thickened. If peripheral nerve trunk involvement is present, usually only one nerve trunk is enlarged. No *M. leprae* are found in the skin and the lepromin test is strongly positive. True tuberculoid leprosy has a good prognosis, many infections resolve without treatment and peripheral nerve trunk damage is limited.

Borderline tuberculoid

The skin lesions are similar to TT leprosy but are larger and more numerous. The margins are less well-defined and there may be satellite lesions. Damage to peripheral nerves is widespread and severe, usually with several thickened nerve

trunks. BT patients are at risk of severe reversal (Type 1) reactions with rapid deterioration in nerve function with consequent deformities.

Borderline leprosy

BB disease is the most unstable part of the spectrum and patients usually downgrade towards LL if they are not treated, or upgrade towards tuberculoid leprosy as part of a reversal reaction. There are numerous skin lesions, which may be macules, papules or plaques and vary in size, shape and distribution. The edges of the lesions may have streaming irregular borders. Annular lesions with a broad irregular edge and a sharply defined punched-out centre are characteristic of BB disease. Nerve damage is common with involvement of several peripheral nerve trunks.

Borderline lepromatous leprosy

Borderline lepromatous leprosy is characterized by widespread, small but variable macules all over the body. With disease progression the macules become infiltrated. Peripheral nerve involvement is widespread and often severe.

Patients with BL leprosy are at risk of both reversal and ENL (Type 2) reactions.

Lepromatous leprosy

The patient with untreated polar lepromatous leprosy may be carrying 10^{11} leprosy bacilli and the characteristic signs of LL are caused by the widespread dissemination of organisms throughout the body. The onset of disease is frequently insidious, the earliest lesions being ill-defined, widely distributed hypopigmented macules. Gradually, the skin becomes infiltrated and thickened and nodules develop. Thickening of facial skin gives rise to the characteristic leonine facies. Hair is lost, especially the lateral third of the eyebrows (madarosis) (Fig. 18.8, see colour plate facing p.00). Dermal nerves are destroyed and there is sensory loss (light touch, pain and temperature) which begins at the hands and feet. Sweating is lost and this can cause profound discomfort in a tropical climate as compensatory sweating occurs in the remaining intact areas. Damage to peripheral nerves is symmetrical and occurs late in disease.

Nasal symptoms (stuffy nose, nose bleeds, loss of sense of smell) can often be elicited early in the disease, and 80% of newly diagnosed lepromatous cases have invasion of the nasal mucosa. Septal perforation may occur. The pathognomonic collapse of the bridge of the nose is secondary to bacillary destruction of the bony nasal spine. Bone involvement is common, with osteoporosis and fractures. Testicular atrophy results from diffuse infiltration and the acute orchitis that occurs with ENL reactions. The consequent loss of testosterone leads to azoospermia and gynaecomastia.

Other forms of leprosy

These include pure neuritic, histoid and Lucio's leprosy, and will not be considered further here.

The eye in leprosy

Blinding complications include lagophthalmos, decreased corneal sensation, acute iritis, chronic iritis and cataract. Patients at risk include those with BL and/or LL disease, those with facial patches, patients experiencing reactions and those with disease of long duration (Table 18.2).

Diagnosis

The diagnosis is made on the clinical findings of one or more of the cardinal signs of leprosy and supported by the finding of acid-fast bacilli on slit skin smears in multibacillary cases. The whole body should be inspected in a good light otherwise lesions may be missed, particularly on the buttocks. Skin lesions should be tested for anaesthesia to light touch, pinprick and temperature. The peripheral nerves should be palpated systematically, examining for thickening and tenderness. Wherever possible the diagnosis should be supported by a skin biopsy, which is essential for accurate classification. Serology is not usually helpful diagnostically because antibodies to the species-specific glycolipid PGL-1 are present in 90% of untreated lepromatous patients but only 40–50% of paucibacillary patients and 5–10% of healthy controls. Polymerase chain reaction for detecting *M. leprae*

EYE DISEASE IN LEPROSY
Conjunctivitis
Infectious/allergic aetiology
Redness throughout and inside eyelids
Treat with tetracycline ointment
Acute iritis
Conjunctiva reddest next to cornea
Eye painful
Sensitive to light
Poor pupil reactivity
Corneal ulcer
Roughness
Discharge and redness
Treat with tetracycline ointment
Chronic iritis
Pupil constricted and irregular
Sluggish reaction
Treat with atropine eye drops

Table 18.2 Key features of eye disease in leprosy.

DNA has not proved sensitive or specific enough for diagnosis.

Slit skin smears

The bacterial load is assessed by making a small incision through the epidermis, scraping dermal material and smearing it evenly onto a glass slide. At least six sites should be sampled (earlobes, eyebrows, edges of active lesions). The smears are then stained and acid-fast bacilli are counted. Scoring is performed on a logarithmic scale per high power field (the bacterial index, BI). A score of 1+ indicates 1–10 bacilli in 100 fields, 6+ indicates > 1000 per field. Smears are useful for confirming the diagnosis and should be carried out annually to monitor response to treatment. Leprosy is essentially a clinical diagnosis.

Differential diagnosis

A wide variety of dermatological conditions might be considered in the differential diagnosis of manifestations of leprosy, which include erythematous macules, hypopigmented macules, papules and nodules. Neurological problems include both mononeuropathies and polyneuropathies. Diabetes is a common cause of peripheral neuropathy and may coexist with leprosy, but does not cause nerve thickening.

Management

Educating a leprosy patient about their disease is the key to successful management. Key issues to be discussed include the low infectivity, the importance of treatment adherence and/or compliance, warnings about reactions, the care of anaesthetic hands and feet and support with social issues.

Chemotherapy

The effectiveness of dapsone against *M. leprae* was discovered in the late 1940s and it was used widely as a single agent. This led to the widespread development of dapsone resistance and in 1982 the WHO proposed a multidrug regimen for the treatment of leprosy. In a multibacillary patient there are three distinguishable types of bacilli: fully drug-sensitive bacteria, drug-resistant mutants and a small population of 'persisters', which are dormant non-multiplying bacilli. Treatment with a multidrug regimen should eliminate nearly all organisms. The first-line antileprotic drugs are rifampicin, dapsone and clofazimine.

Rifampicin

Rifampicin is a potent bactericidal for *M. leprae*. Four days after a single 600 mg dose, bacilli from a previously untreated multibacillary patient are no longer viable. It acts by inhibiting DNA-dependent RNA polymerase, thereby interfering with bacterial RNA synthesis. Rifampicin is well absorbed orally. Hepatotoxicity is rarely a problem. Because *M. leprae* resistance to rifampicin can develop as a one-step process, rifampicin should always be given in combination with other antileprotic drugs.

Dapsone

Dapsone (DDS; 4,4-diaminodiphenylsulphone) acts by blocking folic acid synthesis. It is only weakly bactericidal. Oral absorption is good and it has a long half-life, averaging 28 h. Haemolytic anaemia is the most common side-effect of dapsone treatment and patients with glucose-6-phosphate dehydrogenase (G6PD) deficiency are particularly at risk. The 'DDS syndrome', which is occasionally seen in leprosy, starts 6 weeks after commencing DDS and manifests as exfoliative dermatitis associated with lymphadenopathy, hepatosplenomegaly, fever and hepatitis. Agranulocytosis, hepatitis and cholestatic jaundice occur rarely with DDS therapy.

Clofazimine

Clofazimine is a dye that has a weakly bactericidal action. It also has an anti-inflammatory effect and helps prevent ENL. The major side-effect is skin discoloration, ranging from red to purple-black, the degree of discoloration depending on the dose and the amount of leprous infiltration. The pigmentation usually clears up within 6–12 months of stopping clofazimine, although traces

Fig. 8.2 Patient with massive splenomegaly.

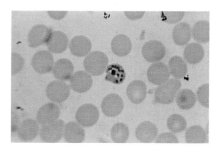

Fig. 9.2 A schizont of *Plasmodium falciparum*.

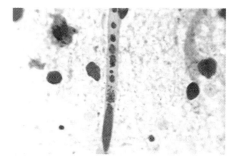

Fig. 9.3 Impression of brain (Giemsa stain) showing a capillary 'blocked' by schizonts in fatal cerebral malaria due to *Plasmodium falciparum*.

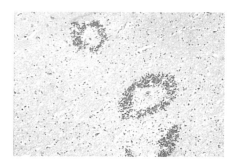

Fig. 9.4 Ring haemorrhages in fatal cerebral malaria: presumably these lesions are irreversible.

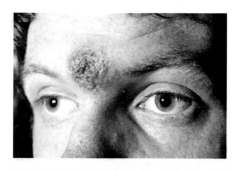

Fig. 11.1 *Leishmania tropica* lesion (Saudi Arabia).

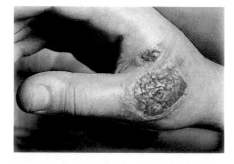

Fig. 11.2 Cutaneous leishmaniasis lesion due to *Leishmania mexicana* (Brazil).

[facing p. 146]

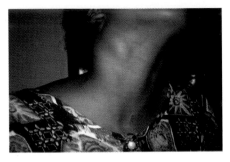

Fig. 12.4 A patient with tuberculous lymphadenitis.

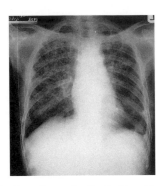

Fig. 12.5 Chest X-ray of patient with miliary tuberculosis.

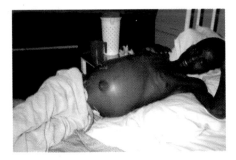

Fig. 13.1 Gross ascites due to tuberculous peritonitis in a patient with AIDS.

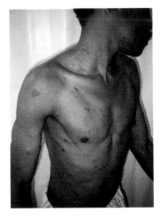

Fig. 13.2 Multiple lesions of Kaposi's sarcoma in a patient with AIDS.

Fig. 14.2 Early onchocerciasis with an itchy papular rash in a student from Cameroon.

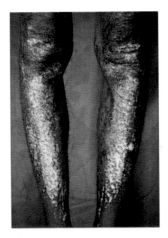

Fig. 14.3 Onchocerciasis: typical depigmentation associated with grossly exaggerated skin fold pattern.

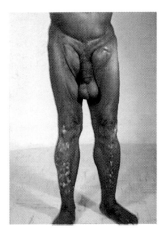

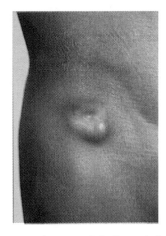

Fig. 14.4 Advanced onchocerciasis (West Nigeria). There is depigmentation of the skin overlying the shins, enlargement of the inguinal lymph glands associated with laxity of the surrounding skin (hanging groins) and generalized presbydermia.

Fig. 14.5 An *Onchocerca* nodule. The typical site in Africa. Many worms may be incarcerated in a multiloculated nodule.

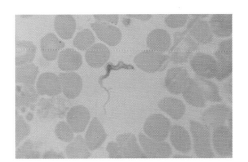

Fig. 15.2 Parasite of *Trypanosoma brucei rhodesiense* in the blood film of a severely ill Zambian adult.

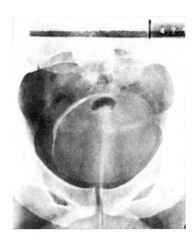

Fig. 17.3 Calcified bladder in *Schistosoma haematobium* infection. Many such bladders are capable of entirely normal function, the calcification involving the eggs rather than the bladder tissues.

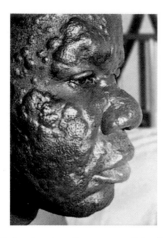

Fig. 18.2 Lepromatous leprosy—nodular form.

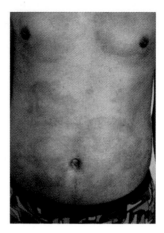

Fig. 18.3 Skin lesions of borderline lepromatous leprosy.

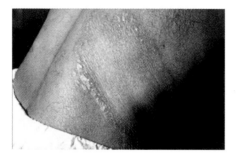

Fig. 18.4 A patient from India with tuberculoid leprosy. The single lesion is anaesthetic, scaly, dry, and has a raised edge.

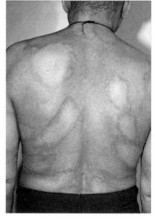

Fig. 18.5 Borderline tuberculoid leprosy with numerous anaesthetic skin lesions.

Fig. 18.6 Visibly thickened posterior auricular nerve. The patient had a 'tuberculoid'-type lesion on the palm of his right hand, and an associated thickening of the dorsal branch of the radial nerve.

Fig. 18.7 An ulnar nerve lesion in a patient from South Africa with lepromatous leprosy.

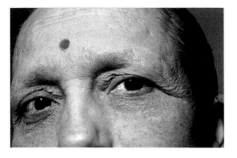

Fig. 18.8 Lateral eyebrow loss or 'madarosis' in a patient with lepromatous leprosy.

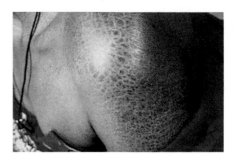

Fig. 18.9 Icthyotic skin changes in a patient on clofazimine treatment.

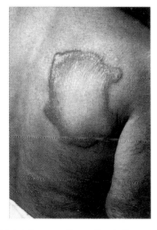

Fig. 18.10 A reversal reaction in leprosy. The previously flat lesions suddenly become hot, painful and raised.

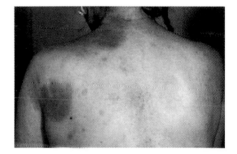

Fig. 18.11 Erythema nodosum leprosum (ENL) in an Asian patient who presented with fever, leucocytosis and painful red skin swellings.

Fig. 18.12 Perforating neuropathic plantar foot ulcer in a patient with lepromatous leprosy and extensive lower limb denervation.

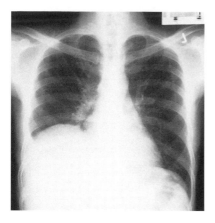

Fig. 19.1 Chest X-ray of a patient with a large amoebic liver abscess, showing a grossly elevated right hemidiaphragm.

Fig. 21.1 Though it looks like urine, this is the 'ricewater' stool from a patient with cholera.

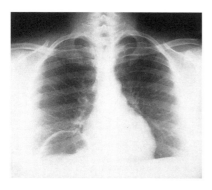

Fig. 27.1 A hydatid cyst at the base of the right lung, found in an African patient 'incidentally'.

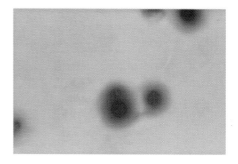

Fig. 32.1 Budding yeast-like organisms of *Cryptococcus neoformans* in the cerebrospinal fluid, stained with Gram's stain.

Fig. 47.1 A chronic tropical ulcer in a poor and malnourished young Nigerian patient.

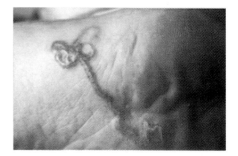

Fig. 50.1 Typical rash of cutaneous larva migrans in a holidaymaker returned from a beach holiday in the Caribbean.

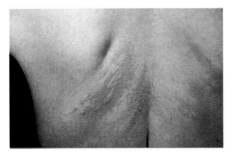

Fig. 52.1 The 'larva currens' rash of strongyloidiasis in a former Second World War prisoner of the Japanese. The serpiginous wheals come and go in a few hours and travel rapidly over the central body areas. The rash is due to tissue larval migration of *Strongyloides stercoralis*.

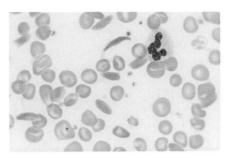

Fig. 56.1 Sickle cells. From Bain B. *Blood Cells: A Practical Guide*, 3rd edn. Oxford: Blackwell Publishing, 2002.

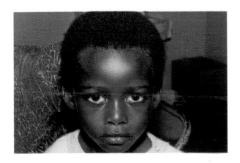

Fig. 56.2 Frontal bossing in a child with sickle cell anaemia.

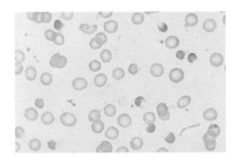

Fig. 56.3 Red cell appearances during acute haemolytic crisis caused by G6PD deficiency. From Bain B. *Blood Cells: A Practical Guide*, 3rd edn. Oxford: Blackwell Publishing, 2002.

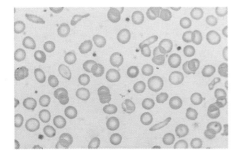

Fig. 57.1 Iron-deficient red cells. The cells are paler and more irregular in shape than normal red cells. From Bain B. *Blood Cells: A Practical Guide*, 3rd edn. Oxford: Blackwell Publishing, 2002.

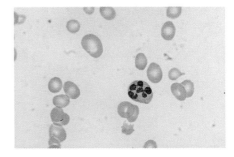

Fig. 57.2 Oval macrocytes and hypersegmented neutrophils in folate deficiency. From Bain B. *Blood Cells: A Practical Guide*, 3rd edn. Oxford: Blackwell Publishing, 2002.

RECOMMENDED MULTIDRUG THERAPY REGIMENS

Type of leprosy	Drug treatment		Duration of treatment
	Monthly supervised	Daily self-administered	
Paucibacillary (PB)	Rifampicin 600 mg	Dapsone 100 mg	6 months
Multibacillary (MB)	Rifampicin 600 mg	Clofazimine 50 mg	24 months
	Clofazimine 300 mg	Dapsone 100 mg	
Paucibacillary single lesion Rifampicin 600 mg, ofloxacin 400 mg, Single dose minocycline 100 mg 'ROM'			

Children
PB: Rifampicin 450 mg monthly and dapsone 50 mg daily
MB: Rifampicin 450 mg and clofazimine 150 mg monthly, clofazimine 50 mg alternate days and
dapsone 50 mg daily

Table 18.3 Modified multidrug therapy regimens recommended by the World Health Organization (WHO).

FIELD USE

Paucibacillary single lesion leprosy	1 skin lesion
Paucibacillary	2–5 skin lesions
Multibacillary	> 5 skin lesions

Table 18.4 World Health Organization classification for field use when slit skin smears are not available.

of discoloration may remain for up to 4 years. Clofazimine also produces a characteristic ichyosis on the shins and forearms (Fig. 18.9, see colour plate facing p. 146). Gastrointestinal side-effects, ranging from mild cramps to diarrhoea and weight loss, may occur as a result of clofazimine crystal deposition in the wall of the small bowel.

Multidrug therapy

This has been used to treat over 10 million people since 1982. Relapse rates are low, ranging from 0% in China and Ethiopia to 2.04 per 100 person-years in India. Patients with high initial bacterial loads are at greater risk of relapse (8 per 100 person-years). So far there has been no reported drug resistance. Toxicity is limited, response is rapid and the duration of therapy is greatly shortened compared to historical treatment regimens. The recommended regimens are summarized in Table 18.3. WHO now recommends treatment of multibacillary (Table 18.4) patients for 12 months only (although there were no controlled trial data to guide this decision). Second-line agents (ofloxacin and minocycline) are used in a single dose treatment regimen with rifampicin ('ROM') for patients

with single lesion paucibacillary disease. If this regimen is used it is imperative to examine the patient carefully to ensure that he or she only has one lesion and does not have any nerve damage. In a trial, ROM was not shown to be as effective as the 6-month WHO-paucibacillary (PB) regimen, but single dose treatment has major operational advantages.

Clarithromycin also has antimycobacterial activity and is sometimes used in tailored regimens.

Reactions and nerve damage

These include Type 1 (reversal) reactions, Type 2 (ENL) reactions and acute neuritis.

Type 1 (reversal) reactions (Fig. 18.10, see colour plate facing p. 146)

These are caused by delayed hypersensitivity towards *M. leprae* antigens in skin and nerve.

Those at risk include all borderline (BT, BB and BL) patients and women in the postpartum. The peak time for reactions is during the first 2 months of treatment. Type I reactions occur in 30% of BL patients. Clinical manifestations include erythema, swelling and tenderness of skin lesions, and pain and tenderness of peripheral nerves with loss of sensory and motor function. Rapid severe nerve damage may occur with Type I reactions, so patients must be warned about symptoms and advised to return for treatment if they develop new weakness or numbness. Nearly all reactions, and especially those with nerve inflammation, must be treated with 40 mg/day prednisolone, reducing by 5 mg/day every month. Physiotherapy will be needed for affected hand, foot and eye muscles.

Erythema nodosum leprosum reaction

This is also known as a Type 2 reaction (Fig. 18.11, see colour plate facing p. 146) and is caused by immune complex deposition, T-cell dysregulation and overproduction of TNF. It affects 20% of LL and 10% of BL patients. There is systemic illness with malaise, fever and raised white cell count and erythrocyte sedimentation rate. Manifestations include widespread erythema nodosum, neuritis, iritis, arthritis, orchitis, lymphadenopathy and renal disease. ENL usually starts in the second year of chemotherapy and may relapse intermittently over several years. It is often difficult to treat.

Mild cases can be treated symptomatically with aspirin or non-steroidal anti-inflammatory drugs (NSAIDs). Antileprosy drugs should be continued and the patient reassured that the reaction will settle. In moderate or severe cases, high-dose prednisolone (60–80 mg/day) should be used for short periods. Thalidomide is a very effective steroid-sparing agent, especially in chronic ENL, and increasing the dosage of clofazimine also helps reduce inflammatory responses. Iritis should be treated with 1% atropine and 1% steroid eye drops.

Neuritis

Neuritis refers to acute and chronic nerve inflammation that may occur without evidence of either a Type 1 or Type 2 reaction. Nerve damage may also occur as silent neuropathy which is defined as 'the development of functional deficit of a major nerve without a manifest neuritis'. Nerve function should be checked carefully during treatment so that silent neuropathy can be detected. Treatment is with 40 mg/day prednisolone as for reversal reactions, reducing slowly over a period of months.

Prevention of disability

Nerve damage produces anaesthesia, dryness and muscle weakness. These three factors lead to misuse of the affected limb with resultant ulceration, infection and, ultimately, severe deformity (Fig. 18.12, see colour plate facing p. 146). Keys to prevention include regular monitoring of nerve function and recording of problems secondary to nerve dysfunction. Patients who need self-care should be identified and their understanding and implementation of this should be monitored (Table 18.5). All patients need general training and social support; some may need surgical referral.

Reconstructive surgery

Reconstructive surgery has a role in both improving function and appearance. Lagophthalmos can be ameliorated by tarsorrhaphy or temporalis muscle transfer. Appropriate tendon transfers can reduce the effects of ulnar and median nerve paralysis and improve drop foot and claw toes. Cosmetic surgery, in particular eyebrow replacement, nasal reconstruction and reduction of gynaecomastia, is important in the rehabilitation of severely deformed patients.

Women and leprosy

Women with leprosy are in double jeopardy. Not only may they develop post-partum nerve damage, but they are at particular risk of social ostracism with rejection by spouses and family. There is little good evidence that pregnancy causes new disease or relapse. However, there is a clear temporal association between parturition and the development of Type I reactions

CARE OF HANDS AND FEET

Self-care training
Inspect for injury
Understand why injuries happen
Soak feet and oil
Remove callus
Exercise hands and feet
Treat injuries promptly

Hands
Problems with heat/pressure/sharp objects

Feet
Plantar ulcers occur at pressure sites
Either walk less or use protective shoe insoles

Footwear
Check shoe fittings
Provide insoles
Arch supports, metatarsal pads
Orthopaedic footwear

Management of plantar ulcers
Clean wound
Bed rest, walking plaster
Check footwear
Find out why it happened

Table 18.5 General care of hands and feet.

and neuritis when CMI returns to pre-pregnancy levels. ENL in pregnancy is associated with early loss of nerve function compared with non-pregnant individuals. Rifampicin, dapsone and clofazimine are safe during pregnancy, but ideally pregnancies should be planned for when leprosy is well controlled. Women may breast-feed while on multidrug therapy but should be warned that low levels of clofazimine are excreted in the breast milk and may cause some skin discoloration in the infant.

Leprosy in childhood

All types of leprosy are seen in childhood, usually after the age of 5 years. Children are at the same risk as adults of developing nerve damage and reactions, and often present with established nerve damage. Reactions should be treated with prednisolone 0.5 mg/kg/day. The WHO has produced separate multidrug thera-py paucibacillary (PB) and multibacillary (MB) blister packs for treating children.

Control and prevention

Vaccines

The substantial cross-reactivity between bacille Calmette–Guérin (BCG) and *M. leprae* has been exploited in attempts to develop a vaccine against leprosy. Trials of BCG as a vaccine against leprosy showed it to confer variable protection, ranging from 80% in Uganda to 20% in Burma. A case–control study in Venezuela showed BCG vaccination to give 56% protection to the household contacts of leprosy patients. Combining BCG and killed *M. leprae* has been tried, but in both a large population-based trial in Malawi and an immunoprophylactic trial in Venezuela there was no advantage for BCG plus *M. leprae* over BCG alone.

Leprosy control programmes

The current strategy of vertical control programmes dedicated to leprosy, with case detection, treatment with the WHO drug regimen and contact examination has been very successful. Efficient treatment is not merely restricted to chemotherapy but also involves good case management with effective monitoring and supervision. An important secondary role of leprosy control programmes is the prevention of disabilities and population education about leprosy. However, the changing prevalence of leprosy means that vertical programmes are unlikely to be sustainable for much longer. New options include combined leprosy and tuberculosis programmes, dermatology programmes and full integration with general health services. Whichever of these models is chosen, it is important that a range of skills should be preserved, from maintaining surveillance and reporting on leprosy indicators through to ensuring that prednisolone is available at health points to providing adequate prevention of disability services. The WHO declared that 'the elimination of leprosy as a public health problem' should be achieved by the year 2005. While this has not been possible for all areas, major progress has been made towards this goal.

Further reading and other resources

LEPRA, Fairfax House, Causton Road, Colchester CO1 1PU, UK. [International leprosy charity, can provide posters, pamphlets, etc.]

Leprosy Review. [The premier leprosy journal, free internet access at www.lepra.org.uk]

Lockwood DNJ. Leprosy elimination: a virtual phenomenon or a reality? Br Med J 2002; 324: 1516–18.

Report of the International Leprosy Association Technical Forum. Evidence-based assessment of leprosy management issues. Lepr Rev 2002; 73 (Suppl.). [Covers everything from epidemiology through diagnosis and treatment to prevention of disability.]

Srinivasan H. Prevention of Disabilities in Patients with Leprosy: A Practical Guide. Geneva: World Health Organization, 1993.

World Health Organization expert committee on Leprosy. Summary of Current Recommendations for the Management of Leprosy in the Field. Geneva: WHO Technical Report Series, 1998. [Current official guidelines.]

PART 3

Other Tropical Diseases

CHAPTER 19

Amoebiasis

Introduction and
    epidemiology, 153
Parasite and life cycle, 153
Pathogenesis, 154
Clinical features, 154

Investigations, 156
Management, 157
Prevention and public health
    aspects, 158

Recent developments, 158
Further reading, 159

Introduction and epidemiology

Entamoeba histolytica is an intestinal protozoan parasite with the ability to invade and cause lysis of cells. *Entamoeba histolytica* occurs worldwide, particularly in situations of poor hygiene and sanitation. Thus, infections are commonly found among people living in developing countries and immigrants or travellers from such countries. In addition, people with learning difficulties in residential institutions, men who have sex with men, and people who are immunosuppressed are also at increased risk.

The most common clinical presentation is amoebic dysentery. Extra-intestinal infections also occur, notably amoebic liver abscess (ALA). It has been shown experimentally that the majority of infections are asymptomatic and only about 20% of people who swallow cysts develop symptoms of dysentery. There are estimated to be between 40 and 50 million cases of symptomatic amoebiasis per year, resulting in 40 000–110 000 deaths.

In the past, the prevalence of infection in developing countries has been estimated to exceed 90% in some communities. This is probably an overestimate as cysts of *E. histolytica* are microscopically identical to cysts of the non-pathogenic *Entamoeba dispar*. *Entamoeba dispar* is about three times as common as *E. histolytica* in developing countries and about 10 times as common in industrialized countries. Other non-pathogenic protozoa found in the human intestine include *Entamoeba moshkovskii* (which also produces cysts identical to those of *E. histolytica/E. dispar*), *Entamoeba coli*, *Entamoeba hartmanni* and *Endolimax nana*.

Other amoebae sometimes associated with disease in humans include *Entamoeba gingivalis* (periodontal disease), *Entamoeba polecki* and *Dientamoeba fragilis* (diarrhoea), *Acanthamoeba* species and *Balamuthia mandrillaris* (acanthamoebic keratitis/granulomatous amoebic encephalitis) and *Naegleria fowleri* (primary amoebic meningoencephalitis).

Parasite and life cycle

Entamoeba histolytica is principally an infection of humans, although some monkeys also harbour the parasite. The four-nucleated cyst is ingested in food or water contaminated by human faeces. The cyst is digested in the gut releasing eight amoebic trophozoites. These live in the colon, normally on the surface of the mucosa, feeding on bacteria and other food residues. The amoebic trophozoite is variable in size, highly motile by means of its pseudopodia and characteristic flowing motion and, when invasive, usually contains ingested red cells.

The amoebae multiply in the gut by simple binary fission. As they move around the colon from right to left, the colonic contents become more solid and the actively motile amoebae stop feeding, empty their food vacuoles, become rounded and secrete a cyst wall.

The cysts are spherical, measuring 10–15 μm, and when mature contain four nuclei and sometimes a glycogen mass and a refractile chromidial bar. Amoebic cysts are passed in the formed stool of people with, usually asymptomatic, amoebiasis. The cysts can survive for prolonged periods in normal environmental conditions; e.g. for more than 12 days in cool faeces and for several weeks in water. They are killed by drying at temperatures above 50°C, freezing below – 5°C and standard treatment of water supplies.

Under normal circumstances amoebic trophozoites are said to be non-infective; however, an epidemic of amoebic dysentery was caused by the introduction of trophozoites via an incorrectly functioning enema machine used by chiropractors in the USA and resulted in several deaths. Person–person transfer and inoculation of trophozoites into skin abrasions or mucous membranes can also occur, resulting in cutaneous amoebiasis.

Pathogenesis

Entamoeba histolytica binds to host intestinal cells by means of a galactose-binding lectin on its surface. Following attachment, the amoeba uses pore-forming molecules called amoebapores and, possibly, phospholipidases, to disrupt the target cell, triggering a process of apoptosis, or 'cell suicide'. The amoeba then phagocytoses the dead cell and in due course the process leads to the development of mucosal ulcers with undermined edges, commonly described as 'flask-shaped' ulcers. However, this 'bind–lyse–eat' model for invasive amoebiasis may be oversimplistic. Invasion also appears to depend on cytoskeleton motility, the secretion of proteases that degrade the extracellular matrix and antibody. The host inflammatory response, including the production of cytokines and inflammatory mediators, accompanied by an influx of neutrophils, are also important in the pathogenesis of invasive amoebiasis.

Clinical features

The incubation period may range from a few days to many years. Amoebiasis is one of several 'tropical' diseases that can have a prolonged latent period and may present more than a decade following exposure.

Intestinal amoebiasis

The clinical spectrum ranges from asymptomatic infections (the majority) to fulminant amoebic colitis. The onset of invasive disease may be precipitated by another gastrointestinal infection or other illness, debility or immunosuppression. Symptoms are usually insidious with abdominal discomfort and loose stools sometimes containing mucus and blood. Patients with mild disease are relatively well with a history of a few loose stools. Investigations may reveal scanty trophozoites in faeces and a few ulcers on endoscopy. Patients with more extensive disease usually remain afebrile and ambulant despite producing 5–15 bloody stools per day containing numerous trophozoites. They are likely to have obvious rectal ulceration on endoscopy. Debilitated or immunosuppressed patients are more likely to present with rapid onset of abdominal pain, vomiting, dysuria, tenesmus and frequent bloody stools. They are likely to be febrile, dehydrated, toxic and anaemic. Abdominal tenderness may be marked and there may be evidence of peritonitis. Endoscopy is contraindicated. The most extreme presentation is that of extensive fulminating necrotizing colitis, which occurs in a minority of, usually immunocompromised, patients and is often fatal.

Complications of intestinal amoebiasis include the following.
• *Peritonitis*—abrupt/insidious onset; may occur after the patient has commenced treatment.
• *Haemorrhage*—resulting in anaemia or shock.
• *Stricture*—especially of the colon and rectum.

• *Post dysenteric ulcerative colitis*—mimicking classical ulcerative colitis; usually resolves slowly without specific treatment; rarely progresses to massive necrosis and toxic megacolon.

• *Skin ulceration*—usually perianal and anogenital regions, but may occur elsewhere, e.g. in surgical wounds and ileostomy/colostomy sites.

• *Amoeboma*—chronic inflammatory mass, single or multiple, most commonly developing in the ileocaecal region, presenting as an acute/subacute obstruction or causing an intussusception.

• *Amoebic abscess*—most commonly in the liver (see below).

The differential diagnosis of amoebic colitis includes the following.

1 Other causes of dysentery or bloody stools such as *Shigella*, typhoid, other *Salmonella*, enteroinvasive and enterohaemorrhagic *Escherichia coli*, schistosomiasis (especially *Schistosoma mansoni* and *S. japonicum*), *Balantidium coli*, *Trichuris trichiura*, tuberculosis, carcinoma, inflammatory bowel disease, ischaemic colitis, arteriovenous malformation and diverticulitis.

2 Any other cause of acute or chronic abdominal pain.

The differential diagnosis of an amoeboma includes tuberculosis, carcinoma, actinomycosis, an 'antibioma' or appendix mass.

Amoebic liver abscess

Trophozoites of *E. histolytica* invade the liver via the portal vein and set about destroying hepatocytes, resulting initially in the formation of microabscesses that subsequently coalesce to form multiple abscesses (25–35% of patients) or, more commonly, a single abscess (65–75% of patients) by the time the diagnosis is made. The surrounding tissue becomes oedematous with a chronic inflammatory infiltrate. Secondary bacterial infection may occur, but is unusual.

Right lobe abscesses are four times more common than those on the left. Amoebic liver abscess is about 10 times more common in males than females. All age groups may be affected, from neonates to the elderly, but ALA is most common in males aged between 20 and 40

years. Fewer than 50% have a history of dysentery within 1 month prior to presentation and many have no history of dysentery at all. A patient may present with an ALA many years after exposure, the transition from latent infection to clinical disease often being precipitated by immunosuppression or debility.

Clinical features

Patients who present in the early precoalescence stage of the development of an amoebic liver abscess may complain of low-grade fever and (usually) right upper quadrant discomfort and tenderness. This stage is sometimes referred to as 'amoebic hepatitis', a term which is perhaps misleading as it is unusual for such patients to have raised aminotransferases or bilirubin.

Most patients with ALA present when the abscess or abscesses are more 'mature' and the clinical symptoms and signs more florid. The history is usually one of gradually increasing but sometimes acute pain in the right upper quadrant of the abdomen. In some cases there is referred pain to the shoulder. Symptoms such as fever, sweats and rigors are common. In some cases the pain is localized to the lower chest wall and may be pleuritic in nature. The patient may also complain of cough and breathlessness and have evidence of a pleural effusion, leading to a mistaken diagnosis of pneumonia. Weight loss, wasting and anaemia occur more frequently in chronic presentations and such patients may be afebrile. Most patients have tender hepatomegaly, sometimes with inflammation and oedema of the overlying tissue. There may be tenderness in the intercostal spaces (Durban's sign). Signs of a pleural effusion may be evident and the apex beat displaced, especially if the left lobe is affected. Chest X-ray frequently reveals a raised hemidiaphragm or a pleural effusion (Fig. 19.1, see colour plate facing p. 146). Jaundice is uncommon and less than half of patients with an established ALA have a raised bilirubin and transaminases, although most have a raised alkaline phosphatase. A neutrophil leucocytosis is present in around 80% of cases.

In patients with a suspicious history but

without obvious tender hepatomegaly, it may be possible to elicit tenderness resulting from a deep-seated abscess by means of a gentle 'thump' over the lower rib cage. This should be regarded as something of a last resort in a situation where more sophisticated diagnostic techniques are unavailable and caution is advised if you do decide to thump your patient: (a) because your patient may thump you back (liver abscesses are usually extremely tender); and (b) worse still, the abscess may rupture.

Complications of ALA include:
• rupture through the skin or into the peritoneum, lung, pleura or pericardium (a particular risk with left lobe abscesses possibly resulting in cardiac tamponade); and
• haematogenous seeding causing abscesses in any organ or tissue, e.g. brain, muscle, kidney or spleen.

The differential diagnosis of an ALA includes pyogenic abscess, hepatocellular carcinoma, liver secondaries, hydatid cyst, hepatitis, tuberculosis, syphilitic gumma and lung pathology.

Investigations

Microscopy

The presence of cysts only on stool microscopy is of little diagnostic value because of the problem in distinguishing between E. histolytica and E. dispar. A diagnosis of amoebic dysentery depends on finding trophozoites of E. histolytica containing ingested red blood cells in a fresh stool sample. Ideally, the stool sample should be examined within 15 min of being produced or should be maintained at body temperature until examined. This is advised because the trophozoites lose their motility and tend to round up as the specimen cools, thus becoming more difficult to identify. Trophozoites of E. histolytica may also be seen in scrapings or biopsies of ulcers identified endoscopically. Non-pathogenic amoebae do not contain ingested red blood cells. In contrast to patients with bacillary dysentery, leucocytes are usually scanty in the faeces of patients with amoebic dysentery. It is rare for trophozoites of E. histolytica to be identified in the faeces of patients with ALA and less than half have cysts.

On aspiration, the pus from an ALA ranges in colour from pink to brown, darkening on exposure to air, and is sometimes described as resembling 'anchovy sauce' (in appearance, not odour). Characteristic trophozoites of E. histolytica can be identified in the pus or, more reliably, in scrapings from the wall of an ALA. Antigen detection is more sensitive. Cysts of E. histolytica are never found in abscesses. Leucocytes are scanty in pus obtained from an amoebic liver abscess unless there is secondary infection.

Antigen detection and polymerase chain reaction

Microscopy remains the principal method of investigation in settings with limited resources. However, stool antigen detection is more sensitive and specific and is being increasingly adopted in more affluent countries. This reliably differentiates between E. histolytica and E. dispar, but a positive test does not guarantee that E. histolytica is responsible for the patient's symptoms. Antigen can also be detected in pus from an ALA. Sensitive polymerase chain reaction (PCR) techniques have recently been developed for detection of E. histolytica in faeces and in ALA, and may be useful in epidemiological studies and in determining the virulence characteristics of different isolates.

Serology

A variety of serological tests have been developed for the diagnosis of invasive amoebiasis. Of these, indirect haemagglutination assay appears to be the most sensitive, particularly in the diagnosis of ALA. However, antibodies may persist for years following a significant infection and 10–35% of people living in developing countries have positive serology. Therefore, caution should be exercised in interpreting the results of serological tests.

Endoscopy

Colonoscopy is helpful in investigating patients

with suspected intestinal amoebiasis in whom stool microscopy or antigen tests are negative or inconclusive. Bowel preparation with enemas or cathartics is not advised because this may interfere with the identification of the parasite. The endoscopic appearance of amoebic colitis resembles that of inflammatory bowel disease and there have been numerous examples of misdiagnosis and consequently disastrous mismanagement. Discrete patchy ulceration with a granular friable mucosa may be seen in acute cases. Larger ulcers with loosely adherent yellowish or grey 'pseudomembranes' tend to occur in more chronic disease. *Entamoeba histolytica* may invade areas of the bowel affected by other pathology, such as a carcinoma, leading to diagnostic confusion. Aspirates, scrapings or superficial biopsies from the ulcer edge should reveal motile erythrophagocytic trophozoites of *E. histolytica* if examined immediately and should also be positive when tested for antigen. The parasites are readily identified by their magenta colour in biopsy specimens using a periodic acid–Schiff stain.

Imaging

Barium enema may demonstrate areas of ulceration, stricture or a filling defect from an amoeboma; however, none of these appearances are specific for amoebiasis, and there is a risk of perforation in patients with severe disease. Ultrasound, computerized tomography (CT) and magnetic resonance imaging (MRI) are very useful in identifying a liver abscess, but cannot reliably differentiate an ALA from a pyogenic abscess. An abscess may not be evident if the patient presents in the early 'precoalescence' stage of disease, and it is worth repeating the scan after a few days if there is a high index of suspicion.

Management

Invasive amoebiasis

Treatment with one of following tissue amoebicides is usually effective:
• *Metronidazole.* Adults 800 mg three times

daily orally for 5–10 days; children 35–50 mg/kg/day in three doses for 5–10 days.
• *Tinidazole.* A single oral dose of 2 g is better tolerated but more expensive. This dose should be continued for 3–6 days in more severe infections; children 50–60 mg/kg/day for 3–5 days.
• *Oral chloroquine* 600 mg base daily for 2 days followed by 300 mg base daily for 2–3 weeks is also effective in the treatment of ALA; children 10 mg/kg/day (max. 300 mg/day base) in 2–4 divided doses for 2–3 weeks.

Eradication of cysts

One of the following luminal amoebicides is usually recommended:
• *Diloxanide furoate.* Adults 500 mg orally three times daily for 10 days; children 20 mg/kg/day in three doses for 10 days.
• *Paromomycin.* Adults and children 25–35 mg/kg/day in three doses for 7 days.
• *Iodoquinol.* Adults 650 mg three times daily for 20 days; children 30–40 mg/kg/day (max. 2 g) in three doses for 20 days.
• *Quinfamide* given in three doses of 100 mg in a single day. This regimen has been used in both adults and children; however, full prescribing information is not yet available.
• *Tetracycline* may also be used as a luminal amoebicide.

Some practical points

A 5-day course of metronidazole is usually sufficient for the treatment of amoebic dysentery and most other forms of invasive amoebiasis. In affluent settings, it is usual to follow on with a course of a luminal amoebicide. If a luminal amoebicide is unavailable, patients with severe infections should be given a 10-day course of metronidazole. Patients treated with tinidazole or chloroquine should also receive a luminal amoebicide.

Parenteral metronidazole is indicated for patients who are severely ill with the addition of gentamicin and a third-generation cephalosporin (if available) or ampicillin to cover secondary sepsis from bowel pathogens. Attention should also be paid to management of fluid and electrolyte disturbances, anaemia, ileus and other complications.

Surgery is recommended in cases of acute colonic perforation in the absence of diffuse colitis and in cases of ruptured amoebic appendicitis. Surgery should be avoided in patients with severe amoebic colitis because the bowel is very friable and difficult to repair or anastomose. However, patients presenting with toxic megacolon or an abdominal abscess should be managed surgically.

Amoebomas usually respond rapidly to medical treatment and failure to do so should raise suspicion of coincidental pathology, such as a carcinoma. Surgery may be indicated in cases of obstruction or intussusception.

Most patients with an uncomplicated ALA will respond to a 5-day course of metronidazole. However, it may be advisable to extend this to 10 days, particularly if a luminal amoebicide is not available. The best guide to the efficacy of treatment is the patient's clinical response. Unless indicated on clinical grounds, there is little point in repeating scans, as these are likely to remain abnormal for several months despite successful treatment.

Indications for aspiration and drainage of an ALA include failure to respond to medical treatment, impending rupture, suspected secondary bacterial infection and diagnostic uncertainty.

Management of asymptomatic individuals passing cysts depends on the clinical context and the availability of resources for diagnosis and treatment. Ideally, one should differentiate between E. histolytica and E. dispar using a stool antigen test and prescribe a luminal amoebicide for those with E. histolytica. This is unlikely to be possible or practical in a resource-poor setting. In these circumstances, there is little point in attempting to identify and treat such individuals, particularly as the majority have E. dispar and all are likely to become reinfected. It is very important to eliminate E. histolytica from the gut of asymptomatic patients prior to immunosuppressive treatment and such patients should receive either a 5-day course of metronidazole (or single dose of tinidazole), followed by a luminal amoebicide, or a 10-day course of metronidazole.

Prevention and public health aspects

Improved hygiene, sanitation and access to safe drinking water are the main issues in preventing infection with E. histolytica. 'Boil it, cook it, peel it or leave it' is the message for travellers.

Recent developments

Nitazoxanide, a thiazolide compound, has been shown to be well tolerated and effective in the treatment of a wide range of gastrointestinal infections in adults and children including E. histolytica, Giardia intestinalis (G. lamblia), Balantidium coli, Ascaris lumbricoides, Strongyloides stercoralis, Trichuris trichiura, Enterobius vermicularis, Taenia saginata, Hymenolepis nana, Fasciola hepatica, Blastocystis hominis, Isospora belli, Dicrocoelium dendriticum, Cryptosporidium parvum, Enterocytozoon bieneusi and Helicobacter pylori. A 3-day course of oral nitazoxanide in the following doses:

- adults and children over 12 years, 500 mg b.d.
- children aged 4–11 years, 200 mg b.d.
- children aged 1–3 years, 100 mg b.d.

has been shown to be safe and effective in treating adults and children; however, full prescribing information is not yet available.

Acquired immunity to recurrent infection with E. histolytica has been shown to be linked to a mucosal immune response against a major virulence factor of the parasite, a Gal/GalNAc lectin responsible for adherence and killing of the host tissue. Small peptides derived from the galactose-binding adhesin administered by the parenteral or oral route have been shown to protect gerbils against experimental amoebic liver abscess. Therefore, the prospects for a vaccine are brightening.

Further reading

Gilles HM, Hoffman PS. Treatment of intestinal parasitic infections: a review of nitazoxanide. *Trends Parasitol* 2002; **18**: 95–7. [Useful summary about this versatile drug.]

Haque R, Huston CD, Hughes M, Houpt E, Petri WA. Amebiasis. *N Engl J Med* 2003; **348**: 1565–73. [Beautifully illustrated state-of-the-art review.]

Petri WA. *Entamoeba histolytica*: clinical update and vaccine prospects. *Curr Infect Dis Rep* 2002; **4**: 124–9. [Summarizes recent findings on *E. histolytica* infection in children, differentiation of *E. histolytica* from *E. dispar*, outcome of *E. histolytica*/HIV coinfection in pregnant women in Tanzania, acquired immunity to *E. histolytica* and prospects in vaccine development.]

Stanley S. Amoebiasis. *Lancet* 2003; **361**: 1025–34. [Comprehensive overview of pathogenesis, clinical and diagnostic features, and treatment.]

CHAPTER 20

Bacillary Dysentery

Microbiology and
   epidemiology, 160
Clinical features, 160

Investigation, 161
Management, 161

Prevention and public health
   aspects, 161
Further reading, 162

The term 'dysentery' is generally used to describe diarrhoea with visible blood and mucus. The term 'bacillary dysentery' is used interchangeably with 'shigellosis' even though numerous other bacteria also cause bloody diarrhoea, including several that are bacilli. Shigellosis occurs worldwide and is associated with poverty, crowding and lack of hygiene and sanitation.

Microbiology and epidemiology

Shigellae are non-motile Gram-negative rod-shaped bacteria, belonging to the family Enterobacteriaceae. According to current criteria for classification on the basis of DNA, Shigellae belong in the genus *Escherichia coli*. However, for historic and clinical reasons, *Shigella* has retained its identity as a separate genus.

Four groups or species are described, all but one of which include several subtypes and serotypes.

Group A: *Shigella dysenteriae* (12 serotypes) tends to cause epidemics (especially *S. dysenteriae* type 1).

Group B: *Shigella flexneri* (13 serotypes) commonly causes endemic dysentery in developing countries.

Group C: *Shigella boydii* (18 serotypes) is common on the Indian Subcontinent.

Group D: *Shigella sonnei* (one serotype) is an important cause of dysentery in the industrialized world.

Studies in animals and epidemiological evidence in humans indicate that *Shigella* infections elicit serotype-specific immunity. Humans are the only important reservoir of infection. People who have asymptomatic infections are important as carriers. Transmission is faecal–oral via flies, food, water and person–person contact, including various sexual practices. Shigellae are notably resistant to gastric acid and a very small ingested dose, as few as 10 bacilli, may cause clinical disease. Over 250 million infections occur each year, resulting in more than 600 000 deaths.

Clinical features

Shigellosis principally affects the colon and, sometimes, the terminal ileum. Clinical manifestations are brought about by a combination of enteroinvasion and toxin production. Organisms invade and multiply in the mucosa causing cell death, inflammation, ulceration, haemorrhage and formation of microabscesses. Shiga toxin, an exotoxin produced by certain strains of *S. dysenteriae* type 1, consists of an enterotoxin causing secretory diarrhoea, a cytotoxin causing cell necrosis, and a neurotoxin that may cause central nervous system (CNS) complications in children. Shiga toxin may also be involved in the pathogenesis of haemolytic uraemic syndrome (HUS).

The incubation period usually ranges from 1 to 8 days with a median of 5 days. The clinical spectrum of shigellosis may range from asymptomatic to fulminant with fatal attacks. Many clinical episodes are mild and self-limiting, with watery diarrhoea without blood or mucus, which resolve spontaneously after a few days.

Shigella sonnei infections are usually milder, but may be severe in infants. *Shigella dysenteriae* and *Sh. flexneri* tend to cause more severe disease with an abrupt onset of bloody, mucoid stools, cramps and tenesmus, often accompanied by fever and, sometimes, dysuria and confusion. Fever, confusion, meningism and convulsions often precede the onset of diarrhoea in young children.

Shigella dysenteriae type 1 may cause a fulminating gangrenous infection with an abrupt onset of fever, chills, rigors, vomiting and toxaemia. The patient can pass 20–60 bloody stools per day, often containing mucus and pus, and sometimes even sloughs of mucosa. Perforation is relatively rare, but severe dehydration, blood loss and sepsis can lead to acute renal failure. HUS occurs in 13% of cases of *Sh. dysenteriae* type 1, usually 1–5 days after the onset of the dysentery. Rarely, a choleraic form may occur with an abrupt onset of profuse watery diarrhoea that later becomes bloody and is associated with a high mortality.

Other complications and sequelae of shigellosis include toxic megacolon, post-dysenteric colitis, strictures, protein-losing enteropathy, granular proctitis, piles, parotitis and rectal prolapse in children. Peripheral neuropathy can also occur, particularly in children. Post-dysenteric Reiter's syndrome and symmetrical arthritis are also well-recognized sequelae.

For the differential diagnosis of bacillary dysentery see Chapter 19 (p. 155).

Investigation

The typical stool of shigellosis is often described as like 'redcurrant jelly'. Microscopically, red blood cells and pus cells are usually numerous whereas bacilli are scanty. Stool culture, even if the sample is fresh, is often difficult, and rectal swabs are more likely to be positive, particularly if directly inoculated onto appropriate culture media at the bedside.

Management

Most cases can be managed supportively with oral rehydration solution. In more severe cases, intravenous fluids, such as normal saline (with or without potassium, depending on renal function) or Ringer's lactate solution will be required. Blood transfusion may be indicated, and patients with HUS may require dialysis.

Antibiotics are indicated in severe infections. The choice depends on local sensitivity. A 5-day course of co-trimoxazole or ampicillin may be effective; however, multidrug resistance is common, especially with *Sh. dysenteriae* type 1, and antibiotics such as nalidixic acid, ciprofloxacin — or another fluoroquinolone — or azithromycin may be a better choice of empirical treatment if available. The possibility of secondary septicaemia from gut anaerobes and other enteropathogens should be considered in severely ill patients.

The use of antibiotics in children with dysentery caused by *E. coli* 0157:H7, which produces a Shiga-like toxin, is associated with an increased risk of development of HUS. However, at the time of writing this association has not been demonstrated among patients with shigellosis.

Vitamin A may reduce the severity of shigellosis in children and is recommended in areas in which dietary deficiency is common.

Prevention and public health aspects

Prevention of shigellosis is very much a matter of basic hygiene and sanitation. Handwashing, preferably using soap, is very important, especially in relation to food preparation and consumption. Food and utensils should be protected from flies. At community level, provision of an adequate quantity of water is more important than the quality of water, although quality is also important, as is sanitary disposal of faeces. Epidemic shigellosis can be devastating in refugee and displaced populations. This is discussed in Chapter 60.

The need for an effective vaccine has become more urgent with the emergence of multidrug-resistant strains of *Shigella*. Several promising candidate vaccines are currently under development but it is likely to be some time before an effective vaccine becomes widely available.

Further reading

Abouhammour W, Burney I. Shigella infection. http://www.emedicine.com [This is an up-to-date summary on shigellosis, which is well worth a visit.]

CHAPTER 21

Cholera

Disease and its mechanism, 163
Changes in fluid and
   electrolytes, 163
Clinical spectrum, 164

Diagnosis, 164
Treatment, 164
Administration of oral
   rehydration solution, 165

Epidemiology, 166
Control of cholera, 166
Further reading, 166

Cholera is a bacterial infection of humans caused by *Vibrio cholerae* 01 (of classical or El Tor biotypes) and *V. cholerae* 0139, which characteristically cause severe diarrhoea and may lead to death—in those severely affected—from water and electrolyte depletion. Spread is directly from person to person by the faecal–oral route, or indirectly by infected food or water. It can spread to any part of the world and may become endemic where standards of environmental sanitation and personal hygiene are low.

Humans are the only animal reservoir of infection. However, *V. cholerae* can survive for several months in aquatic environments and transmission may be maintained from such sources. The El Tor biotype has now largely displaced classical cholera as the major pathogen of public health importance, with 0139 responsible for infection in some areas of South Asia.

onset depends on the size of the infecting inoculum and the natural defences of the host. Gastric acid is a very effective barrier to infection. If it fails, the vibrios multiply very rapidly in the alkaline medium of the small intestine. Disease is caused by the exotoxin released by the vibrios. Cholera toxin comprises two subunits, A and B. The B subunit adheres to the intestinal mucosa and allows entry of the A subunit into the cells. The A subunit 'switches on' cyclic adenosine monophosphate (cAMP) and results in the loss of water, bicarbonate and electrolytes from the cells. There is no significant inflammation of the gut and the epithelium remains morphologically intact. The illness is self-limiting and diarrhoea ceases in a week if the patient survives. Death is usually caused by hypovolaemic shock and its complications, including acute renal failure. With adequate and timely rehydration, mortality rates should not exceed 1%.

Disease and its mechanism

After an incubation period, usually in the range of 1–5 days, there is diarrhoea of rapid onset. The diarrhoea is typically watery, white and flecked with mucus—the infamous 'ricewater stool' (Fig. 21.1, see colour plate facing p. 146). In 80% of cases, vomiting follows soon after the diarrhoea.

Fever is unusual except in children, and short-lived. The temperature is usually subnormal when the patient is first seen. The speed of

Changes in fluid and electrolytes

Dehydration is caused by the profuse diarrhoea, compounded by the usual inability to retain fluids by mouth. The speed with which severe dehydration occurs is greater than in any other disease. Collapse from hypovolaemic shock can occur within a few hours of the onset, and the untreated patient may die within 24 h. Hypoglycaemia is common, especially in children.

Signs of dehydration include loss of turgor in the cheeks leading to a pinched appearance, sunken eyes as a result of orbital dehydration, general loss of skin elasticity detectable by the delay in a pinched skinfold returning to its normal position, and shrivelling of the skin of the fingers—'washerwoman's hands'. Most people are familiar with this appearance as the result of staying in the bath or swimming pool for too long. In severe cases, cerebration is impaired, the voice is weak and husky, and urine output is reduced or ceases altogether.

Cramps of the muscles of the limbs and abdomen are a typical feature of severe cases. The loss of water is relatively more than the loss of electrolytes in children, who often suffer hypertonic dehydration as a result.

Clinical spectrum

The condition described is the severe picture, seen in a minority of infections only. In an outbreak, for every case of severe disease, there will be at least 10 other cases with mild or asymptomatic infections.

Diagnosis

Clinical diagnosis is only possible in severe cases with profuse painless diarrhoea, 'ricewater' stools, gross dehydration and muscle cramps. Only occasionally do other infections produce this picture. A confirmed diagnosis can only be made by isolating the organism.

Direct diagnosis by microscopy

In severe cases, the watery stool contains cholera vibrios in almost pure culture. They can be recognized by dark-field examination of a wet preparation. Identification can be supported by adding specific antiserum, which immobilizes the vibrios immediately. Microscopic identification is much more difficult in milder cases because of the presence of large numbers of normal faecal organisms. In asymptomatic cases, direct microscopy is of no value.

Direct diagnosis by culture

Various media are used for primary isolation. Among the best is thiosulphate–citrate–bile salt–sucrose (TCBS) agar. *Vibrio cholerae* appear as yellow colonies after overnight incubation, which are oxidase-positive, and on microscopy show the characteristic Gram-negative curved bacilli. Small numbers of vibrios can only be detected using an enriched liquid medium such as alkaline peptone water. Specimens taken in the field can be transported to the laboratory in sealed plastic bags, or after inoculation into a holding medium.

Treatment

Initial rehydration

Rehydration is the mainstay of cholera treatment. In severe cases with hypovolaemic shock, the restoration of blood volume is urgently needed, and this can only be achieved rapidly by intravenous infusion (see also p. 7). Because peripheral veins are collapsed in such patients, the initial resuscitative infusion may have to be given via the femoral or subclavian vein in adults or the internal jugular vein or intraosseus route in children.

Fluid in the initial stages is run in as quickly as possible; an initial rate of 4 L/h for the first few litres is the norm in adults. The best guide to success is the return of a palpable arterial pulse. As soon as the systolic blood pressure reaches 90 mmHg, renal function usually returns. Tubular necrosis only usually develops if resuscitation is delayed.

In all patients with hypovolaemic shock, the initial fluid deficit will be at least 10% of the body weight. It is a safe rule of thumb to give one-third of the total estimated deficit in the first 20–30 min.

Choice of rehydration fluid

The type of fluid is less important than an adequate quantity. However, because patients usually have a metabolic acidosis as a result of bicarbonate loss, a deficiency of potassium and a loss of water greater than of salts, a slightly hypotonic alkaline fluid enriched with potassium is the most physiological choice.

The single fluid that meets all these needs, and is suitable both for adults and children, is Ringer's lactate solution. This contains calcium 2 mmol, chloride 111 mmol, lactate 27 mmol, potassium 5 mmol and sodium 131 mmol/L. It is suitable for both initial rehydration and maintenance therapy. The World Health Organization's (WHO) intravenous diarrhoea treatment solution (glucose 10 g, potassium chloride 1 g, sodium acetate 6.5 g and sodium chloride 4 g in 1 L) contains approximately acetate 50 mmol, chloride 80 mmol, potassium 10 mmol and sodium 120 mmol/L. Simpler solutions can be used with almost as good results, certainly in adults.

As soon as the blood volume has been restored and the pulse has returned, the drip can be moved to a more convenient site because the peripheral veins reappear. Vomiting ceases as soon as the acidosis has been corrected, and fluids can be taken by mouth. If potassium was not replaced by infusion, it can now safely be given orally.

Maintenance hydration

When the patient has been resuscitated, careful charting of fluid intake and output must be started. The uncontrollable watery diarrhoea often continues for several days and, once adequately rehydrated, daily losses may exceed 20 L. To measure this accurately, it is best for the patient to be nursed on a special 'cholera cot'—a frame bed covered in rubber sheeting with a hole in the middle to allow the fluid escaping from the anus to be funnelled into a calibrated collecting bucket below. A useful alternative is to insert a Foley (urinary) catheter into the rectum and inflate the balloon to hold it in place. This can then be connected to a urine collection bag.

The urine output must also be charted accurately, and intravenous fluid input should equal the combined volume of stool and urine, plus 500 mL added for insensible losses. Any inaccuracies in the calculations are usually adequately compensated for by the patient's new-found ability to drink. This period of maintenance parenteral fluid therapy can be greatly shortened

by the early resumption of oral rehydration with a sugar–electrolyte solution and tetracycline by mouth.

Oral rehydration with glucose–electrolyte solution

This is used for maintenance hydration in severe cases requiring intravenous therapy for resuscitation, and for all milder cases from the beginning. It is much cheaper than intravenous therapy, requires no special apparatus or skills and is free from the dangers of fluid overload. Its success depends on the fact that the active transport of electrolytes into the mucosal cells is glucose dependant. If glucose is not available, sucrose can be used with almost as good results, as it is rapidly split into glucose and fructose by intestinal enzymes. The secret of successful oral rehydration is to give the fluid frequently, but in small amounts. In developing countries where nursing resources are strained, the task of oral rehydration is often delegated to relatives, especially the mothers of small children. The WHO-recommended solution comprises 1 L of sterile water, dextrose (glucose) 20 g, potassium chloride 1.5 g, sodium bicarbonate 2.5 g and sodium chloride 3.5 g.

If glucose is not available, sucrose can be used instead, but should be increased to 40 g/L, as it generates only half its weight of glucose on hydrolysis. Formulas using locally available carbohydrates, such as rice powder instead of sugar, have been used successfully in many areas. Various prepackaged commercial preparations are available in sachets and have obvious advantages.

Administration of oral rehydration solution

This is either by drinking two to four times an hour or by nasogastric tube. The dosage varies with the calculated deficit and rate of fluid loss. In severe continuing diarrhoea, it is 15 mL/kg/h in frequent divided doses. If this rate of oral administration cannot keep pace with fluid loss (equivalent to 20 L/day for an adult), parenteral

therapy is needed. For mild to moderate diarrhoea, a dosage of 5–10 mL/kg/h is adequate.

Antimicrobial agents in cholera

Tetracycline, doxycycline and furazolidone have
all been shown to reduce the volume and duration of diarrhoea, particularly in those with
severe disease. The normal adult regimens are
tetracycline 500 mg 6-hourly for 3 days, furazolidone 400 mg/day for 3 days, or a single dose
of doxycycline 300 mg. Resistance, particularly
to tetracycline, is frequently reported, and in-
discriminate use of antibiotics for mild cases
should be discouraged.

Epidemiology

The most recent (seventh) pandemic started in
Sulawesi in 1961 and spread relentlessly
through the western Pacific, South East Asia, the
Middle East, Africa, eastern Europe and South
America. The organism is the El Tor biotype,
which has effectively displaced classical cholera
except from a few locations in parts of
Bangladesh. It differs from classical cholera in
two important ways: it more often gives rise to
the chronic carrier state; and relatively fewer of
those infected develop the classical disease. The
chronic carrier state has doubtless facilitated its
spread. Both classical and El Tor cholera aggluti-
nate O1 antisera. In 1993, a new serotype, *V.
cholerae* O139, was reported in southern India,
and has been responsible for outbreaks in eastern India, Bangladesh and Thailand.

Cholera is endemic in many tropical areas
of Asia, Africa and in parts of South America
and transmission is encouraged in areas of
crowding and poor sanitation. Very large outbreaks have occurred in the 1990s in displaced
and refugee communities in central and eastern
Africa. While spread is largely directly by the faecal–oral route, contamination of community
water supplies has been responsible for some
large outbreaks.

Control of cholera

Control involves effective case detection and
management, and emergency public health
measures to reduce spread. Facilities are
required for effective case treatment with
adequate supplies of intravenous replacement
fluids for severe cases and supplies of oral rehydration solution. Longer term control depends
mainly on improving standards of environmental sanitation. The 'old' parenteral killed whole
cell cholera vaccine has no part to play in outbreak control, and is indeed no longer recommended for travellers to endemic areas. It gives
only low individual protection and does not
stop spread from asymptomatic cases.

New oral cholera vaccines are available that
are modified organisms producing immunity to
the B subunit of the toxin. Trials in various endemic areas suggest they may have a role in both
personal protection and possibly in outbreak
prevention.

Further reading

World Health Organization. *Guidelines for Cholera
Control.* Geneva: WHO, 1993. [Standard WHO
document on cholera, its prevention and control.]

CHAPTER 22

Giardiasis and Other Intestinal Protozoal Infections

Giardiasis, 167

Other intestinal protozoa of importance, 169

Further reading, 170

Giardiasis

Epidemiology

Giardiasis occurs worldwide, particularly in areas of poor hygiene and sanitation. Humans are the main reservoir of infection, although beavers have been implicated in outbreaks in North America. It is uncertain whether several species that occur in various domestic pets and other animals actually cause disease in humans. Most infections are acquired through drinking water contaminated with *Giardia* cysts. These are relatively resistant to chlorination and large community outbreaks have occurred from drinking chlorinated but unfiltered water. Cysts may also be ingested on food, particularly salads, and by direct faeco-oral spread, e.g. among preschool children in day-care centres or in other situations of poor hygiene and sanitation. Cysts can survive outside the body for several weeks under favourable conditions.

Parasite and life cycle

Giardia lamblia (also known as *G. intestinalis* or *G. duodenalis*) is a flagellate protozoon that inhabits the upper small bowel. The trophozoite stage of the parasite is a flattened pear-shaped creature about 15 μm long, 9 μm wide and 3 μm thick. It is concave on its ventral surface where it attaches itself, by its sucking disc, to the intestinal mucosa, but does not invade. It has four pairs of flagella for locomotion and multiplies in the gut by binary fission. Large areas of the mucosal surface may be colonized in heavy infections.

Trophozoite adherence disrupts the intestinal brush border and interferes with enzyme activity. Attachment also stimulates an inflammatory cytokine response, resulting in secretion of fluid and electrolytes and damage to enterocytes. Trophozoites usually encyst as they pass distally along the intestine. The cyst is oval, 8–12 μm long by 6–8 μm wide, and contains four small nuclei and a central refractile axoneme. The cysts are infective as soon as passed. When swallowed by a new host, they excyst in the upper gastrointestinal tract and liberate trophozoites.

Occasionally, *Giardia* may colonize the biliary tract and—in patients with achlorhydria—the stomach, usually in association with *Helicobacter pylori*.

Clinical features

The median incubation period is 7–10 days but ranges from 3 days to several months. Susceptibility to infection and disease depends on parasite and host factors. Clinical symptoms may develop after ingesting as few as 10 cysts. Most infections are asymptomatic. Clinical symptoms are more likely to develop, and tend to be more severe, in initial infections and in persons with impaired immunity.

Symptoms are often abrupt with diarrhoea, abdominal cramps, bloating and flatulence. Often there is associated malaise, nausea and belching accompanied by a taste of rotten eggs. The diarrhoea can be variable in character, ranging from watery to greasy, but does not contain blood. Most patients have been symptomatic for several days before seeking medical help, and may have significant weight loss by the time they present. Untreated, the clinical course is variable. Many patients, often after periods of fluctuating symptoms, eventually become asymptomatic. Others continue to have persisting diarrhoea, associated with malabsorption, malnutrition and failure to thrive. In some patients, chronic diarrhoea may be partly related to lactose intolerance, which may persist despite eradication of the *Giardia*.

Differential diagnosis

The differential diagnosis includes a wide range of causes of acute and chronic non-bloody diarrhoea and other causes of malabsorption, including:
• parasitic (fasciolopsiasis, capillariasis, strongyloidiasis, isosporiasis, cryptosporidiosis);
• tropical sprue;
• hypolactasia;
• chronic calcific pancreatitis;
• malnutrition;
• intestinal tuberculosis;
• alpha-chain disease;
• coeliac disease; and
• small bowel lymphoma.

Investigations

The standard method of diagnosis is stool microscopy for the characteristic cysts. Passage of cysts can be intermittent and it may be necessary to examine repeated samples. Motile trophozoites are sometimes seen in saline preparations. Concentration and special staining techniques increase the sensitivity of stool microscopy and it should be possible to diagnose 50–70% of infections on examination of a single stool sample, and more than 90% of infections if three samples are examined.

Various techniques are now available for the rapid detection of *Giardia* antigen in stool samples, e.g. using ELISA and direct fluorescence antibody techniques. These are more sensitive and less time-consuming than stool microscopy, although occasionally it is necessary to test more than one stool sample. A panel enzyme immunoassay (EIA) has been developed for the detection of *G. lamblia*, *Entamoeba histolytica* and *Cryptosporidium parvum* with sensitivities and specificities of over 95% for identification of these organisms. However, antigen testing should not replace stool microscopy because other pathogens may be present and may, in fact, be responsible for the patient's symptoms.

Other methods of diagnosis include duodenal fluid aspiration and microscopy for trophozoites. Duodenal fluid can also be sampled using the 'string test' in which the patient swallows a length of string, one end of which is entwined in a gelatin capsule. The capsule dissolves and the string passes into the duodenum. Having taped the proximal end to the patient's cheek, the string is left *in situ* overnight, or with the patient fasting for 4–6 h. The string is then withdrawn, the duodenal fluid squeezed from the distal end onto a microscope slide and examined for *Giardia* trophozoites. This technique is also useful in diagnosing strongyloidiasis (Chapter 52).

Small bowel biopsy may be helpful in patients in whom an alternative or concurrent diagnosis is being considered, e.g. patients with HIV/AIDS, common variable immunodeficiency or suspected tropical sprue. The typical picture in giardiasis is of villous flattening, deepening of crypts and an increased inflammatory infiltrate in the lamina propria. *Giardia* trophozoites may also be seen in the intervillous spaces.

Management

Most patients respond to oral metronidazole 400 mg three times daily for 5 days, or 2 g/day for 3 days. Paediatric regimens are 15 mg/kg/day in three divided doses, or 40 mg/kg/day for 3 days. Tinidazole is effective as a single oral dose of 2 g for adults and 50 mg/kg (maximum 2 g) for children. Albendazole 400 mg/day for 5 days is also effective. Nitazoxanide is also effective and has proved useful in treating patients who are HIV positive who fail to respond to standard treatment. Other drugs that are sometimes used include quinacrine, furazolidone and paromomycin.

Failure to eradicate the organism following a standard course of treatment may be because of poor compliance, reinfection or, possibly, antimicrobial resistance or underlying immunodeficiency. Persisting symptoms despite eradication of the parasite raises the possibility of continuing lactose intolerance or that *Giardia* was a coincidental finding and the patient's symptoms are attributable to another aetiology.

Prevention and public health

Prevention is all about improving hygiene, sanitation and access to safe water. Cysts of *Giardia* are resistant to standard chlorination of water, therefore flocculation, sedimentation and filtration are of greater importance. Cysts are killed if water is boiled. Micropore filters, with or without iodine resins, are available for personal use and may be handy for travellers.

Other intestinal protozoa of importance

Cryptosporidium parvum
Parasite and life cycle

Cryptosporidium spp. are coccidian protozoans with a worldwide distribution, found in mammals, reptiles, fish and birds. Transmission is faeco-oral and infection most commonly occurs when the oocyst is ingested via contaminated water or food, or following person–person contact. The oocyst releases four sporozoites into the lumen of the small bowel, which invade the epithelial cells where they undergo further stages in a life cycle that, in many ways, resembles that of malaria. *Cryptosporidium* has the ability to produce thin-walled oocysts that maintain its life cycle within the host ('internal autoinfection'), or to produce thick-walled oocysts that are excreted in faeces. The latter are highly resistant to chlorination and small enough to pass through conventional filters. *Cryptosporidium* is notorious in causing epidemics of diarrhoea, even among communities in developed countries with access to treated water supplies.

Clinical features

The incubation period for *Cryptosporidium* has not been clearly established, but usually ranges from 1 to 12 days, with an average of 7 days. *Cryptosporidium* is important in four clinical settings:
- childhood diarrhoea in developing countries;
- travellers' diarrhoea;
- protracted diarrhoea in immunocompromised patients; and
- water-borne outbreaks in developed and developing countries.

Clinical features frequently include watery diarrhoea, abdominal cramps, bloating, weight loss, fever and malaise. Episodes are usually self-limiting but may become chronic or fulminant, particularly in immunocompromised patients, e.g. with HIV/AIDS, and associated biliary tract disease may also occur in this population.

Isospora belli

Isospora belli is a protozoan parasite with a worldwide distribution, usually acquired from faecally contaminated water or food. Disease may occur following ingestion of the mature oocyst and pathology is similar to *Cryptosporidium*. Clinical presentation is usually with watery diarrhoea, sometimes with blood and pus cells, abdominal pain and malabsorption. Infections are usually self-limiting but may become chronic or relapsing in immunocompromised patients.

Cyclospora cayetanensis

Cyclospora is usually acquired from faecally contaminated water, fruit or herbs. Clinically similar to *Cryptosporidium* and *Isospora*, symptoms include prolonged watery diarrhoea, cramps, fever and fatigue.

Microsporidia

Various species of the order Microsporidia are pathogenic in humans and are increasingly recognized to be important, particularly in HIV-infected individuals. The most common is *Enterocytozoon bieneusi*, which occurs in 7–50% of HIV-infected persons with chronic diarrhoea.

Investigations

The diagnosis of cryptosporidiosis is usually made by demonstrating acid-fast oocysts in faeces or luminal aspirates using a modified Kinyoun acid-fast stain. *Cryptosporidium* oocysts appear as round pinkish-red bodies measuring 4–6 μm. A sensitive and specific panel EIA test has recently been developed.

Isospora oocysts are larger and oval, measuring 10×30 μm. They may be visible in an unstained saline preparation and appear red with the modified acid-fast stain. Unusually for a protozoal infection, *Isospora* may cause an eosinophilia.

Cyclospora oocysts are round and measure 8–10 μm. They may be seen in unstained faecal preparations and stain (variably) red with the modified acid-fast stain. They do not stain with iodine. They can be detected as blue fluorescent dots when examined in ultraviolet light.

Modified trichome stains, calcofluor or chemofluorescent stains can be used in expert hands for the diagnosis of *E. bieneusi* and other microsporidia in faeces, and electron microscopy is used in reference laboratories to confirm the identity of the organism. None of these modalities are routinely available in the Tropics and underdiagnosis is the norm.

Management

Most patients with normal immunity recover from these infections spontaneously. Treatment of symptomatic patients with cryptosporidiosis poses problems as few of the available antimicrobials have proven and consistent efficacy. Paromomycin is the agent most commonly recommended for the treatment of *Cryptosporidi-um* associated with HIV. Other agents under investigation include azithromycin and letrazuril, but these appear to be inferior to paromomycin in pilot studies. Hyperimmune bovine colostrum has also been used. HIV-related cryptosporidiosis often improves following the initiation of antiretroviral treatment.

Isospora and *Cyclospora* respond to oral trimethoprim-sulfamethoxazole 160–800 mg four times daily for 7–10 days. HIV-positive patients should then receive a maintenance dose three times weekly, or a weekly dose of Fansidar. Pyrimethamine can be used if the patient is allergic to sulphonamides. Ciprofloxacin is also effective against *Cyclospora*.

Albendazole may produce clinical improvement in patients with *Enterocytozoon bieneusi*, despite persistence of the parasite in stool samples and small bowel biopsies following treatment. It is more effective for treating the less common gut microsporidian *Encephalitozoon intestinalis*. Improvement may also occur with the introduction of antiretroviral treatment.

Nitazoxanide has been shown to be effective against *Cryptosporidium*, *Isospora*, *Cyclospora* and *Enterocytozoon bieneusi* in a number of small studies and in treatment of individual patients. The results of larger clinical trials are awaited.

Further reading

Gardner TB, Hill DR. Treatment of giardiasis. *Clin Microbiol Rev* 2001; **14**: 114–28. [Useful summary including recommendations for treatment during pregnancy.]

CHAPTER 23

Intestinal Cestode Infections (Tapeworms) Including Cysticercosis

Tapeworms, 171
Cysticercosis, 172
Other intestinal cestode
    infections, 172

Management of intestinal
    cestode infections, 173

Prevention and public health
    aspects, 173
Further reading, 173

Tapeworms

Tapeworms are flat segmented hermaphrodites measuring from 10 mm to 20 m. The head (scolex) attaches to the intestinal mucosa by means of suckers or hooklets. All, with the exception of *Hymenolepis nana*, require a secondary intermediate host in which the larvae develop into cysts, usually in muscle. Human infection follows consumption of cysts in undercooked meat or fish. Larval cestode infections can also occur in humans following the ingestion of the egg, the most important being cysticercosis.

Parasites and life cycles

Taenia saginata, the beef tapeworm, is a cosmopolitan infection in which humans harbour the adult worm and cattle harbour the larval stage. Its main importance is in economic losses caused by condemnation of beef carcasses. Human infection is of social importance only. Ethiopia has the highest infection rate in the world. People acquire infection by eating undercooked meat containing cysticerci, the larval stages of the parasite encysted in the muscles of infected herbivores. The cysts evaginate in the intestine, and the head of the worm attaches itself to the mucosa of the upper third of the small intestine by its suckers. Segments called proglottides grow from the head, and new segments are added until the worm contains a chain of 1000–2000. A full-grown tapeworm is often more than 5 m long, and sometimes 10 m.

Proglottides at the tail end of the worm develop fertilized eggs in the uterus and are called gravid segments. When mature, the gravid segments break off the chain (strobila) and leave the anus in the stool or by their own movements. Proglottides sometimes rupture in the intestine, and free eggs are also passed in the stool. The eggs that reach pasture, mainly after disintegration of the mature proglottides, are infective to cattle (and several other herbivores) when swallowed. They hatch in the bovine gut to become oncospheres, and enter the circulation where they are carried to the muscles and encyst as cysticerci. The meat is described as 'measly', and the cysticerci are easily visible to the naked eye.

Taenia saginata cysts can occur in other domestic bovines and a closely related Asian species has been shown to infect pigs, ungulates and monkeys. *Taenia solium*, the pork tapeworm, is a much less common infection than *T. saginata* but far more important because of its ability to cause severe disease in humans. Humans are the definitive host; the pig the normal intermediate host. It is found all over the world where people eat cysts in raw or undercooked pork. For this reason, intestinal infection with *T. solium* is rare in Muslims, Orthodox Jews and vegetarians. *Taenia solium* cysts also occur in dogs and cats.

Ingestion of eggs of *T. solium* can give rise to human cysticercosis, a major cause of epilepsy and other neurological disease in some parts of the world such as Central America and India. Human cysticercosis can occur regardless of religious or dietary affiliation. Human cysticercosis does not occur following the ingestion of eggs of *T. saginata*.

Clinical features of taeniasis

Intestinal infections are usually asymptomatic. The host may only realize that a tapeworm is on board when a proglottid segment appears in faeces or is felt as it passes through the anus. Symptoms may include loss of appetite, nausea or vague abdominal pain. Rarely, complications arise when a proglottid migrates to an unusual site, such as the appendix or pancreatic and bile ducts. Patients who are vomiting profusely for whatever reason may be further distressed by the appearance of several metres of tapeworm in the vomit.

Investigations

The eggs of *T. saginata* are indistinguishable from those of *T. solium* on routine microscopy. To make a specific diagnosis, a mature proglottid is pressed between two microscope slides and the number of lateral branches of the uterus counted. *T. saginata* has 15–20 main branches on each side, *T. solium* has 13 or fewer; but this criterion is not as reliable as once thought. The scolex, measuring about 1 mm, may be found among the smallest immature segments with the aid of a magnifying glass. The presence of hooks distinguishes the scolex of *T. solium* from that of *T. saginata*, which has no hooks.

Serology is sometimes used for epidemiological surveys and may be useful in the diagnosis of cysticercosis. DNA probes have also been developed to differentiate between *T. saginata* and *T. solium*.

Cysticercosis

Tissue cysts of *T. solium* are usually 1–2 cm in size and can be found in many tissues, especially subcutaneous, muscle and brain. During the initial phase of invasion and development there may be pain and swelling, accompanied by eosinophilia. Subsequently, skin nodules can sometimes be felt as movable, small, painless nodules, especially on the arms or chest. Muscle cysts eventually calcify and can be seen as calcified streaks that follow the planes of the fibres of skeletal muscle on X-ray of the forearms, psoas or thigh muscles.

The most important effect of cysticercosis is in the brain. Initially there may be a diffuse encephalitic picture. More usually, the patient presents with single or repeated seizure and neurocysticercosis is the most important cause of epilepsy in many parts of Africa and South America. A small proportion of cases present with features of hydrocephalus. Computed tomography (CT) and magnetic resonance imaging (MRI) are needed to delineate the number and location of cysts in the brain.

There is still controversy about the benefits and drawbacks of active antiparasitic treatment, but expert consensus is that treatment will benefit some patients with neurocysticercosis. The usual regimens are albendazole 15 mg/kg/day for 8 days, or praziquantel 50 mg/kg/day for 15 days. Steroids should be given before and during antiparasitic treatment to reduce the effects of inflammation around damaged cysts. Steroids are also needed (in neurosurgical doses) for short-term management of occasional flare-ups of inflammation and cerebral oedema that occur as cysts degenerate as part of their natural history. Seizures usually respond to first-line anticonvulsant drugs (see Chapter 59).

Other intestinal cestode infections

Diphyllobothriasis

Diphyllobothrium latum is the most common of more than a dozen species of fish tapeworm affecting humans. Human infection follows ingestion of undercooked or raw fish or roe. Infection usually involves a single worm and most are asymptomatic or associated with vague non-specific abdominal symptoms. Megaloblastic anaemia can occur in severe cases.

Hymenolepiasis and dipylidiasis

Hymenolepis nana, the dwarf tapeworm, occurs worldwide, mainly among children. Infections are usually asymptomatic, but abdominal pain, nausea, vomiting, pruritis ani and diarrhoea — sometimes containing blood — have been described in heavy infections. Headache, dizziness, sleep and behavioural disturbances are relatively frequent and convulsions have also been reported. Autoinfection is common.

Hymenolepis diminuta, the rat tapeworm, may affect humans following ingestion of the intermediate host, commonly a weevil, flea or cockroach. Infections are usually asymptomatic and short lived.

Dipylidium caninum may infect humans, usually young infants, following accidental ingestion of a flea, the intermediate host. Infections are usually asymptomatic; however, symptoms may include abdominal pain, diarrhoea, pruritis ani and urticaria.

Management of intestinal cestode infections

A single oral dose of praziquantel (10 mg/kg) is the drug of choice for all of the above intestinal cestode infections. *Hymenolepis nana* requires 25 mg/kg as a single dose. Caution is advised in populations in which cysticercosis is common, because of the risk of precipitating or aggravating neurological symptoms.

Niclosamide, as a single oral dose (500 mg if < 11 kg; 1 g if 11–34 kg; 1.5 g if > 34 kg; 2 g for adults) is also effective. Tablets should be well chewed and swallowed with plenty of water. The routine use of purgatives and antiemetics in patients with *T. solium* prior to cestocidal treatment, in order to prevent retrograde peristalsis of eggs and possible risk of cysticercosis, is not justified on the basis of clinical evidence.

Albendazole, which is used in the treatment of cysticercosis and hydatid cyst, is also effective in treating intestinal taeniasis. Nitazoxanide has also been shown to be effective in infections with *T. saginata*.

Prevention and public health aspects

Control measures for taeniasis are aimed at environmental sanitation, meat inspection and adequate cooking.

Further reading

García HH, Gonzalez AE, Evans CAE *et al. Taenia solium* cysticercosis. *Lancet* 2003; **361**: 547–56. [Comprehensive review with superb illustrations and references; clear discussion of areas of controversy.]

CHAPTER 24

Soil-Transmitted Helminths

Ascariasis, 174

Hookworm, 176

Trichuriasis, 177

Toxocariasis, 178

Further reading, 179

The term soil-transmitted helminths applies to a group of parasites whose life cycle usually depends on a period of development outside the human host, typically in soil. They can be further subclassified according to their mode of infection: direct, modified direct or penetration of the skin. Direct infection occurs when eggs are transmitted from anus to mouth and ingested without ever reaching the soil. Examples include *Enterobius vermicularis* (pinworm or threadworm) and *Trichuris trichiura* (whipworm). Modified direct infection occurs when eggs passed in the faeces only become infectious having spent a period of time — usually in soil — undergoing further development, e.g. *Ascaris lumbricoides* (roundworm).

Penetration of the skin is the method of infection used by the hookworms, *Ancylostoma duodenale* and *Necator americanus*. Under favourable environmental conditions, larvae hatch from eggs deposited in the soil. The larvae lie in wait in the surface layers of the soil until they come into contact with the skin of their unsuspecting host. *Strongyloides stercoralis* employs a similar modus operandi, but instead of eggs, larvae are passed in the stool which either can cause autoreinfection by direct penetration of the perianal skin, or go on to establish an independent life cycle in the soil awaiting the appearance of a new host.

The prevalence and distribution of soil-transmitted helminth infections is a product of lifestyle and life cycle. Of the one billion people infected worldwide, the greatest burden of disease occurs among children, particularly in areas of poor hygiene and sanitation, and has a significant effect on physical and intellectual development. The World Health Organization (WHO) is currently promoting the periodic mass chemotherapy of schoolchildren and women of childbearing age in an attempt to reduce the burden of disease in vulnerable populations.

Ascariasis

Epidemiology
Ascaris lumbricoides affects over 600 million worldwide with a peak prevalence and intensity of infection among children aged 3–8 years. Infection is found wherever conditions of environmental hygiene are poor.

Parasite and life cycle
Ascaris eggs, contaminating vegetables, soil or dust, are swallowed and liberate larvae as they pass through the stomach and small intestine. The larvae penetrate the intestinal mucosa, enter the bloodstream and lymphatics and reach the lungs 4–16 days after infection. The larvae penetrate the alveoli, moult and migrate via the respiratory tract to the oesophagus and on to the small intestine, where they develop into adults, mate and start producing eggs 6–8 weeks after infection. Adults are large cream-coloured worms; males are 15–30 cm long, females 20–40 cm. They live in the small intestine

and obtain nourishment from the intestinal contents. They do not suck blood or damage the mucosa significantly. Females produce up to 200 000 eggs per day. These are excreted in faeces and their ova mature into infective embryos within 1–4 weeks and may remain viable in soil for years.

The morphologically similar *Ascaris* of pigs, *A. suum*, also infects humans.

Clinical features

Ascaris pneumonitis

Fever, cough, dyspnoea, wheeze and urticaria may occur in a proportion of those infected during the migration of larvae through the lungs. Chest pain, cyanosis and haemoptysis occur in more severe cases. Ascaris pneumonitis accompanied by eosinophilia is known as Löffler's syndrome. Symptoms usually resolve spontaneously within 10 days.

Clinical features resulting from adult worms

Intestinal worms are rarely noticed unless passed in the stool. Worms may form a bolus in heavy infections causing intestinal obstruction, volvulus or perforation. Wandering worms can obstruct ducts or diverticula causing biliary colic, cholangitis, liver abscess, pancreatitis or appendicitis, or make an unwelcome appearance in an endotracheal tube during anaesthesia. They have also been known to obstruct nasogastric tubes and even enter the paranasal sinuses. Pneumothorax and pericarditis have also been reported. Even if these nomadic nematodes do not cause an obvious pyogenic peritonitis, they may continue to lay eggs, giving rise to a granulomatous peritonitis.

Investigations

Ascaris pneumonitis is diagnosed on clinical grounds; the presence or absence of eggs in the stools is irrelevant. Larvae and/or eosinophils may be found in the sputum. Chest X-ray findings range from discrete densities to diffuse interstitial—or more confluent— infiltrates. Stool microscopy for eggs is usually adequate for diagnosing established infection,

although stools may be negative if all the worms are male. Worms may also be found by means of barium studies, ultrasonography and endoscopy.

Management

• Albendazole 400 mg as single oral dose clears most infections. Heavy infections may need repeated doses for 2 or 3 days.
• Mebendazole 100 mg orally twice daily for 3 days is effective. Its use in children younger than 2 years is not recommended by the manufacturer. Ectopic migration of *Ascaris* has been reported following the use of mebendazole.
• Piperazine 75 mg/kg (to a maximum of 3.5 g for adults and children > 12 years and a maximum of 2.5 g for children aged 2–12 years). Side-effects are relatively common and can be serious. Therefore, piperazine should only be used if safer alternatives are unavailable.
• Pyrantel pamoate (10 mg/kg up to a maximum of 1 g) can be given as a single dose. Pyrantel and piperazine have antagonistic effects and should never be prescribed concurrently.
• Nitazoxanide is also effective (see p. 158).

Ascaris pneumonitis is managed symptomatically with bronchodilators and steroids if indicated. Symptoms may be exacerbated by larval death, therefore the use of antihelmintics is questionable.

Intestinal obstruction is usually managed conservatively with nasogastric aspiration, intravenous fluids and antispasmodics, followed by an antihelminthic when the obstruction has subsided. Laparotomy may be necessary if this fails or if the patient is seriously ill. It may be possible to manipulate the worms through the ileocaecal valve without having to open the bowel. Surgical or endoscopic removal of single worms blocking bile or pancreatic ducts may be necessary for patients who fail to respond to antihelminthic treatment and those with persisting pain or raised serum amylase.

Prevention and public health aspects

Improved hygiene and sanitation, access to clean water and health education are particularly important. In some communities, human excrement ('night soil') is used to fertilize vegetables and poses an obvious risk. Periodic mass chemotherapy of vulnerable schoolchildren and women of childbearing age is currently being promoted by WHO.

Recent developments

Studies from Thailand indicate that intestinal helminth infections, particularly with A. lumbricoides, appear to be protective against cerebral malaria, even though such infections are also associated with an increased risk of infection with Plasmodium falciparum.

Infection with A. lumbricoides has been shown to dampen the immune response to a recombinant cholera toxin and albendazole treatment of children with ascariasis enhances the vibriocidal antibody response to the live attenuated oral cholera vaccine CVD 103-HgR.

Hookworm

Epidemiology

Ancylostoma duodenale and Necator americanus are widely distributed in the tropics and subtropics, affecting approximately 900 million people worldwide. Necator americanus predominates in the Americas, Australia, sub-Saharan Africa, South Asia and the Pacific islands, whereas A. duodenale is more prevalent in the Middle East, northern Africa, southern Europe, northern India and northern China.

Parasites and life cycle

Hookworms are slender tubes about 1 cm long. They have a mouth and oesophagus at the front, connected by the gut to the anus at the rear. Ancylostoma is larger than Necator. The female body is largely occupied by eggs. The teeth in A. duodenale and the cutting plates in N. americanus are used to pierce the intestinal mucosa. The mouth and pharynx are used to attach the worms to the mucosa by suction.

Hookworm eggs passed in the faeces hatch in warm moist conditions, liberating rhabditiform larvae. These develop into filariform larvae, which inhabit the surface layer of soil. Filariform larvae penetrate human skin via fissures or hair follicles and are carried in the lymphatics and venous circulation to the lungs. Here they enter the alveoli, migrate to the pharynx and then to the small intestine where they mature into adults. Adult hookworms attach themselves to the upper half of the small intestine and feed on blood. An adult A. duodenale may consume between 0.15 and 0.26 mL/day. Necator americanus consumes a relatively modest 0.03 mL/day. Blood loss also occurs at the site of attachment. Loss of plasma proteins may result in hypoproteinaemia. The journey to the intestine takes about a week. Adults are fully grown (approximately 1 cm) in 2–3 weeks and sexually mature in 3–5 weeks, after which eggs begin to appear in the faeces.

Rarely, infection with A. duodenale may occur following ingestion of larvae on contaminated vegetables. Infantile hookworm disease has been described in China and attributed to transmammary transmission, laying infants on contaminated soil, or using nappies made of cloth bags stuffed with infected soil.

Clinical features

Most infections are asymptomatic. Problems arise when dietary iron intake is poor or demands are high, resulting in a gradually worsening iron-deficient anaemia, sometimes associated with hypoalbuminaemia and oedema. Eventually this may lead to cardiac failure. Pregnant women, women with menorrhagia and children are at greatest risk of developing anaemia.

Other symptoms are uncommon and tend to be associated with the early stages of initial infection. 'Ground itch' may occur at the site of larval penetration and, if severe, may be associated with the development of vesicles or pustules. The serpiginous rash of cutaneous larva migrans may be seen in human hookworm infections, although this is more commonly associated with infection with dog or cat hookworm. Larval migration through the lungs may cause a pneumonitis. Occasionally, within a few weeks of a heavy infection, there may be abdominal discomfort, flatulence, anorexia, nausea, vomiting and diarrhoea, sometimes containing blood and mucus. Life-threatening gastrointestinal haemorrhage has been reported as a rare complication in young children with severe primary infections.

Investigations

Eosinophilia is common during tissue migration. Characteristic eggs can be identified by standard faecal microscopy. Concentration methods may be necessary for light infections. Culture techniques similar to those used for *Strongyloides* may also be used. Eggs may hatch in stool samples that are left for a few days before examination, liberating larvae that may be mistaken for those of *Strongyloides stercoralis*, although they are morphologically distinct. Mixed infections of hookworm and *Strongyloides* may also occur.

Humans may be infected with largely non-pathogenic worms whose eggs resemble those of hookworm. The most important is *Ternidens diminutus*, a common parasite in monkeys, baboons and humans in southern Africa. The worms inhabit the large bowel, where they may cause cystic nodules. Because they suck blood, they can cause anaemia in heavy infections. Their eggs closely resemble those of the hookworm, but are larger.

Trichostrongylus worms of many species are natural parasites of herbivores in many parts of the world, and humans can become infected by ingesting the infective larvae on raw vegetables or salads. The adult worms are attached to the small intestine, but cause only slight damage and insignificant blood loss. The eggs of *Trichostrongylus* spp. have more pointed ends than the eggs of true hookworms.

Management

- Mebendazole 100 mg twice daily for 3 days is most commonly prescribed.
- Albendazole 400 mg as a single dose is more effective.
- Pyrantel pamoate 11 mg/kg (maximum 1 g) as a single dose is also effective.

Treatment for iron-deficiency anaemia may be required, preferably with oral iron. Transfusion is rarely necessary.

Prevention and public health aspects

Improved standards of hygiene and sanitation and the wearing of shoes reduce the likelihood of infection. The WHO is now promoting mass chemotherapy of schoolchildren and women of childbearing age in an attempt to reduce the global burden of soil-transmitted helminth infections. Interestingly, lower incidence, prevalence and intensity of infection have been noted among children who have received bacille Calmette–Guérin (BCG).

Trichuriasis

Epidemiology

Trichuris trichiura, the whipworm, has a global distribution, most prevalent in warm humid climates and infects about 900 million people worldwide.

Parasite and life cycle

Infection occurs when eggs contaminating soil, food or fomites are swallowed. Larvae are liberated in the caecum, penetrate the crypts of Lieberkühn and migrate within the mucosa. Mature adult worms are 2–5 cm long, the thinner anterior half of the body being normally partly buried in the mucosa of the large bowel of the host (caecum, colon, rectum). They feed on tissue juices, not blood. Female worms release several thousand eggs per day. After about 2 weeks' development in warm moist soil, the eggs are embryonated and infective.

Clinical features

Most infections are asymptomatic. Heavy infestations may cause severe gastrointestinal symptoms. Bleeding from the friable mucosa may result in iron-deficiency anaemia in children on poor diets. Chronic infection is associated with growth retardation. Severe trichuris dysentery syndrome frequently leads to rectal prolapse.

Investigations

The diagnosis is obvious in children presenting with rectal prolapse when adult worms can be seen attached to the mucosa of prolapsed bowel. In other circumstances, the characteristic eggs may be identified in the stool. Concentration techniques can be used for light infections; however, if eggs cannot be found on direct examination, the infection is unlikely to be of clinical significance. Trichuriasis may cause a significant eosinophilia.

Management

A single oral dose of mebendazole 500 mg is more effective than albendazole 400 mg. Severe infections require either mebendazole 100 mg twice daily for 3 days, or albendazole 400 mg daily for 3 days. Single dose combination treatment using albendazole 400 mg plus ivermectin 200 μg/kg is also highly effective. More recently, nitazoxanide has also been shown to be effective (see p. 158).

Prevention

Prevention consists of simple standard methods of improved hygiene and sanitation.

Toxocariasis

Epidemiology

Young children are at greatest risk of infection with *Toxocara canis* and *T. cati*, parasitic roundworms of dogs and cats. Sandpits in public parks fouled by dog faeces are particularly notorious as sources of infection.

Parasite and life cycle

Infection in humans occurs following ingestion of eggs in sand or soil contaminated by dog or cat faeces. The larva from the ingested egg is released in the intestine and then goes on a prolonged safari through the tissues, lasting 1–2 years. Worms seldom develop beyond the larval stage, so do not reach maturity in the intestine.

Clinical features, investigations and management

Clinical disease is relatively uncommon and depends on the intensity of infection and the organs involved. There are two distinct syndromes: visceral and ocular.

Visceral larva migrans

Visceral larva migrans (VLM) is caused by migrating larvae. Pneumonitis, fever, abdominal pain, myalgia, lymphadenopathy, hepatosplenomegaly, sleep and behavioural disturbances and focal or generalized convulsions can occur. Investigations commonly reveal eosinophilia, anaemia, hypergammaglobulinaemia and elevated titres of blood group isohaemagglutinins. Serological diagnosis may be established using an ELISA.

The treatment of choice for VLM is albendazole 10 mg/kg/day for 5 days. Alternatives include mebendazole 100 mg twice daily for 5 days, tiabendazole 50 mg/kg/day in three divided doses for at least 5 days, or diethylcarbamazine 3 mg/kg three times daily for 21 days. Symptomatic treatment with bronchodilators, steroids or antihistamines may also be indicated.

Ocular larva migrans

Ocular larva migrans (OLM) may occur in lighter infections. A larva invades the eye producing a granulomatous reaction, usually in the retina, resulting in visual disturbance or blindness in the affected eye. This may present as strabismus or go unnoticed. The diagnosis is sometimes made by chance on routine opthalmoscopy. The appearance is usually of chorioidoretinitis with a mass lesion, which may be mistaken for a retinoblastoma. Serology is usually positive. Antibody detection in vitreous fluid is more sensitive and specific. Patients with eye involvement seldom have eosinophilia or other evidence of generalized VLM.

Topical or systemic steroids are indicated in the management of acute OLM. There is no consistent evidence of benefit from the additional use of antihelminthics. Destruction of the larva is possible using laser photocoagulation.

Steroids may also be useful in exacerbations of chronic OLM. Surgery is often required for adhesions.

Prevention and public health measures

In most urban environments, children have to compete with increasing numbers of domestic pets for diminishing open spaces where they can play safely. Pet owners are advised to regularly deworm their dogs and cats and keep them away from children's play areas. In many towns and cities protected play areas are provided for children in public parks. There are also designated areas for pet owners to exercise their animals and pet owners are legally required to clean up if their animals defecate in a public area. Even so, when playing football in most public parks, you are as likely to be fouled by dog faeces as by a member of the opposing team.

Further reading

Awasti S, Bundy DAP, Savioli L. Helminthic infections. *Br Med J* 2003; **327**: 431–3. [Concise review of rationale for global control, and role of helminths in causing cognitive impairment in children.]

World Health Organization. Guidelines for the evaluation of soil-transmitted helminthiasis and schistosomiasis at community level. WHO/CTD/SIP/98.1 http://www.who.int/ctd/intpara/98–1.pdf   [This manual has been compiled to assist health planners at national, regional or district levels in the organization, management and evaluation of surveys on soil-transmitted helminthiasis and schistosomiasis for the development and implementation of control activities.]

CHAPTER 25

Viral Hepatitis

General clinicoepidemiological features, 180

Hepatitis A, 180

Hepatitis B, 181

Hepatitis D, 185

Hepatitis C, 186

Hepatitis E, 187

Hepatocellular carcinoma (hepatoma), 187

Further reading and other resources, 189

The differential diagnosis of jaundice in the tropics includes a variety of causes that are seen less often in developed countries (Table 25.1). Schistosomiasis does not usually cause jaundice, but schistosomal liver damage frequently coexists with chronic viral hepatitis. Many of the infectious causes of jaundice are common in childhood and are less likely in the differential diagnosis of a jaundiced adult in the tropics (e.g. glandular fever group, hepatitis A). The history should always include careful questioning about past and present use of alcohol, 'western' drugs and traditional herbal remedies, many of which can be hepatotoxic. This chapter focuses on the effects and prevention of hepatitis A, B, C, D and E in a tropical context.

General clinicoepidemiological features

The acute syndromes produced by hepatitis A, B, D or E are indistinguishable clinically except that acute hepatitis B patients are a little more likely to experience generalized arthralgia and rashes than patients with the other viruses, and acute hepatitis C rarely causes symptoms severe enough to seek medical treatment. In all cases, a prodrome of malaise, nausea and vomiting, fever and often diarrhoea leads to a phase of jaundice with dark urine and pale faeces. This is often followed by a cholestatic phase, especially in older adults in whom the recovery period can take several months.

No drugs have been shown to alter the course of acute hepatitis caused by these viruses. This includes the ayurvedic remedy 'Liv-52', popular in the Indian Subcontinent, vitamin injections and steroids. Full supportive therapy may be required for patients with severe liver failure, which is characterized by an altered (and falling) level of consciousness, metabolic flap of outstretched hands, ascites, peripheral oedema and a rising international normalized ratio (INR) or prothrombin ratio. Such patients have a high mortality and should be transferred early to a specialist centre if possible.

The water- and food-borne viruses hepatitis A and E have no significant chronic sequelae or carriage state, whereas the parenterally and sexually transmitted viruses hepatitis B, C and D can cause long-term problems in those patients who go on to become chronic carriers. The effects are worse in patients with dual infection (e.g. hepatitis B plus C) and especially with concurrent alcohol abuse, and in the tropics schistosomal hepatic fibrosis commonly coexists with chronic viral hepatitis.

Hepatitis A

Hepatitis A (HAV) is a single-stranded RNA virus, which is primarily spread by the faeco-oral

CAUSES OF JAUNDICE

Prehepatic	Hepatic	Posthepatic
Haemolysis (e.g. favism with G6PD deficiency)	Viral hepatits	Pigment gallstones
	Q fever	Hydatid in biliary tree
Haemoglobinopathies	Drugs	Ectopic ascariasis
Sepsis	Traditional medicines	*Opisthorchis/Clonorchis* infection
Malaria	Leptospirosis	Gall bladder cancer (associated with typhoid carriage)
	Yellow fever	Cholangiocarcinoma (associated with *Opisthorchis/Clonorchis*)

Table 25.1 Some causes of jaundice in the tropics.

route. It is so common in the tropics that almost all individuals in a developing country will have the infection by the age of 10 years, although most will not realize this because young children rarely have significant symptoms. Adults are more likely to become jaundiced, and those over 40 years have a small risk of dying from fulminant hepatitis.

The incubation period is 2–6 weeks and most patients already have detectable anti-HAV immunoglobulin M (IgM) antibodies in their blood by the time they develop symptoms. Over time, the IgM antibody is lost and is replaced by a long-lasting anti-HAV IgG response, with solid immunity (i.e. second infections are very rare). Patients with jaundice excrete virus and are infectious to others who do not wash their hands after contact.

Hepatitis A is mainly a problem for travellers to the developing world. Paradoxically, the older a 'western' person is, the more likely they are to have had hepatitis A in childhood before sanitation was improved. However, the minority of older adults who have no immunity may have substantial morbidity or mortality if they acquire HAV during a visit to the tropics.

Prevention is very effectively provided by active immunization with hepatitis A vaccine, followed by a booster 6–12 months later to produce life-long immunity. Anti-HAV IgG levels are not measured routinely to monitor vaccine response. Passive immunity can be provided by in-tramuscular gammaglobulin made from sera of individuals known or presumed to be immune; however, this is rarely used in the modern era.

Some tropical countries such as Singapore have experienced such an improvement in standard of living that children are no longer exposed to HAV in childhood, but are becoming symptomatically affected at a later age. In such a setting, active immunization of the population might become appropriate but, in general, HAV vaccines have little role in the tropics except for travellers.

Hepatitis B

Hepatitis B (HBV) is a major cause of death worldwide, with more than 2 million deaths per year mainly because of the sequelae of chronic liver disease and hepatocellular carcinoma (HCC). The DNA virus replicates in the hepatocytes which then express antigens such as HBsAg (surface antigen) on their surface, provoking both cell-mediated and humoral responses. These cause liver cell destruction associated with a rise in transaminases and clinical hepatitis, followed by loss of the circulating antigens HBsAg and HBeAg, and a rise in antibodies anti-HBs, anti-HBc and anti-HBe (Fig. 25.1). In some cases, immunity is insufficient to clear all the infected hepatocytes and the patient becomes a carrier, defined as having detectable hepatitis B (usually HBsAg) for over 6 months. Initially, carriers have high levels of circulating virus, analogous to continued acute

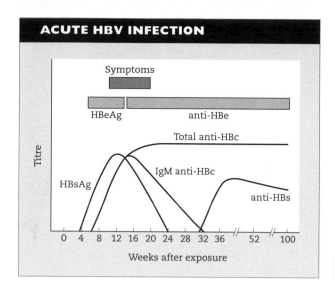

ACUTE HBV INFECTION

Fig. 25.1 Acute hepatitis B (HBV) infection, followed by recovery.

hepatitis, and the peripheral markers for this are detectable HBeAg in blood and high levels of HBV DNA. At some stage, the balance between virus and immunity may alter in favour of the host, in which case there is a clinical 'flare' of hepatitis as most infected hepatocytes are destroyed and the patient loses HBeAg and becomes anti-HBe antibody-positive (Fig. 25.2). These low-grade carriers are less infectious and less likely to develop chronic liver disease or HCC than high-grade carriers. The presence of chronic liver disease (e.g. chronic active hepatitis) can only be determined by clinical examination, by checking liver function tests and by liver biopsy, in addition to serological tests.

Conditions that favour infection progressing to carriage include the neonatal state, chronic illness such as renal failure, and immunosuppression because of HIV or chemotherapy. Immunosuppressed individuals may have high levels of virus but no illness because the destruction of liver cells relies on having a strong immune response. If this is improved (e.g. by successful antiretroviral therapy or by stopping chemotherapy), patients with chronic carriage may develop severe hepatitis.

Infants infected at the time of birth have a 90%

chance of becoming carriers, this risk falling to 10% after infection at 1 year of life and to less than 5% for adults. In the Far East, Polynesia and West Africa, 30–50% of carriers are infected from their mothers at birth and most of the rest are infected by uncertain means in early childhood. This tends to cluster in families with a carrier mother. In such populations, carriage rates in adults are high (≥8%) and carriers tend to have high levels of virus. In other parts of the tropics, most infections are acquired in childhood or infancy and intermediate HBsAg prevalence rates between 2 and 7% are seen in adults (Fig. 25.3, Table 25.2).

In industrialized countries, sexual transmission and shared drug injecting paraphernalia are the major routes of transmission in adulthood. Sexual transmission is important in the tropics but the major threat in this setting is nosocomial and/or iatrogenic hepatitis caused by reuse of needles and infusion sets. Other factors contributing to the spread of HBV in the tropics include traditional practices such as scarification or circumcision using nonsterile instruments, tattooing, acupuncture and barbering practices. Bedbugs (but not mosquitoes) have a small role in transmission in some settings, but

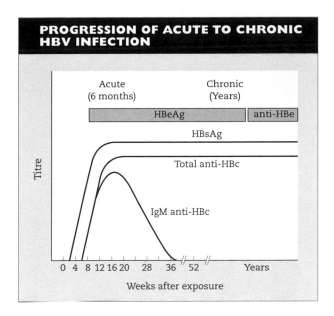

PROGRESSION OF ACUTE TO CHRONIC HBV INFECTION

Fig. 25.2 Progression of acute to chronic hepatitis B (HBV) infection (two patterns).

HEPATITIS B PREVALENCE PATTERNS

Prevalence	Adults with HbsAg (%)	Anti-HBc (%)	Incidence of infection Infant	Child	Adult	Location
High	8–20	70–95	+++	++++	++	China, South East Asia, Polynesia, West Africa
Intermediate	2–7	20–60	++	+++	++	South West Asia, eastern Europe, East and Central Africa, South and Central America
Low	<2	2–6	+	−	+	North, west and central Europe, North America, Australia

Table 25.2 Patterns of hepatitis B prevalence (World Health Organization definitions).

control of bedbugs has little effect on reducing transmission.

Treatment for chronic HBV is only appropriate for high-grade (HBeAg positive) carriers with the intention of converting them to anti-HBe antibody positivity. A small minority subsequently lose HBsAg as well, after a delay of several years. Until recently, the main treatment was high-dose interferon α (IFN-α) injections given three times a week for 3–6 months, limited by side-effects, cost and a success rate of less than 50% even in carefully selected patients. More recently, oral lamivudine (a nucleoside reverse transcriptase inhibitor) has been introduced as long-term suppressive therapy. Such monotherapy is predictably complicated by the emergence of viral resistance and viral escape after 2–3 years, and trials of double and triple

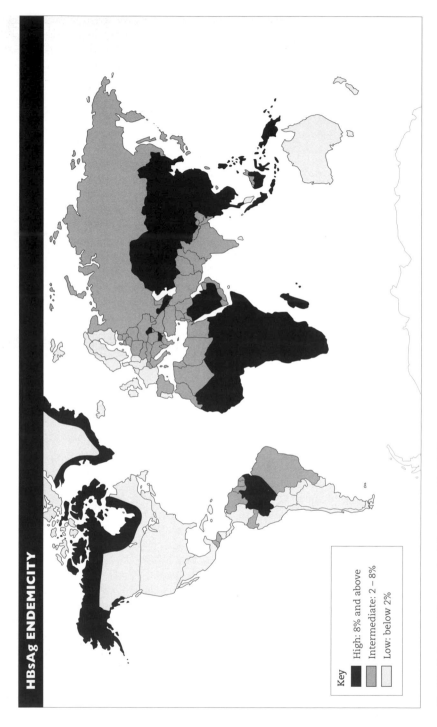

HBsAg ENDEMICITY

Key

High: 8% and above

Intermediate: 2 – 8%

Low: below 2%

Fig. 25.3 Patterns of HBsAg endemicity as defined by WHO.

therapy regimens are in progress. All are limited by cost and side-effects, and require specialist supervision.

Hepatitis B immunization

Very effective vaccines against hepatitis B have been available for over 20 years, containing HBsAg. This may be harvested from the plasma of chronic HBV carriers, and cheap vaccines from this source are available in many parts of the tropics (such as those from the Korean Green Cross). Other vaccines are genetically engineered in yeast culture systems. The vaccine is effective in neonates and three doses of vaccine produce a long-lasting response in 80–90% of children and young adults. Vaccine schedules 'approved' for western licensing are usually 0, 1 and 6 months, or 0, 1, 2 and 12 months (booster fourth dose), or 0, 7 and 21 days followed by a late booster. However, many other strategies are in use and it can be given alongside other Extended Programme on Immunization (EPI) scheduled vaccines in infancy. In western practice it is usual to confirm seroconversion, defined as a rise in anti-HBs antibody > 10 units/L, 4–6 weeks after the third dose of vaccine, and then to give a final booster vaccination 5 years later.

Factors that reduce the efficacy of active immunization include male gender, age > 40 years, smoking tobacco and immunosuppression for any reason. The vaccine must be given into muscle (deltoid or thigh) and inappropriate administration in adipose tissue (e.g. buttocks) is ineffective. Active vaccination can be given at the same time as passive immunization with hyperimmune hepatitis B immunoglobulin (HBIg), a strategy that is used in some countries for extra, early protection of infants born to HBeAg positive (high-grade) carrier mothers. Immunization of infants within 48 h of birth prevents almost all such transmissions, which is usually perinatal rather than intrauterine.

Large community-based studies in different parts of the tropics have shown that infant or childhood vaccination programmes reduce the rate of infection by 90% or more. Most of the few who do become infected have subclinical disease, and very few become carriers. The protective effect in reducing community rates of infection, hepatitis B carriage and hepatocellular carcinoma persist for at least 10 years after such programmes, without further boosters.

Over 140 countries have adopted policies of universal immunization against HBV. This depends on adequate supplies of affordable vaccine, which in turn requires an intact cold chain to maintain vaccine potency. In most tropical settings, the majority of infections occur in infancy or early childhood and universal infant immunization is therefore the most appropriate strategy. Subsequent boosters are not needed in tropical settings. In western settings, in which infections predominantly occur in infants and children of carrier mothers or in adolescence and early adulthood, selective immunization of children of identified carriers may be employed. However, it is more difficult to then arrange for universal or selective adolescent or adult immunization, and most countries have opted for universal immunization in early childhood.

Hepatitis D

Hepatitis D virus (HDV) resembles some plant viruses in that it is incomplete. The outer coat is derived from hepatitis B surface antigen, which means that the virus can only infect individuals who have acute hepatitis B or who are chronic HBV carriers. The virus is transmitted by the same routes as HBV and two epidemiological and clinical patterns are seen. Patients who are infected by hepatitis B and D viruses at the same time have a clinical acute hepatitis that is no more severe than hepatitis B, and are no more or less likely to progress to chronic liver carriage and disease. More importantly, chronic HBV carriers who acquire acute hepatitis D (so called 'superinfection') are likely to develop fulminant hepatitis with a high mortality rate. In western and tropical settings where intravenous drug misuse is a problem, the clue to the arrival and spread of the virus is an epidemic of severe liver disease with high mortality. In tropical settings, similar epidemics are seen in previously asymptomatic HBV carriers of all ages. Such epidemics are regularly reported in South America, examples being epidemics of 'La Brea' or 'Santa Marta' fever. The epidemiological clue is the high proportion of affected patients who develop severe or fatal disease, and would be confirmed by detection of both HBV antigens and anti-HDV antibody responses. However, the latter are rarely available in a tropical setting.

As HDV only affects patients with concurrent acute or chronic HBV, it is prevented by immunization against hepatitis B and by other control measures used to prevent nosocomial or sexual transmission of hepatitis B.

Hepatitis C

Hepatitis C (HCV) is an RNA virus that is predominantly spread by blood–blood contact, and reuse of syringes and infusion equipment is responsible for high rates of hepatitis C endemicity in many parts of the tropics. Other factors encouraging blood–blood contact are also likely to be important in a tropical setting, as for HBV. Sexual and perinatal transmission of HCV is relatively inefficient but accounts for low background prevalence rates. However, the risk of sexual and perinatal transmission rises to about 10% if the 'donor' is also HIV positive. The prevalence of anti-HCV antibodies in Egypt and parts of the Yemen is extraordinarily high, affecting over one-third of adults aged > 40 years. This is because of past well-intentioned mass treatments of populations with tartar emetic for endemic schistosomiasis. The same syringes were used sequentially for many people, resulting in the largest described iatrogenic epidemic of a blood-borne virus in the world.

Hepatitis C rarely causes symptomatic acute hepatitis, but 70–80% of those infected will become chronic carriers, and there is a prolonged 'window' of antibody seronegativity of 2–3 months (ranging up to 6 months) after initial infection. Molecular methods (polymerase chain reaction, PCR) are now routinely used in western settings to detect early infection, and to confirm the presence of viraemia in patients who are anti-HCV antibody positive. Like HIV, HCV has different genotypic groups and a very high rate of viral turnover, leading to frequent mutations of viral quasispecies, so immunity is not solid and patients can be reinfected by HCV after successful treatment.

Chronic HCV is defined by presence of viraemia for more than 6 months, associated with abnormal hepatic histology (on liver biopsy) and/or abnormal liver function tests. Normal liver function tests do not exclude abnormal histology, so liver biopsy is essential for full work-up of patients. The natural history is still poorly understood, and patients who have had single or few exposures to small amounts of virus seem less likely to progress. Overall, the '20' rule applies: 20% of patients initially infected will clear the virus (but still remain antibody positive). Of the remaining 80%, 20% will develop significant liver disease over 20 years. Of those with cirrhosis, up to 20% might develop HCC over a further 20 years. The most important determinant for development of serious liver disease and HCC is concurrent alcohol abuse, and the most important aspect of clinical management is to persuade chronic carriers to reduce or discontinue their alcohol intake.

Other factors that favour more rapid progression (and also resistance to treatment) include older age, male gender, infection with genotypes other than genotypes 2 or 3, and presence of stainable iron in liver biopsies. In counselling patients it is important to emphasize that most will not develop clinically apparent liver disease, and that the main risk to partners and families is sharing of needles and other sharps such as razors, nail scissors, toothbrushes, etc. In tropical settings, concurrent infection with hepatitis B and/or schistosomiasis is common and worsens the prognosis, as does HIV infection. A minority of HIV positive individuals develop fulminant or rapidly progressive HCV for unknown reasons.

Treatment is expensive and rapidly evolving. At the time of writing, treatment regimens incorporating injected IFN-α or pegylated (PEG)

IFN-α, together with oral ribavirin for 6–12 months, produce short-term viral remission in over 50% of cases. This needs careful supervision and costs £10000–20000 per course. There are no immediate prospects for vaccination, and the only means of control is to educate health care workers and others not to reuse any kind of injection or infusion equipment. Intravenous drug misusers (most are anti-HCV antibody positive within 3 years of starting injecting) should not share any kind of injecting equipment, including water and spoons or cookers used to dissolve drugs. Monogamous couples may choose not to start to use condoms (as the risks are small) but HCV carriers should otherwise be encouraged to use condoms with new partners. There is no evidence against breast-feeding.

Hepatitis E

Hepatitis E (HEV) is another virus spread by the faeco-oral route, usually by contamination of water supplies, with a similar incubation period and clinical outcome to HAV infection. The key difference from HAV is that immunity after natural infection is not solid, so that older children and adults can suffer repeated symptomatic infections. Large epidemics that affect people of all ages have been described in the Indian Subcontinent and central Asia for over 50 years, whereas epidemics of HAV primarily affect children. It is also well-recognized in Mexico and North Africa (Fig. 25.4) but the epidemiology in most of Africa is poorly described and probably underestimated. HEV is rare in western countries unless there is a history of travel to the tropics or of contact with a recent traveller, but there is an unexplained increased prevalence of anti-HEV antibodies in some populations at risk of blood-borne virus infections such as renal dialysis patients. Pigs have similar virus infection and it is possible that a zoonotic risk will be confirmed in future.

HEV is endemic in the Indian Subcontinent where it accounts for the majority of sympto-matic acute viral hepatitis admissions to hospital of both adults and children. Like HAV, there are no chronic sequelae.

The second major difference from other forms of viral hepatitis is the high incidence of fulminant hepatitis in pregnant women, with a mortality of 30–50%, particularly in the later stages of pregnancy. Perinatal transmission also occurs with a high mortality in infants who are infected.

Diagnosis is by detection of anti-HEV IgM followed by anti-HEV IgG antibodies. However, assays were not well-standardized and titres of both antibodies decline rapidly (within 1–2 years), so detection of antibodies late after infection is difficult. Passive immunization with pooled immunoglobulin from donors living in endemic areas is not effective. Research on production of a vaccine continues but none is currently available.

Hepatocellular carcinoma (hepatoma)

This aggressive tumour is common in the tropics, and is multifactorial in aetiology. Any patient with chronic active hepatitis is at risk, especially if cirrhosis has developed. Both chronic HBV and HCV are important causes, particularly if patients abuse alcohol. In many parts of the tropics, aflatoxins ingested in food are important epidemiologically as primary or cofactor causes, although it usually impossible to determine this in individual patients.

Patients usually present at a late stage with several months of weight loss, a painful hard irregular mass in an enlarged liver, often with ascites which may be bloodstained. Other signs of chronic liver disease may be obvious, and a 'bruit' or arteriovenous hum can be heard on auscultation over the liver. This is almost pathognomonic of hepatoma, although a 'bruit' can occasionally be heard over the liver in acute alcoholic hepatitis. Ultrasound shows diffuse infiltration of the liver by one or more tumours, often with central necrosis, which may

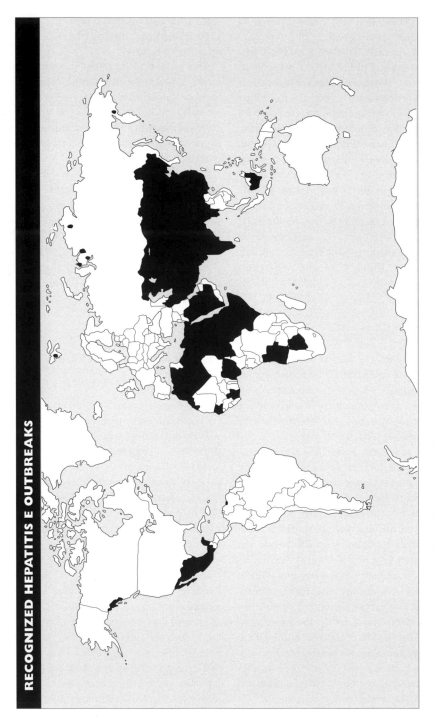

RECOGNIZED HEPATITIS E OUTBREAKS

Fig. 25.4 Countries with recognized hepatitis E outbreaks (after WHO).

resemble metastases from other sites. The other main differential diagnosis is acute amoebic liver abscess, which has a more acute presentation, typical ultrasound appearances, peripheral neutrophil leucocytosis and positive amoebic antibody tests. Most HCC patients have very elevated levels of alpha-fetoprotein in the serum, although this is absent in a minority. By the time that patients present with symptoms, the prognosis is very poor. Surgical removal, embolization and chemoembolization can all be attempted, and liver transplantation may be possible in some cases. In most resource-poor settings the only therapeutic option is palliation.

Prevention is by measures to control HBV and HCV infections, and countries that have initiated mass infant HBV immunization programmes have experienced marked falls in HCC incidence. Control of alcohol abuse is also important, especially in those with chronic HBV or HCV carriage.

Further reading and other resources

Centers for Disease Control—National Center for Infectious Diseases. Viral hepatitis http://www.cdc.gov/ncidod/diseases/hepatitis/index.htm [Western-orientated patient information sheets, detailed statements of best practice, and large downloadable teaching slide sets. First source for hepatitis C.]

Frank C, Mohamed MK, Strickland GT et al. The role of parenteral antischistosomal therapy in the spread of hepatitis C virus in Egypt. Lancet 2000; 355: 887–91. [Describes the largest single iatrogenic blood-borne virus epidemic in the world.]

WHO website http://www.who.int/csr/disease/hepatitis/resources/en/ [Excellent essays, illustrations, maps about all aspects, with extensive electronic links to on-line resources and full scientific referencing for those who want more information. First port of call for hepatitis A, B, D and E.]

CHAPTER 26

Liver Flukes

Liver flukes, 190
Oriental liver flukes, 191

Intestinal flukes, 192

Further reading, 192

Food-borne trematodes include liver, lung and intestinal flukes. One-fifth of the world's population is at risk from these infections, which are endemic in at least 100 countries, half of which are among the poorest in the world. Forty million people are infected and there are at least 10000 deaths per year. Praziquantel is the drug of choice for most food-borne trematode infections. However, a notable exception is *Fasciola hepatica*, for which triclabendazole is emerging as the drug of choice.

Liver flukes

Epidemiology
Fasciola hepatica and *Fasciola gigantica* occur in sheep- and cattle-rearing areas worldwide, notably in the South American Andes, especially Bolivia and Peru.

Parasites and life cycles
In common with other flukes, the life cycles of *F. hepatica* and *F. gigantica* involve certain species of freshwater snail that are infected by miracidia liberated from eggs passed in herbivore (or human) faeces. The snails act as intermediate amplifying hosts, eventually liberating free-swimming cercariae which encyst as metacercariae on water plants.

Human infection occurs when the metacercarial cysts on raw water vegetables (e.g. watercress) are eaten, or are swallowed in contaminated water. These excyst in the

duodenum, releasing larvae which penetrate the intestinal wall, and migrate via the peritoneal cavity to the liver. Having penetrated the liver capsule, they make their way to the bile ducts, where they mature into adults. Maturation in the human host takes 3–4 months. Adult flukes can survive for up to 10 years.

Clinical features
Acute symptoms caused by migrating flukes may develop 6–12 weeks after infection, including fever, malaise, abdominal pain, weight loss, urticaria and respiratory symptoms. Tender hepatomegaly may be evident. Liver enzymes are sometimes mildly elevated. Ectopic flukes can cause granuloma or abscess formation in various organs and migrating erythematous cutaneous nodules, a form of cutaneous larva migrans, may also be seen. Mature flukes in the bile ducts may initially cause fever, anorexia and abdominal pain.

Symptoms usually subside spontaneously once the adult flukes have made themselves at home. Blood loss into the bile resulting in anaemia occurs in heavy infections. Chronic symptoms include recurrent cholangitis or intermittent biliary obstruction in a minority of patients, and fatigue which can persist for more than 10 years.

An acute nasopharyngitis (known as *halzoun* or *marrar*) was previously thought to be caused by pharyngeal attachment of the adult fluke following the consumption of raw liver, but is in fact caused by larval nymphs of *Linguatula serrata*.

Investigations

Eosinophilia is common. Ultrasound is usually normal. Computerized tomography (CT) of the liver may reveal numerous hypodense lesions, and peripheral branched hypodense hepatic lesions, best seen on CT using contrast, are relatively specific for fascioliasis. Serology is helpful in diagnosing *F. hepatica* infections in the acute phase when eggs are unlikely to be present in faeces. Blood spots can be collected on filter paper for later serological testing in large-scale surveys. Serology is less reliable for *F. gigantica*. In established infections, eggs may be present in faeces or in bile aspirate. Concentration techniques may be required. Fasciola excretory–secretory (FES) antigen detection in faeces is useful both in prepatent and in patent infections with *F. hepatica*.

Management

Bithionol 30–50 mg/kg/day in three divided doses on alternate days for 10–15 days has been recommended in the past. Side-effects include mild gastrointestinal upset and pruritus. Triclabendazole, a new benzimidazole, is simpler to use, has few side-effects and is likely to become the drug of choice. A single dose of 10 mg/kg taken with food is usually effective. This may be repeated after 12 h in severe infections. Biliary colic associated with the expulsion of dead or damaged parasites, commonly occurs 3–7 days after treatment and responds well to antispasmodic therapy. Nitazoxanide is also effective against *F. hepatica*, but multiple doses may be required. In contrast to other fluke infections, praziquantel is unreliable in the treatment of fascioliasis.

Prevention and public health aspects

Avoid eating potentially contaminated watercress and other water plants. Treatment of herbivores and snail control measures may sometimes be feasible.

Oriental liver flukes

Epidemiology

Opisthorchis sinensis (also known as *Clonorchis sinensis*) and *Opisthorchis viverrini*, affect about 20 million people in China and South East Asia. *Opisthorchis felineus*, a related species, occurs in eastern Europe and Russia. Animal hosts include domestic dogs and cats. This has important implications for control programmes.

Parasites and life cycles

Eggs passed in human or animal faeces on contact with fresh water release miracidia that infect and multiply in certain species of freshwater snail which act as intermediate amplifying hosts. The snails eventually liberate free-swimming cercariae that, as metacercariae, encyst on susceptible species of freshwater fish.

Human infection occurs when metacercariae are consumed in raw or undercooked fish, or after ingesting metacercariae contaminating cooking surfaces and utensils. Metacercariae excyst in the small bowel, migrate along the common bile duct and colonize the biliary tree where they mature into adults within about 4 weeks. All adult oriental flukes are hermaphrodite creatures of similar appearance, lanceolate in shape, translucent and brownish in colour. In common with other flukes, they possess two suckers. *Opisthorchis sinensis* is about 10–25 mm long by 3–5 mm wide. The other species are about half as big.

Clinical features

Most infections are asymptomatic. Heavy initial infections may present with an illness similar to Katayama fever. Patients with established infections may have vague right upper quadrant abdominal pain that typically occurs in the late afternoon and lasts a few hours. Patients may actually complain of feeling something moving about the liver. Other symptoms include lassitude, anorexia, flatulence, diarrhoea and fever. Hepatomegaly may be evident on examination and heavily infected patients may also be jaundiced. Some patients appear malnourished and are deficient in fat-soluble vitamins.

Recurrent bouts of ascending cholangitis, jaundice and pancreatitis can occur. Biliary cirrhosis and, rarely, cholangiocarcinoma may develop in chronic infections. Cholangiocarcinoma associated with *O. viverrini* is the most common form of liver cancer in north-eastern Thailand, where an estimated 70% of the population are infected with the parasite. In Hong Kong, 15% of all primary liver cancers were found to be cholangiocarcinomas associated with *C. sinensis*. Genetic factors, dietary nitrosamines and aflatoxins have been implicated in the development of this malignancy.

Investigations

Diagnosis is established by identifying characteristic eggs in faeces or in biliary aspirate. Concentration techniques may be required. Immunological tests are generally unavailable and lack specificity. Stool antigen tests show more promise. Ultrasound may reveal abnormalities of the biliary tree and gallstones. Endoscopic retrograde cholangiopancreatography (ERCP) is also useful.

Management

Praziquantel is effective, either 40 mg/kg in a single dose, or 25 mg/kg three times in 24 h after meals. A 3-day course should be used for heavy infections.

Prevention and public health aspects

Health education, improving sanitation and annual treatment with praziquantel can dramatically reduce the prevalence of infection and, on the face of it, infection with oriental liver flukes should be easy to prevent by avoiding raw fish. However, food habits are very difficult to change, even with vigorous health education. It seems that once you have tasted well-prepared raw fish, the cooked item never tastes as good.

Intestinal flukes

Fasciolopsis buski, the most important intestinal fluke in humans, is widely distributed from India to South East Asia, particularly in pig-rearing communities. Metacercariae attached to edible water plants, such as the water caltrop, are ingested, excyst, attach to the mucosa of the duodenum and jejunum and develop into adults causing inflammation and ulceration.

Most infections are asymptomatic. Symptoms are more likely to occur in heavy infections, and tend to be most severe in children. These include epigastric pain, vomiting and diarrhoea, initially alternating with constipation but later becoming persistent. Wasting, oedema and ascites may also occur in severe cases. Characteristic eggs, and sometimes adult flukes, can be identified in faeces and adult flukes sometimes also appear in vomit. Echinostome species, principally found in Asia, can also infect humans causing symptoms similar to *F. buski*. Heterophyids, a group of smaller intestinal flukes affecting humans, cause milder gastrointestinal symptoms than *F. buski*. However, ectopic eggs may be deposited in other organs, e.g. the central nervous system (CNS), presenting as a space-occupying lesion, and the heart, causing myocarditis or valve damage.

Further reading

Gillespie SH, Pearson RD, eds. *Principles and Practice of Clinical Parasitology*. John Wiley and Sons, 2001. Chapters 17 and 24. [This excellent volume on clinical parasitology includes two useful chapters on food-borne trematode infections.]

CHAPTER 27

Hydatid Disease

Echinococcus granulosus (hydatid cyst disease), 193

Echinococcus multilocularis (alveolar hydatid disease), 194

Further reading, 195

Hydatid disease results from the larval stage of a small tapeworm of dogs and other canines developing in humans. The infection is a zoonosis, normally maintained in dogs and sheep or cattle in close association with humans (*Echinococcus granulosus*), or in a wild cycle such as in wild canines and rodents (*E. multilocularis*). Most human infections are with *E. granulosus* and are associated with the rearing of sheep and cattle in climatic conditions varying from tropical to subarctic.

Echinococcus granulosus (hydatid cyst disease)

Life cycle

Infected dogs harbour the 3–6 mm adult tapeworms in their small intestine. The worms possess only three proglottides, the end one being mature. The eggs are liberated either before or after the proglottid escapes in the faeces, and contaminate pasture. When ingested by the normal herbivorous intermediate host, the oncospheres liberated in the gut enter the circulation and are trapped in the capillaries of various viscera, where they develop into cysts. A cyst is composed of a sphere of germinal epithelium containing protruding invaginations (brood capsules) and fluid. From the inner surface of the brood capsules, protoscolices develop, invaginated in much the same way as the cysticerci of *Taenia* spp. The whole structure is a hydatid cyst, and it becomes surrounded by fibrous capsule derived from the host tissue. The cyst may de-velop large daughter cysts in its cavity, each containing more brood capsules. The cyst continues growing for years. Brood capsules that break free from the cyst wall, and individual scolices in the cyst cavity, are called hydatid sand.

Dogs become infected by eating the contents of hydatid cysts in infected carcasses. Sheep or other herbivores become infected by swallowing the *Taenia*-like eggs passed in dog faeces. The strain of *E. granulosus* in the UK that commonly infects horses and has the foxhound as its definitive host is probably not pathogenic for humans. There are several other biological complexes in nature that probably do not pose the risk of human infection.

Clinical features

About 70% of cysts develop in the liver, usually the right lobe, 20% in the lungs and the rest in rarer sites. Cysts may be single or multiple. Symptoms are caused by a mass effect produced by the growing cyst, secondary bacterial infection of the cyst or because of leakage of fluid from the cyst. Hepatic cysts are initially asymptomatic until they become large. A non-tender mass may be evident on examination. If secondary bacterial infection occurs, the cyst may mimic a liver abscess. Spillage or leakage of the cyst fluid, during surgery or following rupture, can precipitate hypersensitivity reactions ranging from urticaria, pruritus and fever to fatal anaphylaxis. Secondary cysts may develop following spillage and seeding in the peritoneal cavity. Leakage from a cyst into the biliary tree may cause colic, urticaria and obstructive jaundice, sometimes complicated by secondary bacterial infection.

Most lung cysts are asymptomatic, found incidentally on a chest X-ray (Fig. 27.1, see colour plate facing p. 146). Symptomatic patients may complain of fever, dyspnoea, chest pain and cough, occasionally with haemoptysis. Secondary infection may result in development of a lung abscess. Pneumothorax, empyema or a hypersensitivity reaction may occur following rupture into the lung. Seeding of pulmonary cysts is uncommon. Rupture of a cyst into a bronchus may cause the patient to cough up clear salty tasting liquid, sometimes followed by the soft white outer membrane of the cyst. A collapsed cyst may have a characteristic 'water lily' appearance on chest X-ray.

Hydatid cysts occur at a variety of other sites including spleen, bone (causing pain and pathological fracture), brain (causing convulsions or a mass effect) and eye (causing proptosis and chemosis).

Investigations
Imaging
Ultrasound is useful for abdominal cysts. X-ray, computerized tomography (CT) or magnetic resonance imaging (MRI) may be useful for detecting cysts elsewhere.

Serology
The specific immunoglobulin G (IgG) enzyme-linked immunoabsorbent assay (ELISA) antigen B-rich fraction (AgB) is the most sensitive serological test. Others include an enzyme-linked immunotransfer blot (EITB) assay and the double diffusion test for arc 5 (DD5). Current serological tests lack sensitivity for extrahepatic cysts. The DD5 may give false-positive results in patients with cysticercosis.

Others
Urine antigen detection tests are promising. Eosinophilia may follow leakage or rupture of a cyst.

Treatment
In the past, surgical removal was the preferred method of managing accessible cysts. Small cysts can be removed intact. Larger cysts should be carefully aspirated and the aspirate replaced with an equivalent volume of a scolicide, such as hypertonic saline, reaspirated after 5–10 min and the procedure repeated. Following reaspiration, the cyst cavity is opened, the membranes are removed and the cavity closed. Percutaneous aspiration of cysts under ultrasound control is now used increasingly as an alternative to surgery. Following initial aspiration, hypertonic saline is injected into the cyst and reaspirated after 20 min.

Patients undergoing surgery or percutaneous aspiration should receive concomitant albendazole, either alone or in combination with praziquantel. Percutaneous aspiration plus an 8-week course of albendazole is more effective than either treatment alone. Laparoscopic treatment of hydatid cysts of the liver and spleen is also effective. Antihelmintic treatment may reduce the need for surgery in patients with uncomplicated pulmonary cysts.

Albendazole is useful for patients with inoperable, widespread or numerous cysts and in patients who are unfit for surgery. The course usually recommended is 400 mg twice daily for adults (5–7.5 mg/kg twice daily for children) for 28 days. This is followed by 14 days' rest and then repeated for 3–12 cycles depending on response. Albendazole absorption is enhanced if taken with fatty meals. Albendazole plus praziquantel has been shown to have greater protoscolicidal activity in animal studies and in vitro compared with either drug alone. Combined therapy has been used successfully in managing inoperable spinal, pelvic, abdominal, thoracic and hepatic hydatidosis and as an adjunct to surgery.

Echinococcus multilocularis (alveolar hydatid disease)

Life cycle
Humans become infected by swallowing the eggs passed by foxes and other Canidae, possibly mainly from contaminated wild ground fruits

such as bilberries and their close relatives, lingonberries and cloudberries, widely eaten in northern Europe. Recent surveys in central Europe have extended the known geographical occurrence of *E. multilocularis* in foxes from four countries at the end of the 1980s to at least 11 countries in 1999. Factors with the potential of enhancing the infection risk for humans in the future include increasing fox populations and parasite prevalences, progressing invasion of cities by foxes, the establishment of urban cycles of the parasite and the spillover of the *E. multilocularis* infection from wild carnivores to domestic dogs and cats. Infection is spreading in North America and Japan. An intense focus of infection has been described in Gansu province, China. Various rodents are the intermediate hosts.

Clinical features

Unlike *E. granulosus*, the cyst produces daughter cysts by external and not internal budding, so it tends to invade progressively into surrounding tissues like a malignant tumour. It is not contained in a well-defined fibrous capsule. It may be 30 years before a patient becomes symptomatic. The primary site of tissue invasion is usually the liver, but metastatic lesions may occur in other tissues. Patients usually present with right upper quadrant pain, hepatomegaly and a palpable mass. Complications arise in around 2% of patients, either because of local invasion or as a result of metastatic lesions involving brain, lung or mediastinum. Untreated, 90% of patients die within 10 years of presentation.

Investigations

Ultrasound, CT and serology are useful in establishing the diagnosis. Histology provides confirmation.

Treatment

Surgical excision is preferred for the primary lesion. Pre- and postoperative treatment with albendazole is recommended. Albendazole 10 mg/kg in cycles of 28 days followed by 14 days' rest is recommended postoperatively and for inoperable patients. Treatment cycles are usually continued for more than 1 year. The optimal duration of treatment remains uncertain. Mebendazole 40–50 mg/kg/day has also been used extensively, sometimes for up to 10 years. Serology may remain positive for several years following successful treatment. A 90% 10-year survival rate is possible with early diagnosis and appropriate treatment.

Further reading

Eckert J, Conraths FJ, Tackmann K. Echinococcosis: an emerging or re-emerging zoonosis? *Int J Parasitol* 2000; **12–13**: 1283–94. [An excellent recent review of the epidemiology and control of human echinococcal infections.]

PAIR: Puncture, Aspiration, Injection, Re-Aspiration. An option for the treatment of cystic Echinococcosis. WHO/CDS/CSR/APH/2001.6 http://www.who.int/emc-documents/zoonoses/whocdscsraph20016.html [Practical guidelines and useful images.]

CHAPTER 28

Pneumonia

Microbiology, 196
Epidemiology, 196
Risk factors for pneumonia, 198

Clinical features, 198
Investigations, 199
Management, 201

Prevention, 204
Future developments, 204
Further reading, 205

Pneumonia is an acute inflammatory disease of lung parenchyma occurring in response to microbial invasion of the distal bronchial tree and alveoli. It is a leading cause of morbidity and mortality in all parts of the world. The incorporation of a chapter on pneumonia in a tropical medicine text underlines the importance of this condition in the developing world—the World Health Organization (WHO) World Health Report 2000 placed lower respiratory tract infections as the leading global cause of ill health. This chapter also serves to highlight the role of specific pathogens, and the difficulties of management when only limited investigational and therapeutic resources are available.

Microbiology

Streptococcus pneumoniae is the most important—by virtue of both frequency and severity—aetiological agent of pneumonia throughout the world and particularly so in the tropics (Table 28.1). Preceding viral upper respiratory tract infections may be important precipitants, although the role of viruses as a primary cause of pneumonia in adults is unclear. Likewise, the contribution of the 'atypical' pneumonia group of bacteria is uncertain, and the infrequency of reporting may simply reflect inadequate surveillance and study. Most importantly, patients with *Mycobacterium tuberculosis* frequently present with an acute pneumonic ill-

ness and can also present with a dual infection with *S. pneumoniae*. In adults, *M. tuberculosis* may be identified in as many as 10% of cases of severe acute pneumonia admitted to hospital in sub-Saharan Africa. Acute presentations of pulmonary tuberculosis in children are also well recognized.

Epidemiology

Pneumonia affects individuals of all ages and the risk of disease exists throughout the year with seasonal variation. Rates are highest during the cooler drier periods, with case numbers beginning to rise towards the end of the rains when malaria is also a significant problem. Pneumonia is invariably reported as a leading cause of admission to hospital in all regions of the tropics, although accurate community-based incidence data are lacking. Attack rates in children are surpassed only by malaria and diarrhoea and account for 18% of the 10 million deaths in children under the age of 5 each year. Equivalent estimates for adults are more uncertain. Rates of pneumonia in HIV-infected adults have been measured at 4000 per 100000 person years. HIV increases pneumonia risk between 5- and 10-fold; thus, between 2.2 and 3.3 million episodes of pneumonia (30–50% with HIV coinfection) can be estimated to occur in adults in Africa each year (assuming 300 million adults, 10% HIV seroprevalence). Pneumonia case

MICROBIAL AGENTS OF PNEUMONIA AND RELEVANT THERAPIES

Microbe	Comment	Therapy
Common		
Streptococcus pneumoniae	Over 80 serotypes, majority of disease caused by just 10 serotypes. Invasive disease in 30% of cases. Most frequent agent of pneumonia	1 Amx/Bzp 2 Ery/Chl
Haemophilus influenzae	Frequently recovered from respiratory tract in pneumonia, rarely invasive. In adults, uncommon and not Pittman type b	1 Amx/Ctm 2 Chl
Mycobacterium tuberculosis	Agent of acute pneumonia in up to 10% of cases in sub-Saharan Africa. Should be considered in all poorly resolving pneumonia	Short course therapy
Influenza, parainfluenza, adenovirus	Believed to be important but scale of problem uncertain in tropics. Likely to be important as precipitants of bacterial infection	Nil
Less common		
Chlamydia pneumoniae *Chlamydia psittaci* *Mycoplasma pneumoniae*	These agents have all been described in the tropics. *Mycoplasma* probably occurs in cyclical epidemics as in industrialized countries and is usually a mild illness. Psittacosis is infrequent. The epidemiology of *C. pneumoniae* is unknown	1 Ery 2 Tet/Cip
Legionella spp.	Reported in South Africa and South Asia, but relative importance from non-selective studies not described	1 Ery 2 Cip
Non-*typhi Salmonella*	Pneumonia often a feature of a disseminated septicaemic infection. A particular problem in children and HIV-infected adults	1 Cip 2 Chl
Salmonella typhi	Typhoidal pneumonia an uncommon late (week 2) complication of untreated disease	1 Cip 2 Chl
Gram-negative bacteria	Unusual primary event. Principally nosocomial and in debilitated/bedridden individuals	1 Gen 2 Cip/Chl
Pneumocystis carinii	Important cause of pneumonia in HIV-infected children under 6 months old. However, an infrequent cause of pneumonia in HIV-infected adults in Africa	1 Ctm
Staphylococcus aureus	Infrequent but serious, classically cavitatory and fulminant	1 Flu & Gen
Bordetella pertussis	Remains an important childhood cause of pneumonia, particularly where vaccine coverage is poor. Often superinfected with *S. pneumoniae* or *H. influenzae*	1 Ery 2 Chl

Continued on p. 198

MICROBIAL AGENTS OF PNEUMONIA AND RELEVANT THERAPIES (*CONTINUED*)

Rare

Q-fever/*Coxiella burnetii*	Zoonotic infection from environment contaminated from infected cattle/sheep. Birth/ abortion products particularly infectious	1 Cip 2 Tet
Bacillus anthracis	Usually rapidly fatal following inhalation of spores after handling and processing infected animal carcasses	1 Bzp 2 Cip/Ery/Chl
Yersinia pestis	Pneumonic disease often appears during outbreaks of plague. Because of human–human potential for spread, cases must be isolated and treated urgently	1 Stp/Cip 2 Chl/Tet
Nocardia, Rhodococcus	Weakly acid-fast saprophytic Actinomycetes that may cause pneumonia in immunocompromised adults. May be misidentified as *Mycobacterium* spp.	1 Ctm/Ery/Cip
Burkholderia pseudomallei	Occasional cause of pneumonia in South East Asia. Particularly associated with diabetes	1 Cfz 2 Chl/Ctm

Abbreviations: Amx, amoxicillin; Bzp, benzylpenicillin; Cfz, ceftazidime; Chl, chloramphenicol; Cip, ciprofloxacin; Ctm, co-trimoxazole; Ery, erythromycin; Flu, flucloxacillin; Gen, gentamicin; Stp, streptomycin; Tet, tetracycline.

Table 28.1 Overview of microbial agents of pneumonia and relevant therapies (1, first choice; 2, second choice).

fatality was 60% in the preantibiotic era. With effective early therapy this may be reduced to 5%. Unfortunately, case fatality rates are still reported at 15–20%, reflecting the problems of late presentation and delays in therapy, common to all resource-limited settings.

Risk factors for pneumonia

• *Age*—small children and elderly adults are at particular risk. In regions with high HIV prevalence this pattern is lost as a consequence of increased pneumonia amongst young adults.
• *Coexistent medical problems*—HIV infection (most important), underlying lung disease, diabetes, nephrotic syndrome, kwashiorkor, marasmus, measles and sickle cell disease/asplenia. Malaria and other tropical parasitic diseases (nematodes, trematodes and leishmania) have

been suggested as predisposing factors for pneumonia by leading to alterations in the humoral immune response and phagocytosis.
• *Social*—overcrowding, migrant labour, refugees, tobacco smoking, alcohol and drug abuse.
• *Environmental*—domestic smoke (from firewood and paraffin), poorly ventilated dwellings, mining-associated dust exposure.

Clinical features

Symptoms
Presentation is typically acute, with a 2–3 day history of cough, fever, dyspnoea and in adults purulent sputum production. A more prolonged presentation may occur if the pneumonia has been partially treated or if there is underlying chronic chest disease, in particular tuberculosis. In addition to fever, cough and dyspnoea, failure to feed is an important feature in infants. Other symptoms may be prominent. Streaky haemoptysis may occur although it will rarely be pro-

fuse. 'Rusty sputum' is almost pathognomonic of S. pneumoniae infection, caused by the pigmentary effect of bacterial exotoxins on haemoglobin, but is infrequent. Headache is often severe and may suggest meningitis, particularly when associated with meningism, although meningitis complicating pneumonia is unusual. Diarrhoea can occasionally be profuse and be the primary presenting complaint and lead to confusion with acute gastroenteritis. Pleuritic chest pain is common, but when referred to the abdomen may lead to confusion with an acute abdomen.

Signs

Patients usually appear unwell and may be tachycardic and febrile. Cyanosis is difficult to assess in individuals with pigmented skin but if found indicates severe disease. Nasal flaring, use of the accessory muscles of respiration and tachypnoea are particularly important to look for in children, and along with lower chest wall indrawing form the basis of the simple diagnostic methods to establish the presence and severity of acute respiratory tract infections in children (Fig. 28.1). Chest signs of consolidation (dullness to percussion, bronchial breathing and aegophony) are classically found on examination. In the earlier phase of pneumonia, coarse inspiratory crackles consistent with retained secretions are more common. Decreased movement of the affected hemithorax may give some clue to the localization of the infection on simple inspection. A pleural rub may be present and does not necessarily predict complicated pleural disease. Presentations with acute psychosis, confusion, hypothermia, jaundice and abdominal pain may also be prominent, but careful examination of the respiratory system should help to clarify the underlying diagnosis.

Differential diagnosis

This may be broad when the presenting symptoms are not predominantly respiratory. As illustrated above, gastroenteritis, meningitis, hepatitis and acute psychiatric illness may all enter the differential. Similarly, diagnostic confusion may occur if the presentation is hyperacute with abrupt onset of fevers and rigors, when malaria, other bacterial septicaemic illness or fulminant viral illness may be considered. Even when symptoms are predominantly respiratory, there are several common misdiagnoses or coinfections that should be considered.

- *Tuberculosis* — this should always be considered in poorly resolving pneumonia and is the prime concern in more prolonged presentations.
- *Asthma* — an increasing problem in urban settings, often coexisting with pneumonia. Low-grade fever and poor air entry (obscuring the characteristic wheeze) are features of life-threatening severe asthma.
- *Diabetic ketoacidosis* — respiratory distress, clouded consciousness and a preceding history of weight loss may be overlooked as a case of HIV infection complicated by pneumonia. The smell of exhaled acetone, a more detailed history to elicit polyuria and polydipsia along with urinalysis should identify the condition.
- *Poisoning* — this should be considered when the respiratory distress is out of proportion to the pulmonary findings and there is evidence of other system involvement, particularly neurological. Important poisons are aspirin, chloroquine, petrochemicals and herbicides either intentionally or accidentally ingested or inhaled.
- *Amoebic liver abscess* — the presence of a right-sided effusion should alert the clinician to consider an amoebic problem. Rarely, only consolidatory changes may be present in the right lower lobe.
- *Pulmonary eosinophilia* — as a consequence of migratory worms may present acutely. Bronchospasm may be present; more importantly, individuals are not unwell.

Investigations

To maximize the use of limited resources, investigations should be reserved for individuals with severe disease or cases where response to therapy has been suboptimal and when the information acquired is necessary for the case management process.

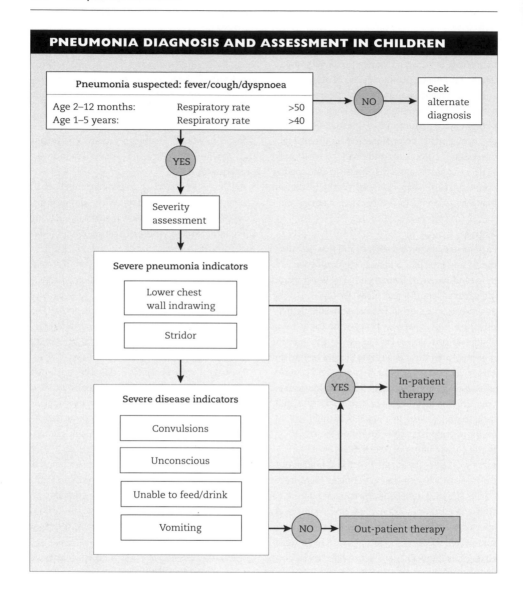

Fig. 28.1 Pneumonia diagnosis and severity assessment in children. In regions with a high prevalence of bacterial pneumonia, use of respiratory rate alone will detect 80% of children who require antibiotics for pneumonia.

Radiology

In severe disease, radiology can be used to identify pneumothoraces, effusions and to rule out alternate diagnoses such as pericardial effusions

or pulmonary oedema which require specific therapy. Bilateral interstitial pneumonia may suggest *Pneumocystis* pneumonia but this radiological pattern is found with both bacterial and tuberculous infections and thus has poor diagnostic precision. When pneumonia is slow to resolve, the presence of mediastinal/hilar lymphadenopathy should stimulate a search for tuberculosis. Cavities should also point towards tuberculosis or, occasionally, anaerobic/aspiration or staphylococcal pneumonia. The presence of pleural effusions in poorly resolving pneumonia should lead to pleural aspiration and examination for empyema.

Sputum Gram stain

This is valuable when performed at presentation on a properly collected sample of sputum. The presence of 10–20 pus cells in a high power (×1000) field and no epithelial cells suggests an adequate specimen. The presence of Gram-positive diplococci (dark purple) in association with pus cells confirms a diagnosis of pneumococcal pneumonia and may be helpful in confirming appropriate management in severe disease (see also p. 13).

Sputum Ziehl–Neelsen stain

This should be carried out on all poorly resolving pneumonia and will identify 60–80% of *M. tuberculosis* infections. Culture for *Mycobacterium*, if available, increases the diagnostic yield. If numerous pus cells are seen on Gram stain in the absence of organisms, Ziehl–Neelsen staining should be performed.

Trans-thoracic lung aspirate

This is a useful technique for recovering lower respiratory tract samples for staining and culture. Aspirates should only be taken from consolidated lung tissue, laterally to avoid the heart. When this is done, pneumothoraces are uncommon (see also p. 13).

Pleural aspiration

Pleural effusions occur in 5–15% of cases of pneumonia. The majority will resolve with antibacterial therapy alone. Sampling of pleural fluid can be helpful in establishing a diagnosis and should always be carried out if empyema is suspected.

Other microbiological tests

Blood cultures have low diagnostic sensitivity and most basic laboratories do not have appropriate facilities, although recovery of an organism from blood is the gold standard (other than lung biopsy) for aetiological diagnosis. Serology and other immune-based diagnostics (e.g. immunofluorescence for *Pneumocystis carinii* pneumonia; PCP) are unlikely to be available and rarely contribute to management during the early crucial phase of pneumonia.

Other investigations

Haemoglobin measurements help to identify individuals who may require blood transfusion. Transcutaneous oxygen saturation measurements, if available, are a useful adjunct to determining the need for oxygen therapy. Biochemical assessment of renal and hepatic function are not essential (except in rare cases of renal failure requiring dialysis) but may provide prognostic information.

In regions of high HIV-prevalence the majority of pneumonia cases will be HIV-infected; rates range from 40 to 90%. If resources are available and national strategies emphasize identification of people with underlying HIV infection (e.g. to receive disease prophylaxis, antiretroviral therapy), pneumonia patients should be offered voluntary counselling and testing. In some instances, HIV testing can aid diagnosis and management (e.g. suspected PCP).

Management

Once pneumonia has been identified or suspected, an assessment of the severity of the condition is rapidly required (Fig. 28.2). The threshold for admission to hospital may be more severe in tropical regions where inpatent facilities are crowded and an assessment of suitability for outpatient oral or inpatient parenteral therapy is required. Indicators of poor outcome in adults are listed in Table 28.2. Vomiting or profuse diarrhoea are relative contraindications to oral therapy as antibiotic absorption may be impaired.

Outpatient therapy

Oral amoxicillin (500 mg t.d.s.) or ampicillin (500 mg q.d.s.) are effective antipneumococcal agents and are a suitable first choice. Alternative agents include co-trimoxazole (promoted as a

PNEUMONIA MANAGEMENT

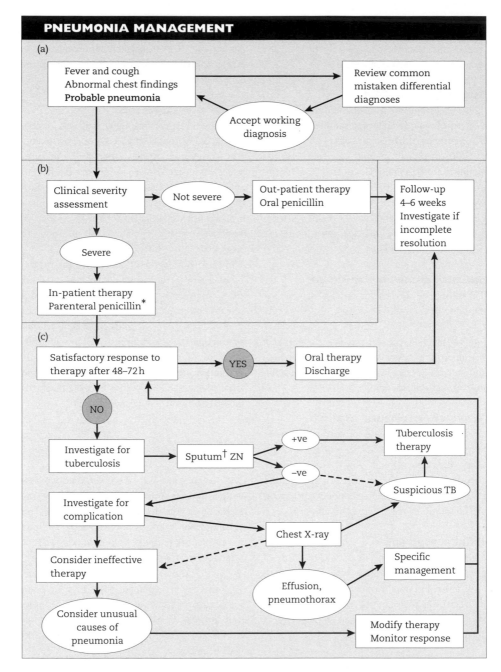

Fig. 28.2 Pneumonia management: (a) establishing a presumptive diagnosis; (b) initiating empirical therapy; and (c) testing and modification of initial diagnosis by assessing response to therapy. * Chloramphenicol if penicillin allergic. † Sputum or pleural aspirate/lung aspirate/regional lymph node aspirate.

INDICATORS OF POOR OUTCOME IN ADULT PNEUMONIA

Severity grading A	Severity grading B
Demographic features	
Age > 55	Increasing distance from health centre
Preceding use of traditional healer	Recent migration/refugee camps
Clinical features	
Diastolic blood pressure < 60 mmHg	Confusion
Respiratory rate > 30 breaths/min	Cyanosis
Pulse rate > 120 beats/min	Extrapulmonary infection
	Jaundice
	Reduced body mass index/wasted
Investigations: available at initial assessment	
White cell count < 4000 cells/mL	Multilobar disease (on X-ray)
	White cell count > 18 000 cells/µL
Investigations: available during therapy	
	Penicillin-resistant pneumococcus
	Coinfection with tuberculosis
	Pneumococcal bacteraemia

Table 28.2 Associations of severe disease and poor outcome in adult pneumonia. The table represents a summary of information from several sources. Factors are graded as A, strongly associated with severe disease and death (five times or greater risk of death if present); and B, moderately associated (less than 5 times or poorly quantified increased risk of death). No validated scheme to determine hospital admission or use of parenteral antibiotics exists; however, the presence of two or more grade A factors should lead to hospital admission if facilities are available. Such an approach will be sensitive for identifying severe disease but will lack specificity.

first choice in children by WHO) and erythromycin. Chloramphenicol can also be used as it is well absorbed and has reasonable activity against *S. pneumoniae*. Individuals managed as outpatients should always be encouraged to reattend to be assessed for tuberculosis if there are persisting respiratory symptoms.

Inpatient therapy

Basic resuscitation (ensuring airway patency and adequate respiratory and circulatory activity) should be rapidly followed by antimicrobial therapy. Parenteral penicillins (benzylpenicillin 1.8 g every 4–6 h or amoxicillin 0.5–1 g every 8 h) remain the agents of first choice. Chloramphenicol is valuable for *H. influenzae* infections or in individuals with penicillin allergy. A macrolide, tetracycline or ciprofloxacin can be added to therapy as oral agents (parenteral preparations are unavailable or expensive) in severe disease if an atypical organism is suspected.

In addition to antimicrobial therapy, other supportive treatment should be initiated. Supplemental oxygen, clearance of respiratory secretions (by changing posture, suctioning and provision of moist air), maintenance of appropriate hydration (with intravenous fluids if necessary) and adequate pain relief (preferably with non-opiate analgesics) may all be required. Assisted ventilation, when available, is necessary in some cases when respiratory muscle fatigue develops, heralded by a rising partial pressure of CO_2 and, more latterly, altered conscious level, blood pressure instability and decreasing respiratory effort.

Complications

An appropriate response to therapy is lysis of fever within 48–72 h of its commencement in association with an improvement in well-being and recommencement of oral intake. Failure of this pattern to emerge requires a reconsideration of the underlying diagnosis, therapy given and/or a search for an early complication — empyema, suppurative pericarditis, lung abscess, pneumothorax or meningitis. The late sequelae of pneumococcal pneumonia (osteomyelitis, arthritis, endocarditis) are infrequent if appropriate antibiotics are given early, although HIV infection can be associated with complications and an increased risk of these metastatic infections. Following discharge from hospital, a follow-up visit in 4–6 weeks is advisable to ensure complete resolution of the pneumonia or further investigation for possible tuberculosis. The chest X-ray changes may take 4–6 weeks to resolve.

Prevention

Active vaccination

Polyvalent pneumococcal polysaccharide vaccines have been available for many years, but uncertainties over their effectiveness and their relative expense ($9 per dose) have limited their use. They are currently recommended for individuals with sickle cell disease but there are no other clear recommendations for use in the tropics. The vaccine is ineffective in HIV-infected adults and children under 2 years of age. The new generation of protein conjugate pneumococcal polysaccharide vaccines are currently under evaluation in children in Africa and the outcomes of these field trials are awaited before further recommendations are made. They have proved extremely effective in the USA at preventing pneumonic and invasive disease. Protection against *Haemophilus influenzae* type b (Hib) pneumonia in children is achievable by vaccination. Attempts to incorporate Hib vaccine in the Extended Programme on Immunization (EPI) regimen are underway.

Chemoprophylaxis

Penicillin should be given on a daily basis to individuals with sickle cell disease or with asplenia to prevent pneumococcal infections. Daily co-trimoxazole is recommended for HIV-infected adults and children in the tropics. This approach is effective in reducing PCP in children under 1 year, but reduction of bacterial pneumonia in adults is less certain.

Future developments

Antibiotic resistance

Penicillin resistance amongst *S. pneumoniae* is on the increase globally and represents a major threat to cheap and effective therapy of lower respiratory tract infections. Alterations in the penicillin-binding proteins of pneumococci lead to incremental increases in resistance to penicillin. At present these changes are probably of little consequence for the treatment of pneumococcal pneumonia — achievable levels of penicillin in blood and pulmonary tissue with standard dosages will be bactericidal for most of the currently 'resistant' pneumococci (unlike the situation with meningitis). However, if pneumococci with higher grade resistance become established, current empirical antibiotic regimens will be ineffective. Moreover, penicillin resistance is associated with resistance to multiple antibiotics including macrolides and co-trimoxazole. The inevitable consequences are therapeutic failures and higher priced therapies. Local information on bacterial sensitivity patterns are essential to plan national and district level policy.

Pneumococcal vaccines

Results of trials of protein conjugate pneumococcal vaccines in Africa and the Philippines will become available from 2002 onwards. There is optimism that these vaccines will prove effective and the price of manufacture will fall to levels to allow them to be incorporated into EPI regimens. In addition to the protein conjugate vaccines, several pneumococcal peptides (pneumolysin, pneumococcal surface proteins) are being studied as vaccine candidates or as components of the conjugate vaccines. These vaccines may be able to overcome the serotype-restricted nature of the polysaccharide vaccines and may be substantially cheaper to manufacture.

Further reading

Allen SC. Lobar pneumonia in northern Zambia: clinical study of 502 adult patients. *Thorax* 1984; **39**: 612–16. [Two hospital case series highlighting the importance of tuberculosis as agents of acute pneumonia in Africa in the AIDS and pre-AIDS era, along with important demographic data.]

Berman S. International conference on acute respiratory infections. *Clin Infect Dis* 1999; **28**: 189–238. [Selection of papers from a variety of authors covering acute respiratory infections with a strong focus on children and the developing world.]

Gilks CF. Royal Society of Tropical Medicine and Hygiene meeting at Manson House, London, 12 December 1996. HIV and pneumococcal infection in Africa. Clinical, epidemiological and preventative aspects. *Trans R Soc Trop Med Hyg* 1997; **91**: 627–31. [Description of HIV and its impact on pneumococcal disease in Africa.]

Greenwood B. The epidemiology of pneumococcal infection in children in the developing world. *Philos Trans R Soc Lond B Biol Sci* 1999; **354**: 777–85. [Summary of the role of pneumococcal disease in childhood illness based primarily on prospective studies in the Gambia.]

Scott JA, Hall AJ, Muyodi C et al. Aetiology, outcome, and risk factors for mortality among adults with acute pneumonia in Kenya. *Lancet* 2000; **355**: 1225–30.

www.who.int/child-adolescent-health [Useful source of practical information for the management of childhood illnesses, with several printable guideline documents for management of pneumonia.]

CHAPTER 29

Lung Flukes

Life cycle, 206   Diagnosis, 208   Further reading, 208
Clinical features, 206   Treatment, 208

Lung flukes of the genus *Paragonimus* are zoonoses but human infection can cause a chronic cough with haemoptysis which can easily be mistaken for tuberculosis. *Paragonimus westermani* is the most common cause of human disease.

Life cycle

Lung flukes are hermaphrodite trematodes. The stout adults, which are about 12 mm long, live as pairs in cavities in the lungs. Large brownish eggs $(85–100 \times 50–60\,\mu m)$ are passed in the sputum or faeces and, if they reach water, develop in about 3 weeks to release a miracidium that infects certain species of freshwater snail where the parasite undergoes asexual multiplication. Cercariae with knob-shaped tails emerge and encyst in freshwater crabs or crayfish.

The definitive hosts of the flukes are carnivores that eat crustacea; humans are incidental hosts infected by ingesting metacercariae in uncooked crab or crayfish meat or their juices. Human infections are most common in Asia, especially Korea where medicinal use of crayfish juice and in parts of China where eating live crabs dipped in rice wine ('drunken crabs') aid transmission (Fig. 29.1). Other species of *Paragonimus* occur in Africa and the Americas. Paragonimiasis was common during the Biafran war in Nigeria and is found in native Indians in Ecuador and Colombia.

Clinical features

Patients with acute infections may present with malaise, shivers, sweats and urticarial skin rash or abdominal pain a few days or weeks after infection.

Lung disease
• Patients present with chronic cough productive of brownish-red sputum, sometimes with haemoptysis.
• Occasionally, there is breathlessness or chest pain.
• Radiology usually shows peripheral nodules or ring shadows; sometimes there is a crescentic shadow of a fluke within the ring. CT scans are useful.
• Pleural effusions or empyema are complications.

Ectopic disease
This is caused by aberrant migration of flukes from the gut. There are many possible presentations but these include:
• abdominal pain and inflammatory masses;
• convulsions;
• cerebral tumour-like presentations;
• mental disturbance; and
• migrating subcutaneous lumps (larva migrans).

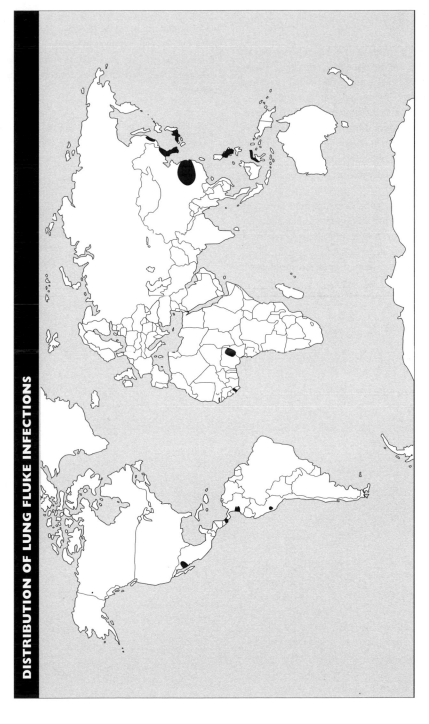

Fig. 29.1 Distribution of lung fluke infections.

Diagnosis

• Microscopy of sputum or stool. Concentrate stool using the formol-ether technique. Examine sputum for brownish flecks that may contain nests of eggs. Mucoid sputum is concentrated by adding 2–3 times the volume of 10% potassium hydroxide for 1 h and then centrifuging for 3 min at 1500 r.p.m. Examine the deposit using the low-power objective.
• Eggs can also be recovered by bronchoalveolar lavage.
• Serological diagnosis by enzyme immunoassay or dot enzyme immunoassay using antigens from flukes or metacercariae is available in some endemic areas.
• Radiological evidence and eosinophilia in early disease may aid diagnosis.

Treatment

Praziquantel given as 75 mg/kg/day in three divided doses for 3 days is highly effective. Triclabendazole is an alternative drug. Treat cautiously if cerebral disease is suspected, as a sudden increase in intracranial pressure is possible. Use dexamethasone to control cerebral oedema.

Further reading

Blair D, Xu ZB, Agatsuma T. Paragonimiasis and the genus *Paragonimus*. *Adv Parasitol* 1999; **42**: 113–222. [Useful review of clinical findings and of recent knowledge of life cycles.]
Calvopina M, Guderian RH, Paredes W, Chico M, Cooper PJ. Treatment of human pulmonary paragonimiasis with triclabendazole: clinical tolerance and drug efficacy. *Trans R Soc Trop Med Hyg* 1998; **92**: 566–9. [Recent drug treatment.]

CHAPTER 30

Tropical Pulmonary Eosinophilia

Clinical features, 209   Treatment, 209   Further reading, 210
Diagnosis, 209

Tropical pulmonary eosinophilia (TPE) is an asthma-like disease caused by a hyperactive immunological response to 'human' filariae, usually *Wuchereria bancrofti* or *Brugia malayi* (see also Chapter 14). Microfilariae are destroyed in the pulmonary capillaries and antigen release recruits large numbers of degranulating eosinophils that release major basic protein and other chemicals. Interleukin (IL) 5 promotes and IL 12 downregulates this process. A granulomatous reaction may eventually progress to fibrosis and permanent lung damage.

Clinical features

• The disease usually occurs in children and young adults, often in young men of Indian descent.
• Patients present with cough, wheeze and shortness of breath, worse at night and often preceded or accompanied by malaise and low-grade fever.
• Signs are similar to those of asthma but there is sometimes lymph node or splenic enlargement and, rarely, seventh nerve palsy or other focal neurological signs.
• The blood shows a marked eosinophilia, usually above 3×10^9/L and sometimes much higher. The erythrocyte sedimentation rate is raised.
• Immunoglobulin E levels are also greatly raised and tests for filarial antibodies show high titres.
• Microfilariae cannot be found in peripheral

blood, even by sensitive filtration methods, but filarial antigen tests are positive.
• The chest X-ray is sometimes normal but more often there is diffuse alveolar mottling with small 1–2 mm nodules in the middle and lower zones. Hilar adenopathy is sometimes seen and, rarely, pleural effusion or areas of hyperacute pneumonitis.
• Lung function tests show restrictive changes and poor gas diffusion; obstructive changes are less common.

Diagnosis

Diagnosis is based on the high eosinophilia with high filarial antibody titres and the response to treatment.

The differential diagnosis includes migrating helminths, especially *Ascaris* in children but also *Strongyloides stercoralis* or hookworms. These usually cause only temporary disability. Severe allergic asthma and bronchopulmonary aspergillosis should also be considered.

Treatment

Diethylcarbamazine (DEC) 6 mg/kg in divided doses given for 12–21 days is the standard treatment. Symptomatic response within days is usual but lung function can remain impaired for months. About 20% respond poorly or relapse. Albendazole 400 mg twice daily for 3 weeks may be added to a second course of DEC in those who do not respond adequately.

Further reading

Cooray JH, Ismail MM. Re-examination of the diagnostic criteria of tropical pulmonary eosinophilia. *Respir Med* 1999; **93**: 655–9. [Cough worse at night, residence in filarial endemic area, eosinophilia > 3300 cells/mm^3 and clinical and haematological response to DEC proved most useful in Sri Lanka.]

CHAPTER 31

Pyogenic Meningitis

Epidemiology, 211
Clinical features, 213
Diagnosis, 214

Management, 215
Epidemic control
(meningococcus), 217

Further reading, 218

Bacterial meningitis is a major cause of morbidity and mortality in the tropics. It has been estimated that 1 in 250 children are affected before the age of 5 in urban West Africa (Dakar, Senegal), with up to 50% mortality. Further south in the 'meningitis belt', 1–2% of people may be affected during the cyclical epidemics of meningococcal disease that occur every 5–7 years. Viral meningitis and viral encephalitis occur in the tropics but will not be considered further here (Chapter 33).

Epidemiology

The majority of cases of pyogenic meningitis are caused by the pneumococcus *Streptococcus pneumoniae* (many serogroups), the meningococcus *Neisseria meningitidis* (serogroups A, B, C, Y, W135) or *Haemophilus influenzae* type b (Hib). The latter rarely affects patients aged over 5 years, and all three are more common in children under 2 years. These young children are an important group of patients because they are least likely to respond to polysaccharide capsule-derived vaccines that have been produced for serogroups of each of the main three bacterial species. At the extremes of age, group B streptococci and staphylococci (both *Staphylococcus aureus* and *Staphylococcus epidermidis*) are important in neonates, and various Gram-negative organisms are important in adults who also become more susceptible to pneumococ-

cal infections as age increases. Antibiotic resistance is common in Hib throughout the world, and in pneumococci in much of the tropics. Clinically, significant resistance has yet to emerge in meningococci.

Meningococcal epidemiology

Epidemics have occurred with regularity in West Africa and the Sudan for at least a century. Lapeysonnie defined the 'meningitis belt' in 1963 as an area with a high incidence and recurring epidemics between latitudes 4° and 16° north with 300–1100 mm annual rainfall south of the Sahara, comprising much of semiarid sub-Saharan Africa including the Sahel. This belt has high levels of seasonal endemicity with large superimposed epidemics of infection of predominantly group A meningococci at irregular intervals. This zone has been extended further south in recent years to include other countries with at least 1 month of reduced humidity (Fig. 31.1). Factors predisposing to meningococcal infection include overcrowding and poor hygiene, so that schools, urban slums, military barracks, prisons and similar large collections of people are at increased risk. Damage to the upper respiratory mucosa by smoke, intercurrent viral infections such as influenza, or external dust (e.g. the annual dry 'Harmattan' wind in Nigeria) facilitates invasion by meningococci. While group A meningococci are usually implicated in West Africa or in previous epidemics related to the Hajj pilgrimage to Mecca, other

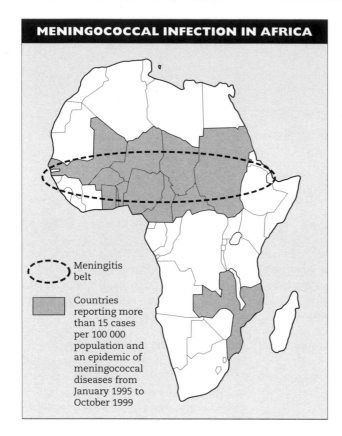

Fig. 31.1 Meningococcal infection in Africa. (Source: WHO)

strains—most recently W135—became prominent after successful vaccination against group A. In South America, epidemics of group C strains have been described but should now be preventable by vaccination. In western countries where immunization against group C has been instituted, group B predominates and there is no effective vaccine against this strain. There does not seem to be increased risk of meningococcal infection in HIV-positive or similarly immunocompromised patients.

Pneumococcal epidemiology

There are many serogroups of pneumococci, the predominance of which varies from country to country, so that sophisticated microbiological surveillance is required in order to plan the relevant mixture to be included in polyvalent

vaccines. Host factors common in the tropics are well-recognized to predispose to pneumococcal infection including haemoglobinopathies such as sickle and sickle–haemoglobin C (SC) disease, splenic dysfunction caused by these or because of removal after trauma, damage to the cribriform plate in the skull by trauma, and HIV infection. HIV has led to a great change in the pattern of meningitis in much of Africa over the past decade. The standard of care for prevention in risk groups is the use of polysaccharide vaccines, although the evidence base is poor and even suggests that pneumococcal vaccine is harmful in some HIV-positive African populations.

Haemophilus influenzae type b epidemiology

Some countries in the tropics have virtually

eliminated invasive Hib disease, including meningitis, by instituting early childhood immunization as practised in the west. This has been shown in the Middle East and in the Gambia.

Vaccines

All three major causes of bacterial meningitis can be prevented by vaccines based on serogroup antigen derived from their polysaccharide capsules. These vaccines require an intact 'cold chain' of refrigeration ($<5°C$) at all stages of transport from manufacturer to delivery at the point of health care. They are less immunogenic in the groups most at risk (under 2 years) and induced immunity is mainly T-cell dependent and relatively short-lived. There is no vaccine for serogroup B meningococci.

New conjugate vaccines are more expensive but also more successful; proven examples being Hib vaccine and the new group C meningococcal vaccine. Similar vaccines are being developed for other serogroups of meningococci and for pneumococci.

DIFFERENTIAL DIAGNOSIS OF PYOGENIC MENINGITIS

- Malaria (cerebral or otherwise)
- Typhoid
- Pneumonia
- Urinary tract infection
- Otitis media/sinusitis
- Severe paediatric gastroenteritis
- Tetanus
- Trypanosomiasis
- Brucellosis
- Rickettsial disease
- Subarachnoid haemorrhage
- Cerebral abscess or other space-occupying lesion, including tuberculoma
- Causes of lymphocytic cerebrospinal fluid (tuberculosis, cryptococcus, viral meningitis, etc.)
- Encephalitis
- Poisoning (alcohol, drugs, etc.) and other cause of coma
- Drug-induced extrapyramidal signs (phenothiazines, antiemetics)

Box 31.1 Differential diagnosis of pyogenic meningitis

Clinical features

The cardinal features of meningitis are no different in the tropics than elsewhere, and symptoms include headache, vomiting, fever, photophobia, loss of consciousness and fits, associated with the clinical signs of fever, neck stiffness and reducing level of consciousness. The index of suspicion is higher in children who may have less obvious features and who often have coexisting malaria parasitaemia. The rash of meningococcal disease develops rapidly and in darker skins may be difficult to see unless the mucosal membranes are involved—always check in the mouth and conjunctivae. In some African groups, the meningococcal rash is slightly raised and can be palpated even if it is difficult to see.

The differential diagnosis in a tropical setting is wide (Box 31.1). The most important clinical decisions are whether the patient has cerebral malaria and/or meningitis, whether there is a space-occupying lesion (often an abscess

related to severe otitis media) or whether the patient has tuberculosis (TB) or cryptococcal meningitis. Examination should focus on clues to underlying predisposing factors, including broken nose, laparotomy scars (for splenectomy) and features of haemoglobinopathy suggesting pneumococcal, or HIV suggesting pneumococcal or cryptococcal meningitis. Optic fundi may show military TB, HIV-related retinopathy (including cytomegalovirus), haemorrhage caused by severe malaria, or tuberculomas as well as features of raised intracranial pressure in long-standing space-occupying lesions. Focal neurological signs suggest a space-occupying lesion, TB or late cryptococcal disease. Most TB meningitis presents with a prolonged history but some cases do present acutely, and patients with cryptococcal meningitis and HIV may have little headache and minimal neck stiffness.

Diagnosis

Lumbar puncture (LP) is mandatory whenever there is suspicion of meningitis. It should only be avoided if there is clear clinical evidence of a space-occupying lesion and, if omitted, the patient should be given antibiotics to cover the possibility of meningitis until it is considered safe to do an LP.

Diagnosis can be achieved with simple biochemical and microscopic tests. The patterns of white cells in cerebrospinal fluid (CSF) vary and there is considerable overlap between the groups (Box 31.2). Abnormalities of cells persist in the CSF for several days, even after antibiotic treatment has been started. Some viral infections, especially mumps, may have neutrophil predominance in the early stages. Biochemical

tests help to distinguish these and a very high protein usually suggests bacterial infection or TB. In TB the CSF protein is sometimes high enough to form a 'spider-web' clot in the tube. The specificity of low CSF glucose for diagnosing bacterial infection is improved by comparing it to simultaneous blood glucose, especially if the patient is diabetic. Urine dipsticks can be used for CSF. Stains of CSF should include Gram's stain for bacteria, which is as sensitive as latex agglutination and other antigen detection systems. Ziehl–Neelson or auramine (for TB) and India-ink stains for cryptococcal infection should always be considered if CSF is lymphocytic or even if there are no CSF white cells in an HIV-positive patient with suggestive symptoms. Cryptococcal antigen tests (if available) are very valuable.

Cultures for all the above organisms should be set up if facilities are available, and large volumes of CSF (10 mL) need to be taken to maximize the chances of growing *Mycobacterium tuberculosis*. Molecular tests such as polymerase chain reaction (PCR) based tests on blood or CSF are valuable in a western setting for diagnosing meningococcal infection, particularly if the patient has already received antibiotics before arrival, but are not available in most tropical settings. Blood cultures should always be taken if facilities are available. Suggestive changes in other tests include neutrophilia in peripheral blood and a high C-reactive protein level, especially in children.

In many tropical settings, culture and more sophisticated tests are not available and Gram's staining will be negative. In such cases the features that suggest pyogenic meningitis, rather than viral meningitis or other organisms, are summarized in Table 31.1. The sensitivity and specificity, and hence usefulness at the bedside, of these features varies from area to area. In a recent study in (mainly HIV-negative) Vietnamese adults, five features were more predictive of TB meningitis compared to bacterial: age, length of history, peripheral white blood cell (WBC) count, total CSF WBC count and CSF neutrophil proportion. In a population of mainly HIV-positive adults in Malawi, patients with

---

CEREBROSPINAL FLUID PATTERNS

Pyogenic — bacterial

Lymphocytic with normal glucose
- most viruses (e.g. polio, enteroviruses, Coxsackie)
- rickettsiae
- HIV
- early TB
- miscellaneous (e.g. endocarditis, neoplastic)

Lymphocytic with low glucose
- partially treated bacterial meningitis
- cerebral abscess
- TB
- some viral (e.g. mumps)
- fungal (e.g. *Cryptococcus, Aspergillus*)
- brucellosis
- syphilis
- leptospirosis
- trypanosomiasis

Eosinophilic
- *Angiostrongylus cantonensis*
- *Taenia solium* (cysticercosis)
- Paragonimiasis

Amoebic (rare)
- *Naegleria* spp.

Box 31.2 Cerebrospinal fluid patterns

CSF CHANGES PREDICTIVE OF BACTERIAL CAUSES

Opening pressure	High
Turbidity	Present
Total WBC	$>2 \times 10^9$/L (2000/mm³)
Neutrophils	>50% if total WBC $> 0.1 \times 10^9$/L (100/mm³)
Glucose	<1 mmol/L (18 mg%)
CSF: blood glucose ratio	<40%
Protein	>2 g/L (200 mg%)
Gram's stain	Positive
Ziehl–Neelsen stain	Negative
India-ink stain	Negative
Culture	Positive
Antigen detection	Positive

Plus peripheral WBC $> 16 \times 10^9$/L

WBC, white blood cell count.

Table 31.1 Cerebrospinal fluid changes predictive of bacterial causes if Gram's stain is negative (see also Table 3.5).

cryptococcal meningitis had longer histories, lower CSF WBC counts and lower proportions of neutrophils than those with bacterial meningitis.

Management

The key features are to assess severity of illness, to make a diagnosis to detect coexisting or underlying disease, and to provide both supportive and specific therapy. For pyogenic meningitis alone, empirical or pathogen-specific antimicrobials should be given as early as possible, before the LP if there is going to be any delay in the procedure. All patients should be reevaluated at least daily, especially when the CSF results become available. Patients with equivocal (lymphocytic) CSF findings are difficult; repeat LP may be required after 2–3 days. The important decision is whether to treat as TB or not, or whether the initial CSF changes were brought about by viral or partially treated bacterial meningitis or missed cryptococcal disease.

Meningococcal disease

This responds to short courses of chloramphenicol or penicillin, and will also respond to third-generation cephalosporins. Uncomplicated meningococcal meningitis has a mortality of < 10% and rarely needs more than 5–7 days of parenteral treatment. Meningococcal septicaemia (with or without meningitis) has a mortality of >40% in most tropical settings, requiring maximal intensive care. Up to 10% of patients experience immune complex disease including uveitis, polyarthritis and pericarditis in the second week of illness. This responds to inflammatory drugs, including short courses of steroids.

Uncomplicated meningitis can also be managed with single doses of Triplopen or similar mixtures of long- and short-acting penicillins, or with Tifomycin, an oily suspension of chloramphenicol administered as 2–3 g once only (but divided into two injections because of volume) for an adult. These regimens are useful for the management of large numbers of patients in an epidemic setting, provided that patients are reviewed daily for evidence of recovery.

Close contacts (family/household) may be given immediate chemoprophylaxis to prevent them from developing illness over the next fortnight. Suitable medications include sulfadiazine (only if the infecting strain is already known to be sensitive) but not penicillins. The alternatives are expensive but include a single dose of ciprofloxacin (15 mg/kg orally in children, 750 mg orally in adults), rifampicin (20 mg/kg twice daily or 600 mg twice daily for adults, for 2 days) or ceftriaxone (50 mg/kg children or 2 g adults intramuscularly once only). 'Ring' vaccination of contacts is more appropriate in an epidemic situation (see below).

Pneumococcal disease

Antimicrobial therapy depends on local susceptibility patterns. In Papua New Guinea, much of sub-Saharan Africa and elsewhere, there may be at least moderate penicillin resistance, so that isolates need to be cultured to direct therapy, and empirical therapy should not be based on penicillin alone. In a western setting, adequate doses of third-generation cephalosporins (cefotaxime or ceftriaxone) are usually adequate, but are beyond the reach of many tropical settings because of cost. Coexistent chloramphenicol resistance is much less common and monotherapy with chloramphenicol is often the only available choice. The commonly recommended mixture of penicillin and chloramphenicol is wasteful of resources and confers no improvement in survival or reduced morbidity in survivors. Meropenem is a safe (but expensive) alternative in proven chloramphenicol and penicillin-resistant cases, or vancomycin can be used. Pneumococcal meningitis has a mortality of over 50% in many tropical settings and in sub-Saharan Africa the majority of these patients are also HIV positive. Up to 40% of survivors will have significant neurological deficits. Treatment should last for at least 10 days for uncomplicated disease and may need to extend beyond 14 days in difficult cases.

Haemophilus influenzae disease

Found mainly in young children, clinically significant resistance to ampicillin and/or amoxicillin and to benzyl penicillin is present in about 50% of cases. Less marked chloramphenicol resistance is present in up to 10% of tropical cases. The treatment of choice is a third generation cephalosporin (if available) or chloramphenicol alone. Empirical ampicillin cannot be used until

the patient's own isolate is known to be sensitive. Chemoprophylaxis of contacts is not usual in tropical settings.

Empirical therapy

Unless there is good epidemiological (current epidemic) and clinical (typical rash) evidence to suggest meningococcal disease, the average adult or child with pyogenic meningitis has to be managed to cover the main three bacterial pathogens, including cover for *Salmonella* sp. in younger children. Monotherapy with chloramphenicol or a third-generation cephalosporin is the treatment of choice. In resource-poor settings with multiresistant pathogens, it has been shown that two doses of Tifomycin 48 h apart are as effective for inpatient treatment as parenteral ampicillin plus chloramphenicol for over a week, in terms of hospital mortality and serious sequelae in survivors.

Patients at extremes of age need extra cover for other pathogens, e.g. antistaphylococcal cover such as flucloxacillin, and/or antipseudomonal cover for neonates (e.g. gentamicin). Patients with possible intracerebral abscess should receive metronidazole to cover anaerobes as well as chloramphenicol.

Use of steroids

Complications of bacterial meningitis are common and severe (Box 31.3). In parts of West Africa they may account for over one-third of cases of deafness. There has been much controversy over the years about using high-dose steroids to prevent this. Previous evidence showed that postmeningitis deafness could be

COMPLICATIONS OF BACTERIAL MENINGITIS

- Cranial nerve palsy
- Hemiplegia
- Deafness
- Subdural empyema
- Abscess
- Late hydrocephalus

Box 31.3 Complications of bacterial meningitis

reduced by dexamethasone given to western children with low mortality from Hib meningitis treated with cephalosporins. The benefit was balanced by a low but appreciable morbidity from gastrointestinal haemorrhage and studies of steroid use in children in the tropics have generally shown harm. One study from Egypt had shown that adults and children with pneumococcal meningitis (not meningococcal or Hib) had reduced mortality and subsequent deafness if given dexamethasone. This has at last been supported by a recent study in northern Europe; again, only patients with pneumococcal meningitis showed benefit from high-dose dexamethasone (40 mg/day intravenously for 4 days), given shortly before the first dose of antibiotics.

High-dose dexamethasone is expensive and extra resources are required to administer it in a tropical setting. There is no evidence to support its use in definite or probable meningococcal disease or in patients who have already received antibiotics of some sort. However, empirical use for other forms of meningitis is likely to become more common in adults.

Epidemic control (meningococcus)

The World Health Organization (WHO) has developed the definition of an 'alert threshold' for an epidemic of meningococcal meningitis as an incidence of >15 cases per 100 000 population for 1 week. Other features that should alert the clinician are a shift in the average age of patients affected, from younger than 5 years to teenagers or older, especially if in a high-risk situation (e.g. refugee camp or in the 'meningitis belt' during the dry season). If more than 3 years have elapsed since the last epidemic, the alert threshold is reduced to 10 cases per 100 000 (Table 31.2).

Early recognition is the key to management (Box 31.4). This should be followed by maximal attempts to confirm the diagnosis by culturing the organism to determine its antimicrobial sensitivities (to guide chemoprophylaxis) and

MENINGOCOCCAL EPIDEMIOLOGY

	Endemic	Epidemic
Incidence/100 000	<10	10–1000
Carrier : case ratio	High	Low
Secondary infections	Rare	Frequent
Peak age	<5	5–15
Usual serogroup	B or C	A, C or W135

Table 31.2 Meningococcal epidemiology.

MANAGEMENT OF MENINGOCOCCAL EPIDEMICS

Early recognition ('threshold' 15 cases/100 000 population/week)
Identify organism
• confirm clinical diagnosis, get CSF from cases
• establish strain and antibiotic sensitivity
Alert peripheral staff
• diagnostic algorithms and case definitions
• treatment algorithms
Alert authorities
• surveillance schemes
• temporary treatment centres
Prevent major outbreak
• mass chemoprophylaxis
• mass vaccination

Box 31.4 Management of meningococcal epidemics

serogroup (to guide vaccination). If facilities to do this are not routinely available, outside assistance is required. This is needed anyway to support enhanced surveillance and to enable provision of adequate supplies of drugs, vaccines, etc. Once an outbreak is declared, a decision must be made on how to get publicity to the affected population and how to conduct surveillance. Health care workers and educated lay persons need to be briefed, using simple case definitions on triaging the worried well as well as the sick at designated assessment centres, and clinical management protocols need to be prepared to guide treatment of patients. The numbers may be large enough to require buildings to be made over specifically for this purpose.

Epidemics of group A disease can be controlled by vaccination — there is ample evidence to support this. The effectiveness depends on both early recognition of an outbreak, and on rapid administration of the appropriate vaccine. Mass chemoprophylaxis can also be used if the organism is sulfa-sensitive, or if a decision is made to use fluoroquinolone, but this is often not practical in a large population.

Further reading

Cuevas LE, Kazembe P, Mughogho GK, Tillotson GS, Hart CA. Eradication of nasopharyngeal carriage of *Neisseria meningitidis* in children and adults in rural Africa: a comparison of ciprofloxacin and rifampicin. *J Infect Dis* 1995; **171**: 728–30. [A single dose of ciprofloxacin (15 mg/kg orally) was as effective and safe as rifampicin 20 mg/kg orally twice daily for 2 days or ceftriaxone 50 mg/kg intramuscularly (children <2 years only) in eradicating meningococcal carriage in children in Malawi. Adult doses are 750 mg, 600 mg twice daily or 2 g, respectively.]

de Gans J, van de Beek D. Dexamethasone in adults with bacterial meningitis. *N Engl J Med* 2002; **347**: 1549–56. [A European study in immunocompetent adults showing a definite benefit of high-dose dexamethasone for those with pneumococcal meningitis.]

Gordon SB, Walsh AL, Chaponda M *et al.* Bacterial meningitis in Malawian adults: pneumococcal disease is common, severe and seasonal. *Clin Infect Dis* 2000; **31**: 53–57. [Change in local epidemiology probably because of HIV, but seasonality is still present. Adult mortality is 61%.]

Hasbun R, Abrahams J, Jekel J, Quagliarello VJ. Computed tomography of the head before lumbar puncture in adults with suspected meningitis. *N Engl J Med* 2001; **345**: 1727–33. [Simple clinical features predict the absence of normality of computerized tomography of the head in adults in a Western setting.]

Santaniello-Newton A, Hunter PR. Management of an outbreak of meningococcal meningitis in a Sudanese refugee camp in Northern Uganda. *Epidemiol Infect* 2000; **124**: 75–81. [Case fatality rate of 13% and attack rate of 0.3% (group A meningococcus). They had better experience than the Cuevas *et al.* group and estimated vaccine protective effect to be approximately 83%. The biavalent (A + C) vaccine they used had to be given subcutaneously. They emphasize the need for early epidemic recognition and immunization campaign.]

Thwaites GE, Chau TTH, Stepniewska K *et al.* Diagnosis of tuberculous meningitis by use of clinical and laboratory features. *Lancet* 2002; **360**: 1287–92. [Source reference on diagnostic methods and development of simple formula for bedside use in adults in Vietnam: age, length of history, white blood cell (WBC) count, total cerebrospinal fluid (CSF) WBC count and CSF neutrophil proportion are five important variables that combine to distinguish TBM from bacterial meningitis.]

http://www.who.int [World Health Organization website. An excellent source of latest country-specific information and guidelines for diagnosis and treatment.]

CHAPTER 32

Cryptococcal Meningitis

Organism and
  epidemiology, 219

Clinical features, 219

Diagnosis, 219

Treatment, 220

Organism and epidemiology

Cryptococcal disease is caused by the yeast-like fungus *Cryptococcus neoformans*. There are two varieties: var *gattii* and var *neoformans*. Var *neoformans* occurs worldwide and is found in the environment related to avian droppings. Var *gattii* occurs predominantly in the tropics and is much harder to find in the environment, but occurs in association with eucalypt trees.

Cryptococcal disease occurs in both immunocompetent and immunocompromised hosts. Infection is acquired by inhalation and predominantly causes pulmonary disease or cryptococcal meningitis (CM). In the tropics prior to the AIDS epidemic, var *gattii* caused CM in immunocompetent individuals. However, most cases of cryptococcal disease in the tropics now occur in patients with HIV infection and are caused by var *neoformans*. CM occurs in HIV-infected individuals thoughout the tropics, predominantly in those with CD4 cell counts of less than 100×10^6/L. It is one of the most common identified causes of death in HIV-infected Africans.

Clinical features

Cryptococcosis presents as pneumonia, cryptococcal meningitis or disseminated disease. Pneumonia is less common, often asymptomatic or mild and may resolve spontaneously. A small number of immunocompromised patients have a rapidly progressive pulmonary illness. CM presents as a subacute or chronic meningitis with headache, fever and alteration in mental state, often of several weeks' duration. The disease mimics tuberculous meningitis and may be indistinguishable clinically. Clinical signs include fever, cranial nerve palsies and visual disturbance associated with papilloedema; neck stiffness is relatively uncommon. Disseminated infection occurs in advanced immunosuppression, often presenting as fever. Skin lesions are common and may contain the organism.

Diagnosis

A high index of suspicion should be maintained in those with HIV infection and unexplained fever or headache. Lumbar puncture usually demonstrates a raised cerebrospinal fluid (CSF) protein and white cell count with a predominant lymphocytosis, although the cell count can be normal in early CM or if the patient is very immunosuppressed. The lumbar opening pressure is commonly raised. Cryptococci can be seen after Gram staining or simply demonstrated in the CSF by the addition of a few drops of India ink (Fig. 32.1, see colour plate facing p. 146). This outlines the capsule of the organism and makes it easier to distinguish the organism from white cells. India ink staining is positive in up to 70% of patients. The organism can also be cultured from the CSF, blood or, occasionally, skin lesions. Latex agglutination tests can detect cryptococcal polysaccharide antigen with a high sensitivity in the CSF or the blood and may be very useful if culture facilities are not available.

Treatment

Standard treatment of CM is amphotericin B (0.3–0.7 mg/kg/day) in combination with flucytosine (75–150 mg/kg/day). Treatment may need to be continued for several weeks. However, there are often practical difficulties in the administration of amphotericin in resource-poor settings because of its renal toxicity and the need to monitor renal function. Flucytosine may also cause marrow suppression, which may be problematic in some HIV patients. Fluconazole (400 mg or more daily) is also effective in the treatment of cryptococcal meningitis and has fewer side-effects than amphotericin and flucytosine, although it may take longer to sterilize the CSF.

Relapse is common in HIV patients following successful initial treatment. Secondary prophylaxis with fluconazole (200 mg/day) is effective in preventing relapse and needs to be continued for life.

Many patients with CM have raised intracranial pressure, which is associated with a poor prognosis. There is no proven method of reducing this but many clinicians advocate repeated lumbar punctures to reduce the intracranial pressure. In the absence of treatment of CM, the disease is uniformly fatal. Even with treatment, case fatality rates can be as high as 70% in some parts of the tropics.

CHAPTER 33

Encephalitis

Causes of encephalitis, 221
Arboviral encephalitis, 221
Japanese encephalitis, 221
West Nile virus, 224
St Louis encephalitis virus, 224

Tick-borne encephalitis
  virus, 224
Murray valley encephalitis
  virus, 225
Equine encephalitis viruses, 225

La Crosse virus, 225
Management of patients with
  encephalitis, 225
Further reading, 225

Encephalitis (inflammation of the brain parenchyma) is strictly speaking a pathological diagnosis that should only be made with histological evidence at autopsy or from brain biopsy. Because of the obvious practical limitations of this, clinical definitions are often used. Most patients present with the triad of fever, headache and encephalopathy (reduced level of consciousness). Many also have focal neurological signs and seizures. However, patients occasionally present simply with abnormal behaviour, which may be mistaken for psychiatric illness.

Causes of encephalitis

Encephalitis can be caused by many viruses, other organisms and autoimmune processes (Table 33.1) However, viruses transmitted by insects (arboviruses; Chapter 40) make encephalitis especially common in the tropics.

Arboviral encephalitis

The arboviruses that cause neurological disease in humans come principally from three viral families (see Fig. 40.1). The most important are the flaviviruses, especially Japanese encephalitis virus. Flaviviruses exist in enzootic cycles, being transmitted by mosquitoes (in warm climates) or ticks (in cooler northern climates); most

humans are coincidentally infected 'dead end' hosts. In addition, alphaviruses and bunyaviruses cause central nervous system (CNS) disease in the Americas.

Japanese encephalitis

Japanese encephalitis is the most important cause of epidemic encephalitis worldwide. There are an estimated 35 000–50 000 cases annually with 10 000 deaths. More than half the survivors have severe neurological sequelae.

Epidemiology

The virus is transmitted naturally between birds by *Culex* mosquitoes, especially *Culex tritaeniorhynchus*, which breeds in rice paddy fields. Peri-domestic animals (especially pigs) act as amplifying hosts, and subsequently humans become infected (Fig. 33.1). In southern tropical areas the disease is endemic, in northern temperate zones it occurs in summer epidemics (Fig. 33.2). The geographical area affected is expanding, possibly because of increasing irrigation projects, and spread by birds. In endemic areas of rural Asia the virus is ubiquitous and almost all children are infected, but only a small proportion develops disease.

Clinical features

Following a non-specific febrile illness, which

INFECTIOUS CAUSES OF ENCEPHALITIS

Arthropod-borne viruses (often epidemic and geographically localized)
Flaviviruses (Japanese encephalitis, West Nile)
Alphaviruses (Venezuelan and Eastern equine encephalitis)
Bunyaviruses (La Crosse encephalitis)

Non-arthropod-borne viruses (mostly sporadic and non-geographically localized)
Herpes viruses (Herpes simplex virus types 1 and 2, varicella zoster virus, Epstein–Barr virus)
Enteroviruses (Coxsackie, echovirus, enterovirus type 71)
Paramyxoviruses (measles, mumps, Nipah)
Rabies (Chapter 36)
Human immunodeficiency virus

Acute disseminated encephalomyelitis
Occurs several weeks after an acute infection (often viral) or vaccination

Other infectious causes of encephalitis
Usually distinguishable from viral encephalitis, either because of their slower onset, or because they
    are associated with other features (e.g. multiorgan failure, rash)
Trypanosomiasis, especially *Trypanosoma brucei rhodesiensis*
Toxoplasma occasionally presents with diffuse encephalitis
Amoebic meningoencephalitis, especially *Naegleria fowleri*
Typhus, especially African tick typhus (*Rickettsia africae*) and scrub typhus in Asia (*R. tsutsugamushi*)
Secondary syphilis, and other spirochaetes (e.g. relapsing fevers)

Table 33.1 Infectious causes of encephalitis.

JAPANESE ENCEPHALITIS VIRUS

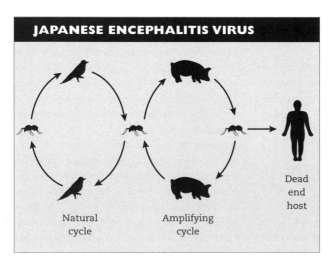

Dead
end
host

Natural   Amplifying
cycle   cycle

Fig. 33.1 The transmission cycle of Japanese encephalitis virus. The virus is transmitted naturally between aquatic birds by *Culex* mosquitoes. During the rainy season, when there is an increase in the number of mosquitoes, the virus 'overflows' into pigs and other peri-domestic animals, and then into humans, from whom it is not transmitted further.

DISTRIBUTION OF ARBOVIRUSES

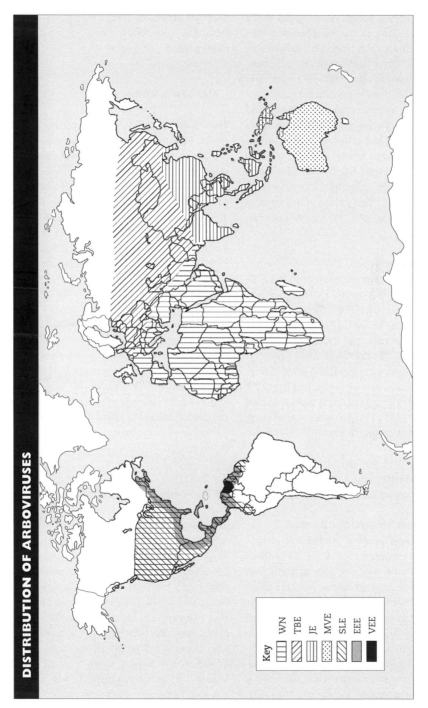

Fig. 33.2 Map showing the approximate global distribution of CNS arboviruses. In North America Western equine encephalitis has a similar distribution to St Louis encephalitis (SLE), and La Crosse has a similar distribution to West-Nile (WN). EEE, Eastern equine encephalitis; JE, Japanese encephalitis; MVE, Murray valley encephalitis; TBE, tick-borne encephalitis; VEE, Venezuelan equine encephalitis.

Key

WN
TBE
JE
MVE
SLE
EEE
VEE

may include coryza, cough, vomiting, diarrhoea and headache, patients develop a reduced level of consciousness. This may be heralded by convulsions; status epilepticus is common. Focal neurological signs include upper and lower motor neurone signs; a 'Parkinsonian syndrome' with tremor, cogwheel rigidity and mask-like facies; rigidity spasms, flexor and extensor posturing, and other signs of raised intracranial pressure. Other patients present with aseptic meningitis, febrile convulsions or a polio-like acute flaccid paralysis. The differential diagnosis includes other causes of encephalitis and non-viral causes of CNS disease (see Table 3.1).

Investigations

Typically, lymphocytes are present in the CSF (5–100/mm^3), with a normal glucose level and slightly elevated protein (50–100 mg/dL). However, the CSF may be normal if taken too early or too late in the disease. IgM antibodies appear in the serum and CSF after the first few days of illness and can be detected by ELISA or rapid diagnostic kits. Virus is sometimes isolated from CSF or brain tissue at postmortem. Computerized tomography (CT) or magnetic resonance imaging (MRI) may show characteristic midbrain changes.

Management

This is discussed at the end of this chapter.

Prevention and public health

Vaccines against Japanese encephalitis include a formalin-inactivated vaccine (requires three doses; occasional mild side-effects) and a new Chinese live attenuated vaccine (safe and effective; single dose may be effective). There is some evidence that vector control may be useful, including treating rice paddy fields with neem cake (a natural insecticide) and intermittent irrigation of rice paddies to prevent *Culex* breeding. Individual protective measures include avoiding mosquito bites (easier for visitors than residents) and using DEET-containing spray, bed nets and protective clothing.

Future developments

Infectious copy DNA clone vaccines are in development.

West Nile virus

This is a flavivirus widely distributed throughout Africa, Asia, the Middle East, southern Europe, and has recently spread to North America. It is transmitted between water fowl by *Culex pipiens*. West Nile virus was previously considered to be a cause of fever–arthralgia–rash syndrome, with occasional CNS disease. There have been recent epidemics of CNS disease in Romania (in 1996, 600 people) and America (in 1999, 60 people and in 2002, 3500 people). Vaccines are in development.

St Louis encephalitis virus

This is a flavivirus found in the Americas. It previously caused large epidemics (3000 cases/year) but now only around 100 cases/year. It may be a good example of how arboviral disease can be controlled if there is enough money for sentinel surveillance and vector control (spraying).

Tick-borne encephalitis virus

The disease caused by this virus has many synonyms, e.g. Russian spring–summer encephalitis. The organism is a flavivirus transmitted between rodents and other small mammals by *Ixodes* ticks. It is also transmitted by ingesting infected goat's milk. The disease is found in a wide area from eastern Europe to the Far East and particularly affects forestry workers and hikers, who present 1–2 weeks after a tick bite. It is a biphasic illness with a high fever for 1 week followed by an afebrile period, then a meningoencephalitis or myelitis with upper limb, respiratory and bulbar flaccid paralysis. The CSF often shows neutrophils, and a peripheral leucocytosis and elevated erythrocyte sedimentation rate (ESR) may mimic bacterial meningitis. A formalin-inactivated vaccine is available.

Murray valley encephalitis virus

This is a flavivirus causing encephalitis in Australia and Papua New Guinea.

Equine encephalitis viruses

Venezuelan, Eastern and Western equine encephalitis viruses are mosquito-borne alphaviruses transmitted between birds and/or small mammals by *Culex, Aedes* and *Culiseta* mosquitoes. They cause epidemics of encephalitis in horses and humans in the Americas.

La Crosse virus

A bunyavirus transmitted between chipmunks and squirrels, principally by *Aedes* mosquitoes. It causes up to 200 cases of encephalitis in the USA annually but has a low case fatality rate.

Management of patients with encephalitis

There is no specific antiviral treatment for most forms of viral encephalitis. Aciclovir is effective for herpes simplex virus type I and related viruses. In the West it is therefore given as soon as encephalitis is suspected. In the tropics, where other causes are more common, aciclovir is often reserved for cases that are atypical for the common local arboviral cause (e.g. wrong age, wrong season) or for cases with typical radiological changes of herpes. Ideally, encephalitis patients in a coma should be sedated and ventilated on an intensive care unit. This allows airway protection, maximum medication to control seizures and hyperventilation to reduce raised intracranial pressure but this is often not possible.

Whatever the viral cause, attention must be paid to the complications of encephalitis.
• *Pneumonia* — often caused by aspiration. Treat with broad-spectrum antibiotics.
• *Seizures* — sometimes these may be subtle motor seizures: look for twitching of a digit, the mouth or eye. Confirm with electroencephalogram if possible. Treat seizures with diazepam. Treat status epilepticus with phenytoin (using a cardiac monitor) or phenobarbital (see Chapter 59).
• *Raised intracranial pressure* — there may be elevated CSF opening pressure at lumbar puncture. Look for clinical signs of brainstem herniation syndromes (Chapter 3). Nurse patients at 30° with the neck straight, give osmotic diuretics (e.g. mannitol), hyperventilate.
• *Malnutrition* — despite nasogastric feeding, very common in patients who are ill for more than a few days.
• *Bedsores* — minimized by good nursing care. Placing rubber gloves (inflated with water) between the knees and below the heels can help prevent some sores.
• *Contractures* — encourage the family to keep joints supple; use splints (improvised if necessary) to keep joints in position.

Further reading

Davis LE. Acute viral meningitis and encephalitis. In: Kennedy PGE, Johnson RT, eds. *Infections of the Nervous System.* London: Butterworths, 1987: 156–76. [General review of viral meningoencephalitis.]

Solomon T. Exotic and emerging viral encephalitides. *Curr Opin Neurol* 2003; **16**: 411–18. [Newer infections.]

Solomon T, Dung NM, Kneen R *et al.* Neurological aspects of tropical diseases: Japanese encephalitis. *J Neurol Neurosurg Psychiatry* 2000; **68**: 405–15.

Whitley RJ, Grann JW. Viral encephalitis: familiar infections and emerging pathogens. *Lancet* 2002; **359**: 507–13. [Useful update on West Nile encephalitis.]

CHAPTER 34

Acute Flaccid Paralysis

Pathophysiology and clinical
    presentations, 226

Further reading, 228

Acute flaccid paralysis is defined as weakness in one or more limbs, or the respiratory or bulbar muscles, resulting from damaged lower motor neurones. Poliomyelitis was the most important cause, but since it has declined other causes have become more important.

Classically, in acute flaccid paralysis there is weakness with reduced tone (flaccid weakness) and reduced or absent reflexes. Differentiating from upper motor neurone weakness is usually straightforward, but it should be remembered that acute spinal shock (e.g. caused by trauma) can initially cause flaccid paralysis before spasticity develops.

Pathophysiology and clinical presentations

Broadly speaking, there are two pathophysiological processes that cause acute flaccid paralysis (Fig. 34.1). These are direct viral damage of lower motor neurone cell bodies in the anterior horn of the spinal cord (e.g. polio, other enteroviruses, flaviviruses); and a para- or post-infectious immunologically mediated process damaging the motor nerves, and often sensory nerves (e.g. Guillain–Barré syndrome), sometimes caused by antibodies directed against the gangliosides (glycolipids in the nerve cell membranes). Recognizing the clinical features of these two patterns helps in determining the likely cause (Table 34.1).

Anterior horn cell damage causing acute flaccid paralysis

Polio

Infection with this enterovirus can be asymptomatic, can cause a mild non-specific febrile illness, viral meningitis or paralytic poliomyelitis, which can be spinal or bulbar. Paralytic poliomyelitis is biphasic, with a non-specific fever followed by a brief afebrile period before the central nervous system (CNS) is invaded. This is heralded by further fever and an acute-onset asymmetrical flaccid paralysis of one or more limbs, which may be painful. Since the World Health Organization campaign to eradicate polio using the oral polio vaccine, the number of cases has dropped from more than 350 000 cases in 1988 to approximately 1900 cases in 2002. Most of these came from South Asia (India, Pakistan and Afghanistan), West Africa (mainly Nigeria) and Central Africa (mainly Democratic Republic of Congo).

Enterovirus 71

Enterovirus 71 has caused epidemics of acute flaccid paralysis in recent years (especially in Asia), often in association with hand, foot and mouth disease. Many other enterovirus, Coxsackie virus and echovirus serotypes occasionally cause acute flaccid paralysis.

Japanese encephalitis virus

Japanese encephalitis virus, West Nile and other flaviviruses typically cause meningoencephalitis,

CAUSES OF ACUTE FLACCID PARALYSIS

	Direct viral damage to anterior horn cells (e.g. polio)	Immune-mediated damage to peripheral nerves (e.g. Guillain–Barré syndrome)
Paralysis onset	During (or straight after) febrile illness	Several weeks after febrile illness
Pattern of paralysis	Asymmetrical	Symmetrical
Time to reach maximum weakness	Short (e.g. 2–3 days)	Long (e.g. 7–14 days)
Sensory involvement	No	Often (depending on exact disease)
CSF	Increased lymphocytes (e.g. 100/mm³)	Increased protein (e.g. 100 mg/dL, especially late in the disease)
Pain	Often limb muscle pain	Often back pain

Table 34.1 Clinical features to distinguish causes of acute flaccid paralysis.

ACUTE FLACCID PARALYSIS

Immune-mediated damage (GBS)
to myelin (AIDP)
to axons (AMAN)

Direct viral damage of anterior horn cells
Polio, enteroviruses, echovirus,
Coxsackie virus
JEV, other flaviviruses

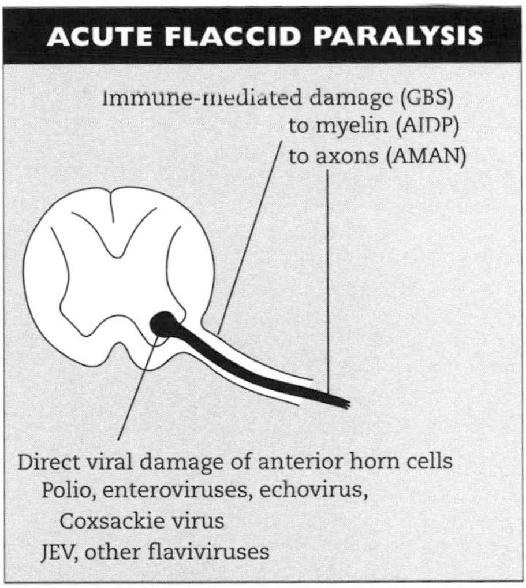

Fig. 34.1 Pathophysiology of acute flaccid paralysis. Immune-mediated Guillain–Barré syndrome (GBS) occurs in two forms: in acute inflammatory demyelinating polyneuropathy (AIDP) the myelin is damaged; in acute motor axonal neuropathy (AMAN) the motor axons are targeted. Viruses such as polio and Japanese encephalitis virus (JEV) cause paralysis by directly attacking the lower motor neurones (the anterior horn cells).

but flaviviruses can also present with a pure flaccid paralysis that can be clinically similar to polio (Chapter 33).

Immune-mediated causes of acute flaccid paralysis

Guillain–Barré syndrome is now recognized as a group of disorders classified according to the predominant type of nerve injury (axonal or demyelinating) and the main nerve fibres involved (motor, sensory, cranial). Different antiganglioside antibodies are associated with different diseases.

Acute inflammatory demyelinating polyneuropathy (AIDP, or 'classical' Guillain–Barré syndrome)

This typically presents several weeks after a febrile illness with back pain, then symmetrical ascending flaccid paralysis and sensory changes. Recovery is usual. Treat rapidly progressing symptoms with intravenous immunoglobulin if available.

Acute motor axonal neuropathy (AMAN, or Chinese paralytic syndrome)

This typically follows diarrhoea caused by Campylobacter jejuni. Symmetrical weakness is present with no sensory changes. Residual weakness is common. Occurs in summer epidemics in China.

Rare causes

Consider rare causes of paralysis if the diagnosis is not apparent.

• Any exposure to toxins?
• Any tick bites (tick paralysis is a slowly ascending paralysis that recovers when the tick is removed)?
• Exposure to rabid animal (paralytic rabies)?
• History of a severe sore throat with neck swelling (diphtheritic neuropathy)?
• Consumption of poorly preserved food (botulinum toxin)?

Nerve conduction studies

Where available, nerve conduction studies may help distinguish further:

• *Anterior horn cell damage* — motor amplitude is reduced because motor cell bodies have been damaged.
• *Classical Guillain–Barré syndrome* (autoimmune demyelinating polyneuropathy) — motor and sensory nerves have reduced conduction velocities and delayed distal latencies because demyelinated nerves conduct more slowly.
• *AMAN* (Chinese paralytic syndrome) — motor amplitudes are reduced because motor axons have been damaged.

Further reading

Anonymous. Progress toward global poliomyelitis eradication, 1999. *Morbid Mortal Week Rep* 2000; **49**: 349–54.
Bolton CF. The changing concepts of Guillain–Barré syndrome. *New Engl J Med* 1995; **333**: 1415–16. [Good explanation of acute motor axonal neuropathy as a form of Guillain–Barré syndrome.]
Solomon T, Willison H. Infectious causes of acute flaccid paralysis. *Curr Opin Infect Dis* 2003; **16**: 375–9. [Annotated review of main infectious causes.]

CHAPTER 35

Spastic Paralysis

Causes and anatomy, 229

Assessment of the patient with
spastic paralysis, 229

Important tropical causes of
spastic paralysis, 229

Spastic paralysis is caused by damage to the upper motor neurones, and is characterized by weakness with increased tone, brisk reflexes and extensor plantars.

Causes and anatomy

Some of the important causes of spastic paralysis are shown in Fig. 35.1. The upper motor neurones can be affected anywhere along the corticospinal tract, but the associated features give a clue as to the site of damage as follows.

• A pure spastic paraparesis with no sensory changes is usually caused by damage in the brain where sensory and motor pathways are far apart (e.g. spastic diplegia in cerebral palsy, or frontal meningioma).

• In the spinal cord, sensory pathways lie close to the motor pathways and so are often also affected by any pathology. Look for dorsal column signs (loss of light touch, vibration and joint position sensation) and a sensory level.

• Intrinsic cord lesions are usually painless.

• Extrinsic lesions causing spinal cord compression often also press on the sensory roots as they leave the spinal cord, and thus cause pain in the distribution of those roots (radicular pain).

Assessment of the patient with spastic paralysis

Important features to determine include the following.

• Speed of onset—rapid in vascular disease, more prolonged in inflammatory, infectious and compressive disease.

• Past and current medical problems—cerebral palsy, tuberculosis, HIV, macrocytic anaemia.

• Urinary hesitancy, frequency or retention (the latter is a late feature).

• Constipation, incontinence, reduced anal tone and sensation.

• Presence of pain—extrinsic lesions compressing spinal roots cause radicular pain (e.g. tumours); abscesses give back pain and local tenderness.

• Examine for a gibbus of Pott's disease (see below), or naevus/hairy patch of spinal dysraphism (spina bifida occulta).

• Sensory level—nipples are T4, umbilicus is T10.

Important tropical causes of spastic paralysis

Tuberculosis

Tuberculosis (TB) causes spastic paralysis in one of three ways:

1 TB of the vertebral bones leads to collapse (Pott's disease) and secondary cord compression (a bony prominent sharp kyphosis caused by a collapsed vertebrum is known as a gibbus).

2 Chronic TB meningitis leads to secondary arteritis and cord infarction.

3 A tuberculoma compresses the cord directly.

Many patients have a personal or family

CAUSES OF SPASTIC PARALYSIS

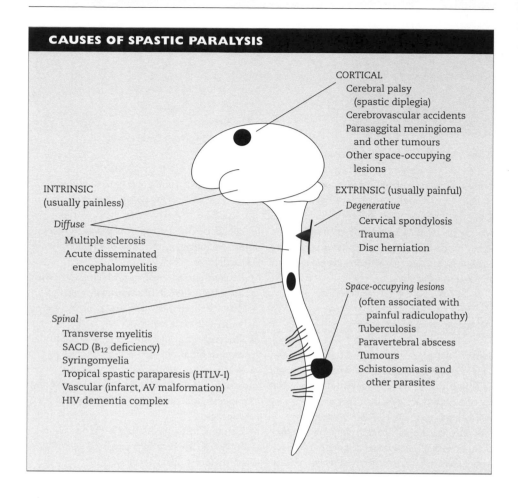

CORTICAL
Cerebral palsy
(spastic diplegia)
Cerebrovascular accidents
Parasaggital meningioma
and other tumours
Other space-occupying
lesions

INTRINSIC
(usually painless)

Diffuse
Multiple sclerosis
Acute disseminated
encephalomyelitis

EXTRINSIC (usually painful)
Degenerative
Cervical spondylosis
Trauma
Disc herniation

Spinal
Transverse myelitis
SACD (B$_{12}$ deficiency)
Syringomyelia
Tropical spastic paraparesis (HTLV-I)
Vascular (infarct, AV malformation)
HIV dementia complex

Space-occupying lesions
(often associated with
painful radiculopathy)
Tuberculosis
Paravertebral abscess
Tumours
Schistosomiasis and
other parasites

Fig. 35.1 Causes of spastic paralysis. AV, arteriovenous; SACD, subacute combined degeneration of the spinal cord.

history of TB or evidence of the disease elsewhere. Investigate with a tuberculin test, erythrocyte sedimentation rate (ESR), chest X-ray, spinal X-ray, computerized tomography (CT) or magnetic resonance imaging (MRI) scanning if possible. Treatment consists of antituberculous therapy, plus surgery if appropriate.

Spinal epidural abscess

Spinal epidural abscess is most frequently caused by *Staphylococcus aureus*. Patients present with the triad of fever, backache/tenderness and radicular pain, followed by rapidly progressive spastic paraparesis, sensory loss and bowel and bladder dysfunction. There is usually a peripheral leucocytosis, elevated ESR and mild cerebrospinal fluid (CSF) pleocytosis with an increased protein level. An X-ray may show soft tissue changes and associated vertebral osteomyelitis. On myelography, the flow of contrast medium is blocked. MRI is the investigation of choice. Patients should be treated with emergency surgical drainage and antibiotics.

HIV myelopathy

HIV myelopathy is a progressive myelopathy and is a frequent finding in patients with the AIDS–dementia complex. There is usually spasticity with increased or decreased reflexes, ataxia, incontinence and dorsal column signs.

The MRI scan is usually normal, but at autopsy there is vacuolation of the spinal cord white matter, especially in the posterior and lateral columns of the thoracic cord. HIV myelopathy is thought to be caused directly by HIV-1 invasion. Other causes of myelopathy in AIDS include lymphoma, *Cryptococcus* and herpes viruses.

Subacute combined degeneration of the spinal cord (SACD)

This is caused by vitamin B_{12} deficiency resulting from poor diet or impaired absorption secondary to tropical sprue, *Diphyllobothrium latum* infection (fish tapeworm), gastrointestinal surgery or as part of pernicious anaemia. The disease leads to a mixture of upper motor neurone deficit (corticospinal tract damage), sensory deficit caused by dorsal column involvement and peripheral neuropathy, some-times with optic neuropathy and dementia. Classical findings are extensor plantars with absent knee jerks. Neck flexion causes shooting pains down the arms (Lhermitte's sign). There is usually a macrocytic anaemia. The disease is treated with intramuscular vitamin B_{12} injections 1000 µg/day for 6 days, then reducing to a maintenance dose (Chapter 57).

Tropical spastic paraparesis

Tropical spastic paraparesis is found in the Caribbean, Seychelles, equatorial Africa, Japan and parts of the Americas. Most cases are caused by human T lymphotrophic virus type 1 (HTLV-1), transmitted sexually, by exposure to blood products or by breast milk. The disease presents with progressive spastic paraparesis, impaired vibration and joint position sensation and bowel and bladder dysfunction.

CHAPTER 36

Rabies

Rabies in the dog, 232
Rabies in other animals, 232
Clinical features in humans, 232
Diagnosis, 233

Treatment, 233
Postexposure treatment, 234
Postexposure vaccine
  regimens, 234

Pre-exposure
  immunization, 236
Prevention of rabies, 236
Further reading, 236

The rabies virus is a bullet-shaped RNA rhabdovirus that is widely prevalent among warm-blooded animals in many tropical countries. Worldwide, dogs are the main transmitters of rabies to humans: bats are involved in North America, certain South American countries, some areas of Europe and, more recently, Australia. Between 35 000 and 50 000 people die from rabies each year, with most deaths occurring in Asia.

On reaching the brain of an infected animal, the rabies virus spreads along peripheral nerves to reach the skin and the lachrymal and salivary glands. An animal bite with virus-laden saliva is therefore the usual mode of infection, although the disease can also be acquired by salivary contamination of cuts, abrasions and mucous membranes, and occasionally by inhalation. Rarely, rabies has been transmitted by corneal transplants taken from donors dying from unrecognized rabies. The risk of rabies in a person bitten by a rabid animal varies widely up to 50%, depending on the site and nature of the bites and the animal species.

Rabies in the dog

This is usually furious: the dog rushes around emitting a high-pitched bark, and biting not only people but also objects. The animal may paw at its mouth, as if trying to dislodge a foreign body and salivate excessively. It nearly always dies within 15 days of becoming infective. This fact is made use of when a dog that has bitten someone is impounded. If the dog is alive 15 days after the bite, it is very unlikely to have been infective when it bit.

Dogs can be given a high degree of protection against rabies by an attenuated live vaccine.

Rabies in other animals

Dogs, foxes, wolves and jackals are the major reservoirs of rabies in most parts of the world, and cats can also transmit infection, usually themselves suffering from a paralytic illness typical of 'dumb rabies'. In South America and the Caribbean, bats are very important vectors and cause enormous economic losses because of death of cattle from paralytic rabies. Bats can also cause human disease, not only from their bites but also from the inhalation of aerosolized bat secretions by speleologists exploring caves, mainly in the Americas.

Clinical features in humans

Once inoculated, the rabies virus travels centripetally along peripheral nerves to reach the spinal cord and brain. The incubation period is related to the inoculum size and the time taken to reach the brain: generally it is 2–8 weeks but may range between 9 days and a year or more.

The incubation period is proportional to the distance the virus has to travel and therefore tends to be shorter in children, and after bites on the face and neck.

The disease may be heralded by pain, paraesthesiae or pruritus in the bitten area; these occur in 30–80% of cases. The illness proper usually begins abruptly with fever, insomnia, anxiety and other psychological disturbances. In the encephalitic (furious) form of the disease the patient suffers from intermittent episodes of confusion, agitation and aggression verging on mania, with intervening periods of calm and lucidity. Copious secretion of ropy saliva is characteristic and the patient may literally 'froth at the mouth'.

Painful spasms of the throat muscles, often precipitated by attempts to swallow, are accompanied by 'hydrophobia', an overwhelming terror of drinking. Spasms often become more widespread, to involve the diaphragm and respiratory muscles: they may be accompanied by spitting, grimacing, vomiting, opisthotonos and seizures. Spasms may also be precipitated by air blowing on to the face (aerophobia). Death usually occurs within a week, with respiratory and bulbar paralysis.

Some 20% of patients suffer from a paralytic form of rabies. There is an ascending sensorimotor neuropathy with ocular, cranial and laryngeal palsies and sphincter disturbances: fasciculation may be seen in muscles. Hydrophobia is rare.

Diagnosis

Diagnosis during life

In most cases the diagnosis will be made on the characteristic clinical picture: laboratory tests for virus detection may not be available and currently lack sensitivity. Immunofluorescence of corneal impression smears or skin biopsies can detect viral antigen, especially late in the illness. Skin biopsies are taken from the nape of the neck so as to contain hair follicles and peripheral nerves. Attempts to culture virus from saliva, cerebrospinal fluid (CSF) or biopsies may not succeed in the later stages of illness because of the presence of neutralizing antibodies. Diagnostic antibody tests do not become positive until about the eighth day and are difficult to interpret in vaccinated patients.

Rapid diagnostic methods based on nucleic acid amplification are being developed and look very promising: viral RNA can be detected in saliva and CSF as early as the second day of illness.

Postmortem diagnosis

Because of cross-infection hazards a postmortem should only be performed if absolutely necessary. Staff should be immunized and must wear protective clothing and visors.

Immunofluorescence of fresh brain is the technique most widely used to demonstrate viral antigen. Rabies virus can be detected using tissue culture and mouse inoculation. Polymerase chain reaction (PCR) methods can also be applied. Histology, which is not routinely performed, shows a diffuse meningoencephalitis with extensive neuronal destruction and the presence of Negri inclusion bodies within brain cells.

Treatment

Rabies is almost invariably fatal and the disease causes great suffering. The main aim of treatment is to relieve symptoms with heavy sedation such as a mixture of a phenothiazine, a barbiturate and diamorphine. Intensive care may, in rare instances, have saved one or two lives.

Precautions with rabies patients

The patient's saliva is potentially infective. Everyone in contact with the patient should be protected as for barrier nursing, with the addition of visors. Staff should receive rabies immunization (four intradermal injections of 0.1 mL of human diploid cell vaccine, each given into a different limb on the same day).

ASSESSING THE RISK OF RABIES

1 Was the person bitten, or licked on an open wound or mucous membranes by an animal?
2 Where on the body was the bite/lick?
3 Where did the incident take place and on what date?
4 What species was the animal?
5 Is rabies known or suspected in the species? In the area?
6 Is there a known and contactable owner?
7 (a) Was the animal behaving normally at the time?
   (b) Had it been vaccinated?
   (c) Is the animal being held under observation?
8 If the animal was a dog or a cat did it become ill while under observation?
9 If the animal has died, does laboratory examination of the animal's brain confirm rabies?
10 Has the bitten person previously received rabies vaccine? How much does the person weigh (relevant to rabies immune globulin dosage)?

Table 36.1 Ten questions to ask the patient.

Postexposure treatment

The aim is to eliminate or neutralize rabies virus during the incubation period. Management is aided by asking the patient 10 key questions (Table 36.1).

First aid treatment

Wash and flush the wound vigorously with soap and water, detergent, or water alone. Then apply ethanol (700 mL/L) tincture or aqueous solution of iodine, or 0.1% quaternary ammonium compounds (with the latter, remove all traces of soap first).

Immediate treatment by or under the direction of a physician

1 Treat as in first aid treatment above, then:
2 If indicated (Table 36.2), use topical rabies immune globulin around the wound (see below).
3 Postpone suturing of wound; if suturing is necessary, use immune globulin locally.
4 Where indicated, institute antitetanus procedures and administer antibiotics and drugs to control infections other than rabies.

Postexposure immunization
(see Table 36.2)

Active immunization
Purified inactivated cell culture or duck embryo vaccines are recommended as they are potent and safe. Costs can be reduced by intradermal administration. Whichever vaccine is used it must be started as early as possible after exposure and should never be withheld whatever time has elapsed.

Passive immunization
Rabies immune globulin (RIG) should be given for high-risk exposures (Table 36.2) unless the patient has been previously fully immunized. The dose is 20 IU/kg body weight for human RIG and 40 IU/kg for equine RIG. Up to half the dose is given by infiltration in and around the wound and the rest given by intramuscular injection at a site separate from the first dose of vaccine. If RIG is initially unobtainable, it may be given up to day 7. Equine RIG may cause allergic reactions and the usual precautions against anaphylaxis must be taken. Serum sickness occurs in 1–6% of patients, usually 7–10 days after injection. At the time of writing this preparation is not being manufactured.

Postexposure vaccine regimens

The following vaccines are recommended by the World Health Organization (WHO).
• Human diploid cell vaccine (HDCV).
• Purified vero cell vaccine (PVRV).
• Purified primary chick embryo cell vaccine (PCECV).
• Purified duck embryo vaccine (PDEV).

TREATMENT OF RABIES

Nature of exposure	Status of biting animal (irrespective of previous vaccination)		Recommended treatment
	At time of exposure	During 15 days*	
1 Contact, but no lesions; indirect contact; no contact	Rabid		None
2 Licks of the skin; scratches or abrasions; minor bites (covered areas of arms, trunk and legs)	Suspected as rabid	Healthy	Start vaccine. Stop treatment if animal remains healthy for 15 days*
		Rabid	Start vaccine; administer rabies immune globulin if appropriate upon positive diagnosis and complete the course of vaccine
	Rabid; wild animal, or animal unavailable for observation		Vaccine + rabies immune globulin, according to previous immunization history
3 Licks of mucosa; major bites (multiple or on face, head, finger or neck)	Suspect or rabid domestic or wild animal, or animal unavailable for observation		Vaccine + rabies immune globulin, according to previous immunization history. Stop treatment if animal remains healthy for 15 days*

* Observation period applies to dogs and cats. Ten days recommended by WHO.

Table 36.2 Treatment according to nature of exposure. (Source: UK Department of Health, 2000.)

WHO recommended schedules

1 Intramuscular administration (into the deltoid, never the buttock). For use with all recommended vaccines. One dose of vaccine on days 0, 3, 7, 14 and 28. Alternatively, two doses on day 0 (one into each deltoid), one dose on day 7 and one on day 21.

2 Two-site intradermal method ('2-2-2-0-1-1'). For use with PVRV, PCECV and PDEV. Days 0, 3 and 7: one intradermal dose given at each of two sites over the deltoid. Days 28 and 90: one intradermal dose given at one site on the upper arm.

The intradermal dose is one-fifth of the intramuscular dose.

3 Eight-site intradermal method ('8-0-4-0-1-1'). For use with HDCV and PCECV. Day 0: 0.1 mL of vaccine at each of eight sites (deltoid, lateral thigh, suprascapular region and lower quadrant of abdomen). Day 7: 0.1 mL of vaccine at each of four sites (deltoid and thighs). Days 28 and 90: 0.1 mL of vaccine at one site (deltoid).

Nerve tissue vaccines

These are less potent and are more likely to cause neuroparalytic complications. Suckling mouse brain (SMB) vaccine is often used in Latin America. Subcutaneous doses are usually given daily for 7 days, with booster doses at 10, 20 and 90 days. Vaccines prepared in fixed sheep and goat brains (e.g. Semple vaccine) carry an appreciable risk of postvaccinal encephalitis or polyneuritis.

Need for flexibility

These vaccination recommendations are intended only as a guide. Modification of standard procedures may be justifiable in areas of low rabies endemicity or where there is no indication of rabies infection in the animal species involved. Local expert advice should be obtained whenever possible.

Pre-exposure immunization

This is recommended for travellers to endemic areas and those at special risk (e.g. veterinarians and animal handlers). Three doses of a cell culture or PDEV are given on days 0, 7 and 28. The vaccines may be given intramuscularly or intradermally; doses according to manufacturers' instructions. The antibody response may be impaired by concomitant chloroquine administration, particularly if the intradermal route is used.

Prevention of rabies

In endemic areas this depends upon the mass immunization of dogs, import controls and the elimination of strays. Unvaccinated dogs (those not wearing a collar with a vaccination tag) should be regularly caught and destroyed. The best results are obtained by professional dogcatchers using bait. Attempts to control dogs by soldiers armed with assault rifles are invariably unsuccessful and dangerous.

Further reading

Anon. Human rabies prevention: United States, 1999. Recommendations of the Advisory Committee on Immunization Practices (ACIP). *Morbid Mortal Week Rep* 1999; **48** (RR–1): 1–21. [Largely concerned with USA practice, this discusses rabies epidemiology in bats, terrestrial carnivores and other wild animals. Extensively referenced.]

Department of Health. *Memorandum on Rabies Prevention and Control*. London: Department of Health, 2000. [Contains useful information on all aspects of rabies control: immunization, patient management, postmortem precautions and veterinary action.]

World Health Organization. *WHO Recommendations on Rabies Postexposure Treatment and the Correct Technique of Intradermal Immunization Against Rabies*. World Health Organization WHO/EMC/ZOO.96.6, 1997. [Essential reading, this gives straightforward advice on first aid, pre- and postexposure immunization and the use of rabies immune globulin.]

CHAPTER 37

Tetanus

Bacteriology and
  pathogenesis, 237
Clinical manifestations, 237

Diagnosis, 238
Treatment, 238

Epidemiology and
  prevention, 238
Further reading, 239

Bacteriology and pathogenesis

Tetanus is caused by *Clostridium tetani*, a Gram-positive obligate anaerobe with a terminal spore, which is ubiquitous in the environment, particularly in soil. Spores are highly resistant to light and temperature; clinical disease occurs when spores are inoculated into wounds. Most cases of tetanus are related to acute injuries. Spores germinate under anaerobic conditions at the site of a wound and the growing bacteria produce two toxins: tetanolysin and tetanospasmin. Tetanospasmin undergoes proteolytic cleavage and binds and enters the presynaptic terminal. There, tetanospasmin cleaves the protein that allows fusion of synaptic vesicle with the membrane and thus prevents transmitter release. Tetanospasmin is able to travel retrogradely via axons to cell bodies and cross synapses, thus reaching the spinal cord, brain and autonomic nervous system. It primarily affects inhibitory glycine or γ-aminobutyric acid (GABA) neurones, leading to increased firing and lack of normal relaxation and causing the classical spasms of tetanus.

Clinical manifestations

The incubation period varies according to the site of injury and is shorter in severe disease, with an average incubation of 8 days for severe disease. There are several classical clinical subtypes, which reflect the main site of action of toxin.

Generalized tetanus

This is the most common clinical form. It often commences with trismus ('lock jaw') in which the patient is unable to open their mouth or risus sardonicus, a grimace caused by spasm of facial muscles. The predominant feature is of repeated spasms which may involve the neck, thorax, abdomen or extremities. Generalized spasms with opisthotonos also occur. Spasms are precipitated by external or internal stimuli, may last for minutes and are painful as full consciousness is retained. Respiratory compromise may occur because of involvement of the glottis or diaphragm. The disease may continue to progress for up to 10 days after the first symptoms.

In severe tetanus, autonomic dysfunction may occur after several days. Hypertension, tachycardia, arrhythmias and hyperpyrexia may all occur and can be extremely difficult to manage. Recovery may take up to 4 weeks: case fatality rates can reach 60%, with death usually occurring because of respiratory or autonomic involvement.

Neonatal tetanus

This is usually caused by infection of the umbilical stump. The risk of infection is related to the length of stump, the care and cleanliness with which the cord is ligated and cut and the cleanliness of the environment. Neonatal tetanus only occurs in the children of non-immune mothers. Symptoms and signs occur 1–10 days postpartum. Initially, generalized weakness and floppiness of the baby are noticed, with irritability and an inability to suck and feed. Subsequently, spasms, opisthotonos and hypersympathetic states occur. Up to 90% of affected infants die and mental retardation is common in survivors.

Localized tetanus

This is usually a mild form in which rigidity is limited to muscles near the site of injury. Weakness of the muscles may also occur because of the action of the toxin at the neuromuscular junction. Symptoms may be mild and persist for months. If the diagnosis is not made, progression to the generalized form may occur.

Cephalic tetanus

This is the rarest form and occurs in head injuries or with middle ear infection. The incubation period is normally 1–2 days and the major clinical manifestations are caused by involvement of cranial nerves with facial paresis, dysphagia and extraocular palsies. This form can also progress to generalized tetanus.

Diagnosis

The diagnosis is usually made clinically. Bacteriology is of little help, the organism is often not found and a positive wound culture for *Cl. tetani* does not prove that the organism is toxin-producing and causing disease. Blood and cerebrospinal fluid (CSF) findings are usually normal. The differential diagnosis is limited but includes strychnine poisoning, dystonic reactions, hypocalcaemia and seizures in adults and metabolic or neurological causes of posturing in neonates.

Treatment

Tetanus should be treated by the administration of tetanus immunoglobulin (human tetanus Ig 500–5000 IU i.m. or equine tetanus Ig 10^4–10^6 IU i.m.). Intrathecal administration of tetanus Ig gives no additional benefit. Wounds should be débrided to prevent further germination of spores. Metronidazole or benzylpenicillin should be given to prevent multiplication of bacteria. Much of the care is supportive. External stimulation should be reduced to prevent precipitation of spasms: patients should be nursed in a quiet, dim environment. The airway should be protected; endotracheal intubation or tracheostomy is sometimes necessary. Spasms can be treated by the use of high doses of benzodiazepines; baclofen is also effective. Some patients require paralysis with neuromuscular junction blockers. The treatment of autonomic instability is often difficult as manifestations can alter quite rapidly. Intravenous labetolol is useful for the management of hypertension; atropine or pacing may be needed for bradycardias and sympathomimetics and fluids are sometimes necessary to treat hypotension.

Neonatal tetanus is treated in a similar fashion to generalized tetanus.

Epidemiology and prevention

Although tetanus occurs worldwide, it is predominantly a problem of tropical and developing countries, being particularly common in the Philippines, Vietnam, the Asian Subcontinent, Indonesia and Brazil. The annual estimated number of cases globally is 500 000 and neonatal tetanus now accounts for about half of cases. However, there has been a significant reduction in the number of cases through the use of vaccination, particularly maternal vaccination, and better obstetric practice and care of the cord.

Tetanus is a vaccine-preventable disease. Immunization with tetanus toxoid is very effective. Children should receive vaccination with tetanus toxoid as part of the routine diphtheria, tetanus and pertussis (DTP) immunization given at 6, 10 and 14 weeks. A single dose of tetanus toxoid in pregnancy leads to protective titres in mother and neonate: standard recommendations are that non-immunized mothers should receive two doses during pregnancy. Routine booster doses should be given at 10 yearly intervals, or for tetanus-prone injuries if not immunized within the last 5 years. There is probably no need to give more than five doses in a lifetime.

Further reading

Farrar JJ, Yen LM, Cook T et al. Tetanus. J Neurol Neurosurgery Psychiatry 2000; **69**: 292–301. [A useful review of the subject.]

CHAPTER 38

Brucellosis

Epidemiology, 240
Clinical features, 240
Diagnosis, 241

Treatment, 242
Follow-up, 243

Public health aspects, 243
Further reading, 244

Brucellosis (Malta fever, Rock fever) is one of the classical zoonoses (infections of animals transmitted to humans). It is an important cause of fever in many parts of the world and is often underdiagnosed because of lack of laboratory facilities.

Epidemiology

Brucellae are Gram-negative coccobacilli. At least six species infect a wide variety of land-based mammals and new species have recently been described in marine mammals such as whales and seals. Three species are responsible for most human infections:

1 *Brucella abortus*, usually a disease of cattle, is prevalent in Africa, the Indian Subcontinent and temperate zones;

2 *Brucella melitensis*, whose normal ruminant host is sheep and goats but is also found in camels, is particularly prevalent in countries around the Mediterranean, the Middle East and Central and South America; and

3 *Brucella suis*, whose natural host is pigs, is still a problem in the USA.

Brucella canis (natural host dogs) can rarely infect humans. The organisms are intracellular and can remain hidden in the reticuloendothelial system so that clinical incubation periods after infection range from several weeks to months. Despite this, brucellosis does not appear to be more common or more severe in patients with HIV. In animals, they are important causes of epididymitis, abortion and infertility, but host animals may appear symptomless.

Humans acquire infection from ingesting milk or dairy products such as laban, lassi, buttermilk and cheeses that have not been pasteurized. The products of abortion and placentae from infected animals are highly infectious and farmers and veterinarians can easily become infected by aerosol transmission from the products of conception. Rarely, human brucellosis can be acquired via breast milk, sexual transmission or transfusion of blood products. Veterinarians and farmers sometimes have localized skin disease caused by direct contact with infected animal products. Brucellosis is not transmitted by eating the meat of infected animals unless it is eaten raw and has been externally contaminated.

Clinical features

The symptoms of brucellosis are of recurrent prolonged bouts of fever. If specific treatment is not given, undulating patterns of fever may last for several weeks, followed by an afebrile period and then relapse. Approximately half of all cases are associated with focal musculoskeletal symptoms, which may be the only clinical clue that differentiates brucellosis from other causes of fever such as typhoid, Q fever, malaria, etc. In an endemic area, it is the first clinical diagnosis for any patient who presents with fever and difficulty in walking. Fever is worse at night and may be associated with profuse sweating. Patients are depressed, anorexic and lethargic, although the onset of these symptoms is often insidious. A small proportion present with more pronounced neuropsychiatric disorder or low-grade meningoencephalitis, and 5–10% of men have orchitis which must be distinguished from mumps. Patients often have a dry cough, mimicking the presentation of typhoid. Epistaxis is an unusual but well-recognized presentation because of associated thrombocytopaenia, but other features of bleeding disorder such as haematemesis or malaena are very unusual.

The overall pattern of presentation varies with the age of the patient and the infecting species. *B. abortus* infections have a more insidious onset, are more likely to affect the axial skeleton and to become chronic. *B. melitensis* tends to have a more acute onset and is more likely to affect peripheral joints as well as the vertebrae. Children often present with fever and a single clinically affected joint, typically the hip or knee, and this may be mistaken for rheumatic fever or septic arthritis. *B. suis* infections have an acute presentation complicated by focal deep tissue abscesses.

Patients look unwell and are lethargic but do not look as toxic as those with enteric fever. The temperature is almost invariably raised but often returns to normal during a 24-h cycle. Up to 10% have cervical or other lymphadenopathy, which must be differentiated from glandular fever, HIV or tuberculosis (TB) adenitis. One-quarter have mild to moderate splenomegaly. The chest is usually clear, even if the patient has a cough. Individual joints may show signs typical of septic arthritis with swelling, heat, tenderness and effusions. There may be local tenderness, especially on movement of vertebrae or sacroiliac joints, but deformity of the back or long tract neurological signs are very unusual and suggest TB rather than brucellosis. Brucellosis is rarely fatal unless complicated by endocarditis (~ 1% of cases), but causes prolonged debilitation and loss of productivity.

Diagnosis

The full blood count shows low white blood cells (WBCs) with lymphopaenia and mild thrombocytopaenia. Occasionally, there is more pronounced reduction of platelets and haemoglobin. Mild elevation of alkaline phosphatase and transaminases is common. Blood culture is the most reliable method of confirming the diagnosis, but will only be positive in about two-thirds of *B. melitensis* cases and less than one-third of cases caused by *B. abortus*. If modern 'signalling' blood culture systems are used, they usually become positive within a week but culture should be prolonged to 3 weeks to detect late positives. If basic culture facilities are all that are available, cultures should be prolonged to at least 6 weeks, with most of the positives occurring between days 7 and 21. Laboratory staff must be told that brucellosis is a possibility, both so that cultures are prolonged and so that they are aware of the significant hazard that *Brucella* poses to laboratory workers because of the risk of aerosol spread.

A single bone marrow culture has a better yield than three sets of blood cultures and is occasionally useful in patients with pyrexia of unknown origin (PUO) who have been given antibiotics. Synovial fluid should be cultured for *Brucellae* in any case of septic arthritis in an endemic area, and aspirates or biopsies from abnormal tissues such as lymph nodes or liver should also be cultured. Cerebrospinal fluid (CSF) usually shows mild elevations of lymphocytes and proteins, and organisms may be cultured.

Serological tests are still based on the old (Wright's) standard agglutination test (SAT). *Brucella* antigen supplied with the kit is added to successive dilutions of patient serum, and if visible agglutination occurs the test is positive. These tests are notoriously affected by the 'prozone phenomenon', which causes false-negative results. This occurs because patients with brucellosis have immunoglobulin A (IgA) antibodies, which interfere with agglutination at low dilutions, and the blocking effect is only overcome at increasing serum dilutions. Thus, the result might be negative at dilutions of 1/40, 1/80, 1/160 and 1/320 and positive only at 1/640.

Many inexperienced laboratories will only dilute serum to 1/160 and therefore miss the true positives.

As with all serological tests, a fourfold rise in titre in acute and convalescent samples (10–14 days later) is strongly suggestive of brucellosis, but this result is too delayed to guide the immediate management of patients with fever. In endemic areas, many patients have had previous exposure to brucellosis and have low titres of antibodies already, so the diagnostic 'cut-offs' for a single sample to be positive have to be set higher, typically at 1/160 or 1/320. In a non-endemic area, or for an expatriate who has recently been exposed for the first time in an endemic area, a titre of 1/80 would be strongly predictive of brucellosis. About 10% of blood culture-positive patients have negative serological results at first presentation, so a negative result does not entirely rule out brucellosis.

The SAT is affected by the antigen used and many different commercial kits are available. Some provide antigens for both B. abortus and B. melitensis but there is much cross-reaction and one cannot reliably distinguish these infections on the basis of serology alone. Mercaptoethanol can be added to patient serum to dissociate IgM and therefore indicate if IgG predominates (more suggestive of chronic infection), but this is only moderately reliable. The SAT has been adapted in some reference laboratories to be performed in microtitre trays, the so-called MAT. All these tests cross-react with some other Gram-negative bacteria (Yersinia, cholera) and recent cholera immunization.

Enzyme-linked immunoabsorbent assay (ELISA) tests have also been developed and are used in some laboratories, but are not internationally standardized and are more expensive. Rose Bengal serological tests, designed for animal diagnosis, are used as screening tests on human sera in many tropical settings, but their specificity and sensitivity have not been validated for this purpose. Urinary dipsticks to detect antibodies are sensitive but not widely used, and assays for circulatory or urinary antigen remain experimental. Molecular techniques (polymerase chain reaction; PCR) to detect DNA are sensitive but not routinely available.

Tissues such as bone, lymph node or liver contain non-caseating granulomas but the distinction from tuberculous granulomas is not always easy. Radiological bone changes are also seen later; typically, erosions at the edge of joints or the end-plates of vertebrae, with associated sclerosis. Marked bony destruction is unusual and is more suggestive of TB. Isotope bone scans show hotspots in affected bones and joints and frequently reveal further foci of infection that are asymptomatic. The clinical and radiological features that discriminate between spine involvement with TB or brucellosis are shown in Table 38.1.

Treatment

Three questions guide management, once a presumptive or definite diagnosis has been made.
1 Is the disease acute (duration < 1 month) or relapsing or chronic (> 6 months)?
2 Is there focal disease of bone or joints?
3 Has tuberculosis definitely been excluded?

Adults with acute non-focal disease should be treated for a minimum of 6 weeks. Patients with focal disease and/or chronic disease require 3 months of treatment. Monotherapy should not be used because, although clinical illness responds in the short term, early relapse occurs in more than 30% of cases. At least two antibiotics should be used for all cases. Patients in whom tuberculosis has not been excluded have to be treated for both infections simultaneously or should be given antimicrobials to which only brucellosis responds (i.e. streptomycin or rifampicin should not be used).

The time-honoured combination of an oral tetracycline for 6–12 weeks plus 1 g/day streptomycin intramuscularly for 2–3 weeks is still the best regimen. The preferred form of tetracycline is 100 mg doxycycline twice daily as it is easier to take and less likely to cause renal toxicity. It is likely that modern aminoglycosides such as gentamicin (5 mg/kg/day for 7 days) could be substituted for streptomycin, but this has not yet been confirmed. An alternative regimen is doxycycline with rifampicin, both given for 6 weeks or 3 months. The relapse rate after

SPINE RADIOLOGY: DIFFERENCES FROM TUBERCULOSIS

	Brucellosis	Tuberculosis
Site	Lumbar + others	Dorsolumbar
Vertebrae	Multiple or contiguous	Contiguous
Discitis	Late	Early
Body	Intact until late	Morphology lost early
Canal compression	Rare	Common
Epiphysitis	Anterosuperior (Pom's sign)	General: upper + lower disc region, centre, subperiosteal
Osteophyte	Anterolateral (parrot beak)	Unusual
Deformity	Wedging uncommon	Anterior wedge (gibbus)
Recovery	Sclerosis of whole body	Variable
Paravertebral abscess	Small, well localized	Common + discrete loss transverse process
Psoas abscess	Rare	More likely

Table 38.1 Radiology of spine: differences from tuberculosis.

this regimen is ~10% compared to ~5% with doxycycline/streptomycin, and some national programmes discourage use of rifampicin for this purpose, reserving it for tuberculosis and leprosy treatment. Co-trimoxazole in high doses (three tablets twice a day for large adults) can be used but can cause anaemia and drug rashes, and should be supplemented with daily folic acid. It provides a good alternative to tetracyclines in young children when given with a second antibiotic. There is reasonable evidence that children (< 12 years) are adequately treated by 3 weeks rather than 6 weeks of therapy.

Pregnant women should not receive tetracyclines and are usually given rifampicin alone or with 2 tablets co-trimoxazole twice daily (avoid or use folate supplements in the first trimester). Patients with endocarditis or meningitis should receive three drugs (e.g. addition of ceftriaxone) and endocarditis often necessitates valve replacement as well. Fluoroquinolones have been disappointing for routine treatment but some physicians add them as a third drug in difficult cases.

Follow-up

Patients should be seen at 3 and 6 weeks to en-courage adherence to antibiotic therapy. The most useful features are improvement in general mood and health, with return of appetite and weight. Serology is not very useful as it follows a variable pattern for months to years after successful treatment and does not predictably rise to warn that relapse is imminent. Relapse is conventionally defined as a further episode of brucellosis occurring less than 6 months after the first. This is usually a result of failure to take adequate antibiotics for long enough rather than being due to drug resistance, and should be treated with a further 3-month course of two antibiotics as for a first episode; some would insist on including streptomycin for retreatment in order to be sure the drugs have been taken. Chronic brucellosis is difficult to define serologically and difficult to distinguish from chronic fatigue syndrome, depression or malingering. Immunity after brucellosis is not solid in humans, who may suffer from repeated infections. No vaccine is available for human use.

Public health aspects

Brucellosis is controlled by simple measures that require political commitment. The first is education of the public to eat or drink only pasteurized milk and dairy products, but this is often difficult to achieve in the face of tradition. Animal herds can be protected by administration of live vaccines, and such control has also been shown to reduce the incidence of human infection. Control of infected herds or flocks is usually based on 'test and slaughter' (i.e. if any animal in the herd tests positive, the whole herd is slaughtered). The public will only comply with such measures if they are offered adequate financial compensation.

Further reading

Ariza J. Brucellosis. *Curr Opin Infect Dis* 1996; **9**: 126–31. [Concise overview of most human aspects.]

Centers for Disease Control. Brucellosis: frequently asked questions. http://www.cdc.gov/ncidod/dbmd/diseaseinfo/brucellosis_g.htm [Useful general detail on the web.]

Hoover DL, Friedlander AM. Brucellosis. In: Sidell FR, Takafuji ET, Franz DR, eds. *Medical Aspects of Chemical and Biological Warfare: Textbook of Military Medicine*. Office of the Surgeon General, Dept of the Army, USA: 1997, 513–21. Available on-line free at http://www.nbc-med.org/SiteContent/HomePage/WhatsNew/MedAspects/contents.html. [Convenient free textbook chapter.]

Khan MY, Mah MW, Memish ZA. Brucellosis in pregnant women. *Clin Infect Dis* 2000; **32**: 1172–7. [Confirms the risk of humans aborting as well as animals, also therapeutic successes with rifampicin and co-trimoxazole.]

Madkour MM. *Madkour's Brucellosis*, 2nd edn. Berlin & Heidelberg: Springer-Verlag, 2000.

CHAPTER 39

Typhoid and Paratyphoid Fevers

Organisms, 245
Mode of infection, 245
Typhoid fever, 245
Diagnosis, 247

Treatment, 248
Carrier state, 250
Typhoid vaccine, 250

Paratyphoid A and B, 250
Paratyphoid C, 251
Further reading, 251

Typhoid and paratyphoid fevers are illnesses caused by *Salmonella typhi* and *Salmonella paratyphi* A, B and C. They cause a systemic septicaemic illness which is also called enteric fever. The many zoonotic salmonellas, which usually cause gastroenteritis, occasionally also cause an enteric fever-like illness, especially severe in patients with HIV infection.

Typhoid and paratyphoid are most common where standards of personal and environmental hygiene are low and only to this extent are these diseases tropical. There are estimated to be 20 million cases of typhoid fever worldwide each year with more than 200 000 deaths. The incidence is more than 100/100 000 population/ year in the Indian Subcontinent and South East Asia, and 10–100/100 000 population/year in other resource-poor countries in Asia, Africa, the Caribbean, Central and South America. In endemic areas, the disease is most common in children and young adults (aged 2–35 years).

Organisms

The organisms are Gram-negative bacilli of the Enterobacteriacae. All possess somatic (O) and flagellar (H) antigens. *S. typhi* and *S. paratyphi* C possess a surface (Vi) antigen that coats the O antigen and potentially protects it from antibody attack. *S. typhi* and *S. paratyphi* A and B infect only humans. *S. paratyphi* C may also affect a variety of animals.

Mode of infection

Infection is usually by ingestion, with transmission in water (mainly *S. typhi*) and food. Ingestion of 10^5 *S. typhi* organisms may cause a relatively low attack rate with a long incubation period. Increasing the infecting dose to 10^9 organisms raises the attack rate to 95% and greatly shortens the incubation period. Conditions causing low gastric acidity allow a lower inoculum to cause infection. The most important reservoirs of infection are asymptomatic convalescent or chronic human carriers. Food-handlers, who are also carriers, are a potentially important source of transmission.

Typhoid fever

After ingestion, the organisms attach and then penetrate the small intestinal mucosa, and are transported by the lymphatics to mesenteric lymph glands. There they multiply, and enter the bloodstream via the thoracic duct and are carried to the bone marrow, spleen, liver and gallbladder. At these sites the bacilli are able to survive and multiply inside macrophages. Eventually, the bacteria are rereleased into the bloodstream and this second bacteraemia corresponds to the onset of symptoms.

There is a secondary invasion of the bowel via the infected bile. Macrophages collect in large

245

numbers in the intestinal lymph follicles, particularly the Peyer's patches in the ileum. The strong inflammatory response in the Peyer's patches may lead to hyperplasia, necrosis and ulceration in 7–10 days if the inflammation does not resolve. Involvement of blood vessels may lead to bleeding and, if the whole thickness of the bowel is involved, perforation follows.

Elsewhere in the body, foci of inflammation with macrophages and lymphocytes, so-called typhoid nodules, are scattered in various organs, especially the liver, spleen, marrow and lymph glands. More diffuse organ involvement also occurs, affecting the myocardium, kidney and lung. Late in the disease there may be abscess formation, most often affecting bone, brain, liver or spleen. Serious disease of the brain, lung and kidneys is not invariably accompanied by typhoid nodule formation, and the assumption is that some unidentified toxin must be the cause.

The natural course of the untreated disease is very variable. In a classical case, fever has returned to normal at the end of the third week and repair processes then begin. However, in some cases, fever and symptoms last for only a few days (particularly in preschool children) and in others may continue for many weeks. Death most commonly results from perforation, haemodynamic shock associated either with intestinal haemorrhage or severe toxaemia, and occasionally from other complications such as meningitis.

Clinical picture

The average incubation period is about 14 days, but can vary from less than a week to more than 3 weeks. The only almost constant symptom is fever. The onset is usually gradual, and rigors are unusual. Fever increases day by day in the first week, often with an evening rise. A high and sustained fever (39–40°C) then continues for another week or more, falling by lysis in the third or fourth week.

Patients with typhoid usually feel very unwell in general, with malaise, generalized aches and pains, and anorexia. Abdominal pain or discomfort, headache, diarrhoea or constipa-

tion and a non-productive cough are common symptoms.

Physical signs

These depend not only on the severity of the illness but on the length of time the patient has been ill. In patients who seek medical aid early, there has usually been no significant dehydration from diarrhoea, and the patient often looks relatively well and is mentally alert. In contrast, the patient who presents after 2 weeks of illness is often very toxic, mentally stuporose and gravely dehydrated. The high fever may be accompanied by hepatomegaly, splenomegaly (often tender), mental changes including apathy, signs of bronchitis and meningism.

Rose spots are usually only seen in fair-skinned patients. They are found from day 7 onwards and take the form of pink macules, usually scanty and found mainly on the trunk. They fade on pressure from a glass slide. In occasional patients, the pulse rate is disproportionately slow compared with the fever, and may not reach 100 b.p.m. even when the temperature is 40°C (so-called relative bradycardia).

Complications

These may develop as the illness progresses, and can follow a clinically mild attack. The clinician must remember that typhoid patients may present with the complication rather than with the symptoms of typhoid fever, and these patients are often difficult to diagnose. The most common and important complications are as follow.

• *Perforation*. This typically occurs in the third week. Toxic patients may show few signs of peritonitis, except for abdominal distension, increasing toxaemia and a rising pulse. Surgery is preferable to conservative management, and excision or segmental resection is safer than simple suturing, because the gut wall immediately surrounding the perforation may be too friable to hold sutures. Antibiotic therapy should be broadened to cover gastrointestinal organisms contaminating the peritoneum.

• *Haemorrhage*. Patients may have repeated small bleeds that resolve without specific treatment. Massive bleeding is typically a complication of the third week. Surgery is seldom needed provided that blood transfusion is available.

• *Severe toxaemia.* Some patients have severe disease characterized by delerium, obtundation, stupor, coma or haemodynamic shock (not caused by gastrointestinal haemorrhage). For unexplained reasons, this complication has been reported more frequently in Indonesia, Papua New Guinea and West Africa than in other countries.

• *Haemolytic anaemia.* This may occur in patients with glucose-6-phosphate dehydrogenase (G6PD) deficiency; typhoid depresses G6PD levels in normal as well as in deficient patients.

• *Typhoid lobar pneumonia.* This is a rare complication of the second and third week. Rusty sputum is not produced.

• *Meningitis.* This may be the only obvious manifestation of typhoid, when it resembles any other pyogenic meningitis. It usually occurs in young children.

• *Renal disease.* This may present as renal failure or an acute nephrotic syndrome, and is probably an immune-complex nephritis. Recovery after successful chemotherapy is usual.

• *Typhoid abscess.* This is a late complication that can occur almost anywhere, especially in the spleen, liver, brain, breast and skeletal system.

• *Skeletal complications.* These are mainly suppurative arthritis and osteomyelitis. Both may be greatly delayed in onset. Zenker's degeneration of muscle or polymyositis may occur.

• *Other complications or sequelae.* Many are described including suppurative parotitis, acute cholecystitis, deep venous thrombosis, psychiatric disturbance and Guillain–Barré syndrome.

Diagnosis

Culture

Culture of the organism is the mainstay of diagnosis. Unfortunately, this technology is often lacking in those hospitals in developing countries that most need it. Blood culture is the most useful, particularly in the first and second week. The average number of bacteria in the blood is low, so an adequate volume of blood should be taken for culture to increase the likelihood of a positive result. Bone marrow culture gives a higher culture positive rate, probably because the concentration of organisms is 10-fold higher than in the blood, and may even yield a positive culture after chemotherapy has been started. A string capsule used to sample duodenal contents can yield positive cultures, but in practice this method is not widely used. Aspirates from rose spots, cerebrospinal fluid (CSF) or pus from abscesses may also yield positive cultures. Stool culture does not confirm the diagnosis, as the patient may be a chronic carrier. Faecal and urine cultures are mainly of value for the detection of carriers.

Serodiagnosis

The Widal test, which measures agglutinating antibodies to the somatic (O) and flagellar (H) antigens, is widely used. Although the test is relatively cheap and straightforward to perform, it lacks specificity and sensitivity.

1 In endemic areas low levels of antibodies are detectable in the healthy population, presumably because of prior exposure to the organisms.

2 Numerous non-typhoid salmonellae share O and H antigens with *S. typhi.*

3 H antibody titres remain high for a long time after typhoid immunization.

4 In typhoid patients, titres often rise before the clinical onset, making it very difficult to demonstrate the diagnostic fourfold rise between initial and subsequent specimens.

5 A significant number of culture-positive patients develop no rise in titre at all.

However, if the test is interpreted intelligently, bearing all these facts in mind, a significant number of patients will be correctly diagnosed by the Widal test, when all other methods have failed. Interpretation of the result is helped by knowledge of the background levels of antibodies in the local healthy population.

New antibody tests that detect different antigens to those used in the Widal test are being developed and look promising.

Other laboratory findings

The white blood cell (WBC) count is usually within the normal range, as is the differential count, but there may be leucopenia or leucocytosis and relative lymphocytosis is common. Biochemical tests usually show only minor changes, such as slight elevation of transaminases and bilirubin. A considerable elevation of indirect bilirubin is often associated with haemolytic anaemia in patients with G6PD deficiency and in children. In severe cases, albuminuria is almost invariable. There may be evidence of disseminated intravascular coagulopathy (DIC), although this rarely causes a clinical problem.

Treatment

Severely ill patients who may be mentally uncooperative require good nursing care and careful supportive medical care, including attention to fluid and electrolyte balance. The mainstay of treatment is effective antimicrobial chemotherapy.

Chemotherapy (Table 39.1)

Chloramphenicol used to be acknowledged everywhere as the drug of choice, with amoxicillin or co-trimoxazole as effective alternatives. However, in recent years, multidrug-resistant (MDR) isolates of S. typhi and S. paratyphi A resistant to all three antibiotics have been widely reported in the Indian Subcontinent and countries of South East Asia.

The fluoroquinolone antibiotics, third-generation cephalosporins and azithromycin have proved effective alternatives for treating MDR infections. Unfortunately, these antibiotics are expensive, in particular the cephalosporins. Widespread use has led to the emergence of strains with reduced susceptibility and resistance to the fluoroquinolones. In some countries in Asia (Indonesia, Papua New Guinea), Africa (except Kenya) and South and Central America, most strains remain susceptible to chloramphenicol.

Relapses after chemotherapy occur in a variable proportion of patients (2–10%), are usually rather less severe than the initial illness and respond to the same chemotherapy.

Fluoroquinolones

A 5–7-day course of ciprofloxacin or ofloxacin (Table 39.1) has proved extremely effective for the treatment of fully susceptible isolates, with rapid resolution of fever and symptoms. There have been concerns about the use of fluoroquinolones in children, because of evidence of damage to the cartilage in the growing joints of animals. However, there has been widespread compassionate use without problems of fluoroquinolones in children with multiresistant Gram-negative infections, where alternatives were unavailable.

Strains with reduced susceptibility to ciprofloxacin and ofloxacin have appeared, particularly in India and Vietnam. Infections with these strains may fail to respond to fluoroquinolone therapy. Microbiology laboratories may have difficulty in detecting these strains with the currently recommended methods. However, resistance to the related antibiotic nalidixic acid is a useful marker. Patients with enteric fever who are still sick after 7 days of an adequate dose of fluoroquinolone are likely to be infected with a nalidixic acid-resistant strain and should be changed to an alternative antibiotic. Fully fluoroquinolone-resistant strains are also beginning to appear in India.

Third-generation cephalosporins

Third-generation cephalosporins such as ceftriaxone, cefotaxime and cefixime are effective in treating MDR strains. However, the response to treatment is frequently slow, with the fever and symptoms taking 7–10 days to resolve.

Azithromycin

There is emerging evidence that azithromycin is another effective alternative for MDR typhoid in adults and children. It has not been used yet in severe disease.

Chloramphenicol

The drug is bacteriostatic only. A fairly

ANTIBIOTICS TO TREAT TYPHOID FEVER

Antibiotic	Dosage	Frequency	Route	Duration (days) Non-severe	Duration (days) Severe or complicated	Side-effects
Chloramphenicol	50–100 mg/kg/day Reduce dose 30 mg/kg/day when fever ceases	4	o (im/iv)	14–21	14–21	Bone marrow depression
Amoxicillin	75–100 mg/kg/day	3	o/im/iv	14	14	
Co-trimoxazole (trimethoprim-sulfamethoxazole)	8 mg/kg/day trimethoprim + 40 mg/kg/day sulfamethoxazole	2–3	o/im/iv	14	14	Nephrotoxic Allergy Not for children <2 years
Ciprofloxacin*	20 mg/kg/day	2	o/iv	5–7	10–14	
Ofloxacin*	10–15 mg/kg/day	2	o/iv	5–7	10–14	
Pefloxacin*	800 mg	2	o/iv	5–7	10–14	
Ceftriaxone	50–80 mg/kg/day	1–2	im/iv	7–10	10–14	
Cefotaxime	100–150 mg/kg/day	3–4	im/iv	7–10	10–14	
Cefixime	20–30 mg/kg/day	2	o	7–10	Not recommended	
Azithromycin	8–10 mg/kg/day	1	o	7	Not recommended	

im, intramuscularly; iv, intravenously; o, orally.
*Nalidixic acid-resistant isolates may not respond.

Table 39.1 Choice of antibiotic to treat typhoid fever.

prolonged course must be given to prevent re-lapse, such as a total of 14 days, or 12 days after fever has abated. It commonly takes 48 h before the fever shows a response, and 5 days or more until the patient becomes completely afebrile in severe cases. A Herxheimer-type reaction is sometimes seen early in treatment, and should be treated with steroids.

Amoxicillin
This is more expensive than chloramphenicol, but at least as effective if given in high doses. Ampicillin is inferior to chloramphenicol.

Co-trimoxazole
The clinical response is at least as rapid as with chloramphenicol.

Steroids
Adults and children with severe typhoid characterized by delerium, obtundation, coma or shock were shown to benefit in a study from Indonesia from the prompt administration of dexamethasone. The dosage given was 3 mg/kg by slow intravenous infusion over a period of 30 min followed by 1 mg/kg given at the same rate every 6 h for eight additional doses. Hydrocortisone given at a lower dose was not effective.

Carrier state

This commonly persists for some months into convalescence, and when it terminates spontaneously such patients are called convalescent carriers. They are an obvious source of infection to others, but even more important are chronic carriers (1–3% of cases) in which a persisting focus of infection smoulders on in the gallbladder (faecal carriers) or urinary tract (urinary carriers) for more than 1 year. In most endemic areas, few carriers are identified because culture facilities do not exist. Persistent elevation of Vi antibodies often accompanies the carrier state.

The excretion of organisms by carriers is variable and erratic. Chronic faecal carriers may have chronic cholecystitis with or without gall-stones, or pathological abnormalities in the urinary tract, including *Schistosoma haematobium* infection in chronic urinary carriers. *Schistosoma mansoni* may be associated with a relapsing non-typhoid *Salmonella* septicaemia.

Treatment of chronic carriers
Ciprofloxacin 750 mg twice daily for 28 days has proved effective. If ciprofloxacin is unavailable and the strains are susceptible, two tablets of co-trimoxazole twice a day for 3 months, or 100 mg/kg/day amoxicillin combined with 30 mg/kg/day probenecid, both for 3 months, may also be effective. Faecal carriers with gallstones only respond temporarily to chemotherapy and cholecystectomy is needed to terminate the carrier state in such cases.

If the patient is intelligent and conscientious, and not a food-handler, the carrier state need not be treated at all, for the fastidious maintenance of high standards of environmental and personal hygiene will prevent transmission of the infection to others.

Typhoid vaccine

Two vaccines are currently available. The live attenuated oral vaccine (Ty21a) requires three doses over 5 days with a booster recommended every 5 years. This vaccine is not recommended for children under the age of 6 years. The second, a purified Vi antigen vaccine, is given as a single dose intramuscularly; boosters are recommended every 3 years. It is not recommended for children under the age of 2 years. A new modified conjugate Vi vaccine is now under development that was 92% effective in a study in Vietnam. This vaccine has the potential to be effective in very young children. The disadvantage of all these vaccines is their cost.

If typhoid does develop in a vaccinated subject, it is no less severe than in the unvaccinated.

Paratyphoid A and B

These usually infect via contaminated foods in

which the organisms have multiplied. For this reason, diarrhoea and vomiting may precede septicaemia. Many mild cases occur. Treatment is as for typhoid.

Paratyphoid C

This commonly produces septicaemia without involvement of the gut, and abscess formation is common.

Further reading

Parry C, Hien TT, Dougan G, White NJ, Farrar JJ. Typhoid fever. N Engl J Med 2002; **347**: 1770–82. [Comprehensive review and source reference.]

CHAPTER 40

Arboviruses

Vectors and hosts, 252   Clinical syndromes, 252   Further reading, 253

Arbovirus (short for arthropod-borne virus) is an ecological description for viruses that are transmitted between vertebrate hosts by insects — principally mosquitoes, ticks, sandflies or midges. There are more than 500 arboviruses, in four viral families, but only a small number are medically important (Fig. 40.1). Some arboviruses are named after the disease they cause (e.g. Yellow fever, or O'nyong nyong — 'joint weakening' in a Ugandan dialect), some after their insect vector (e.g. phleboviruses after 'phlebotomus' — sandflies) and some after the geographical area where the disease first occurred (e.g. Japanese encephalitis).

Vectors and hosts

Following infection with an arbovirus, most animals develop life-long immunity to that virus. An arbovirus therefore needs a ready supply of immunologically naïve hosts. A few arboviruses (notably dengue) have evolved to use humans as the 'natural host'; however, most use small mammals or birds because of their high reproductive rate. For these 'enzootic' viruses, humans are coincidentally infected 'dead-end hosts' and do not transmit the disease. In some situations an 'amplifying host' increases the amount of circulating virus and acts as a link to human infection.

Clinical syndromes

The majority of human infections with arboviruses are asymptomatic or cause a mild non-specific febrile illness. When an arbovirus causes disease, it usually leads to one of three clinical syndromes.
1 Fever–arthralgia–rash (FAR).
2 Viral haemorrhagic fever.
3 Central nervous system (CNS) infection.

Most viruses cause a single syndrome, but there can be overlap; e.g. dengue viruses can present with a FAR syndrome (dengue fever), a haemorrhagic syndrome (dengue haemorrhagic fever) and even, occasionally, CNS disease. The most important haemorrhagic fevers are considered in Chapter 41, and CNS arboviruses are discussed in Chapter 33. FAR arboviruses are summarized below.

Fever–arthralgia–rash arboviruses
Chikungunya
Chikungunya occurs in Africa, India and South-East Asia. Humans and primates are the natural host and *Aedes* and *Culex* mosquitoes transmit the disease.

O'nyong nyong
O'nyong nyong occurs in Africa and is the only arbovirus transmitted by *Anopheles* mosquitoes. Humans are the only hosts and a common clinical feature is conjunctivitis.

Ross River
This occurs in Australia and is transmitted by *Aedes* and *Culex* mosquitoes. It can cause 'epidemic polyarthritis'.

ARBOVIRUS GROUPS

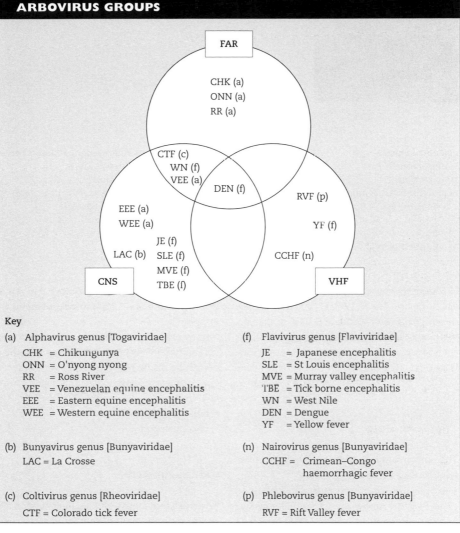

Key

(a) Alphavirus genus [Togaviridae]

    CHK = Chikungunya
    ONN = O'nyong nyong
    RR = Ross River
    VEE = Venezuelan equine encephalitis
    EEE = Eastern equine encephalitis
    WEE = Western equine encephalitis

(b) Bunyavirus genus [Bunyaviridae]

    LAC = La Crosse

(c) Coltivirus genus [Rheoviridae]

    CTF = Colorado tick fever

(f) Flavivirus genus [Flaviviridae]

    JE   = Japanese encephalitis
    SLE = St Louis encephalitis
    MVE = Murray valley encephalitis
    TBE = Tick borne encephalitis
    WN   = West Nile
    DEN = Dengue
    YF   = Yellow fever

(n) Nairovirus genus [Bunyaviridae]

    CCHF = Crimean–Congo
            haemorrhagic fever

(p) Phlebovirus genus [Bunyaviridae]

    RVF = Rift Valley fever

Fig. 40.1 Medically important arboviruses grouped according to disease syndrome (top), and listed by genus (bottom). CNS, central nervous system; FAR, fever–arthralgia–rash; VHF, viral haemorrhagic fever. Viral families are indicated by [ ].

Colorado tick fever

This is found in the Rocky Mountain states of the USA and is transmitted among small mammals by *Dermacentor* ticks. It causes CNS disease in 10% of children; haemorrhagic disease is rarer. It is easily confused with the rickettsial disease, Rocky Mountain spotted fever.

Dengue virus

This is the most common FAR arbovirus and is discussed more fully in Chapter 41.

Further reading

Solomon T. Arboviruses. In: Dawood R, ed. *Travellers' Health*, 3rd edn, pp. 151–70. London: Oxford University Press, 2002. [A leisurely discussion of concepts and diseases.]

Viral Haemorrhagic Fevers

Epidemiology, 254
Pathogenesis, 254
Management, 254
Identifying viral haemorrhagic
    fever in the febrile patient,
    254
Diagnosis, 256
Management, 258

Lassa fever, 259
South American haemorrhagic
    fevers, 259
Ebola and Marburg
    haemorrhagic fevers, 259
Haemorrhagic fever with renal
    syndrome, 260

Crimean–Congo haemorrhagic
    fever, 260
Rift Valley fever, 260
Dengue haemorrhagic fever
    and yellow fever, 260
Further reading, 260

Few diseases cause as much terror as the viral haemorrhagic fevers (VHFs), but this is often out of proportion to the actual harm they do. VHFs are caused by a diverse group of viruses from four viral families: the Arenaviridae, Filoviridae, Bunyaviridae and Flaviviridae (Table 41.1).

Epidemiology

The epidemiology can be simplified by considering three questions (Fig. 41.1).
1 How is the virus transmitted in its natural cycle—via arthropods, directly or unknown?
2 How do human index cases become infected—via insects, directly or unknown?
3 Is there direct transmission between humans to cause nosocomial spread?

Pathogenesis

The pathogenesis varies according to virus, but usually includes a combination of vascular damage, coagulopathy, immunological impairment and end-organ damage. These lead to:
• increased vascular permeability—the major pathophysiological process for most VHFs,

which allows plasma to leak from the vessels into the tissue and causes shock, oedema and effusions;
• haemorrhagic manifestations, which are sometimes relatively minor (e.g. petechiae) or can be major (e.g. gastrointestinal bleeding in Crimean–Congo haemorrhagic fever; CCHF);
• hepatic and renal failure; and
• encephalopathy.

Management

The management of VHFs includes the identification and treatment of suspected cases, limiting further spread (for the directly transmissible VHFs) and identifying others who may have been infected.

Identifying viral haemorrhagic fever in the febrile patient

Most patients with suspected VHF turn out to have malaria, typhoid or another non-transmissible disease. Unnecessary alarm can be avoided, and attention focused on likely cases of VHF, by considering the following (Fig. 41.2).

MAJOR VIRAL HAEMORRHAGIC FEVERS

Virus	Genus, family	Geographical area	Natural cycle	Human disease
Lassa	Arenavirus, Arenaviridae	Western Africa	Mastomys rodent	Human–human spread occurs. 2–15% mortality. Treat with ribavirin
Ebola and Marburg	Filovirus, Filoviridae	Sub-Saharan Africa	Unknown	Nosocomial spread common. 25–90% mortality. No antiviral treatment
Hantaan, and others (Haemorrhagic fever with renal syndrome)	Hantavirus, Bunyaviridae	Far East, Europe	Various rural rodents	No human–human spread. 1–15% mortality depending on virus. Treat severe disease with ribavirin
Crimean–Congo haemorrhagic fever	Nairovirus, Bunyaviridae	Eastern Europe, Asia, Africa	Hyalomma ticks, and livestock	Human–human spread. 15–30% mortality. Treat with ribavirin
Rift Valley fever	Phlebovirus, Bunyaviridae	Africa, Middle East	Aedes and other mosquitoes, and livestock	Human–human spread not documented, but possible. Most infections asymptomatic. 50% mortality for VHF. Treat with ribavirin
Dengue	Flavivirus, Flaviviridae	Tropics and subtropics worldwide	Aedes mosquitoes and humans	No human–human spread. Mortality < 1% with adequate fluid treatment. No antivirals
Yellow fever	Flavivirus, Flaviviridae	Africa, South America	Various mosquitoes and monkeys	No human–human spread. 20–50% mortality. No antivirals

Table 41.1 Overview of the major viral haemorrhagic fevers.

VIRAL HAEMORRHAGIC FEVERS

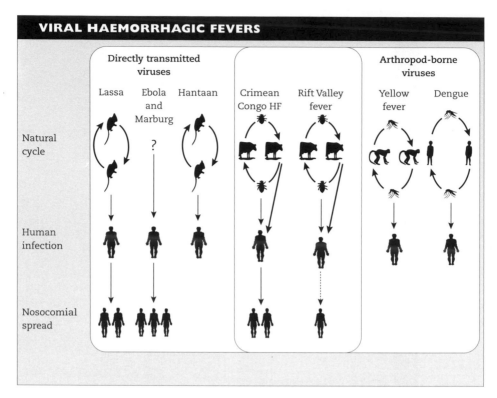

Fig. 41.1 Ecological overview of viral haemorrhagic fevers showing natural cycle, transmission to humans and potential for nosocomial spread. Note the distinction between directly transmissible viruses (Lassa, Ebola, Marburg and Hantaan), arboviruses (yellow fever and dengue), and those transmitted by both routes (Crimean Congo haemorrhagic fever and Rift Valley fever). (Modified from Solomon 2002, with permission from Elsevier Science.)

- Most VHFs are acquired in rural, rather than urban areas.
- Travel history should include details of activities that may have caused exposure.
- An interval of 3 weeks between possible exposure and onset rules out VHF.
- Most early symptoms are non-specific but certain features should ring alarm bells; e.g. pharyngitis with ulcers or causing difficulty swallowing, retrosternal chest pain, conjunctival injection and prostration.
- Haemorrhagic manifestations may not be obvious—look for petechiae in the skin folds and axillae, gum bleeding and microscopic haematuria, and perform a tourniquet test (see p. 263).
- Look—repeatedly if necessary—for a rising haematocrit (caused by plasma leakage), pleural effusions on decubitus chest X-ray, leucopenia, thrombocytopenia and proteinuria.

Diagnosis

The differential diagnosis of VHFs includes many causes of fever in the tropics (Table 41.2). Laboratory diagnosis early in the illness is by virus isolation, reverse transcriptase polymerase chain reaction (PCR) or antigen capture enzyme-linked immunoabsorbent assays (ELISAs). Subsequently, IgM and IgG ELISAs are used. Because of the infectious nature of the directly transmissible VHFs, these tests must be carried

ALGORITHM FOR IDENTIFYING VHF PATIENTS

	Directly transmissible VHF*					Non-directly transmissible VHF		
	Ebola/ Marburg	Lassa	South American VHFs	CCHF	RVF†	HFRS	DHF	Yellow fever

1. Obtain a travel history:

	Ebola/ Marburg	Lassa	South American VHFs	CCHF	RVF	HFRS	DHF	Yellow fever
Africa	+	+		+	+		+	+
Middle East				+	+		+	
Asian subcontinent				+			+	
Europe				+		+		
Far East						+	+	
Americas			+				+	+

2. Ask about activities that may have caused exposure to virus:

	Directly transmissible					Non-directly transmissible		
Exposure to human cases	Recent contact (<3 weeks) with any sick individual with unexplained fever and bleeding					–	–	–
Exposure to animal reservoir	?Monkeys ?Bats	Rodent excreta (urban) (rural)		Livestock		Rodent excreta	–	Monkeys
Activities undertaken	Jungle visits, caving	Cleaning basements, etc.	Farming, harvesting	Farming, abattoir work, rural activities		Rural, agricultural work	Urban mosquito exposure	Jungle mosquito exposure

3. Look for suggestive clinical features:

Early features	Pharyngitis	Rash	Facial oedema
	Conjunctival injection	Venepuncture oozing	Small pleural effusions
	Retrosternal chest pain	Petechial haemorrhages	Abdominal pain
	Prostration	Mucosal bleeding	Tender hepatomegaly

Late features	Shock	Haematemesis	Renal failure
	Pleural effusions	DIC	Encephalopathy
	Ascites	Hepatic failure	Acidosis
	Pericardial effusions		

4. Consider investigative findings common in VHFs:

Leucopenia	Proteinuria	Prolonged TT, APTT
Thrombocytopenia	Haematuria	Elevated transaminases
Rising haematocrit	Renal impairment	

5. If malaria film and other tests negative, and patient deteriorating despite presumptive treatment, suspect VHF:

For a directly transmissible VHF, begin isolation procedure; alert medical, nursing, laboratory, cleaning and laundry staff, public health officials	For non-transmissible VHF, ensure standard safe practices are being followed. Inform public health authorities

6. Start intravenous ribavirin if one of the following suspected:

–	+	+	+	+	+	–	–
Ebola/ Marburg	Lassa	South American VHFs	CCHF	RVF	HFRS	DHF	YF
Directly transmissible VHF					Non-directly transmissible VHF		

Fig. 41.2 Algorithm for identifying VHF patients. (Modified from Solomon 2002, with permission from Elsevier Science.) * Directly transmissible between humans. † Patients with VHF caused by RVF should be treated as infectious, although direct transmission between humans has not yet been shown. CCHF, Crimean Congo haemorrhagic fever; DHF, dengue haemorrhagic fever; HFRS, haemorrhagic fever with renal syndrome; RVF, Rift Valley fever; YF, yellow fever.

DIFFERENTIAL DIAGNOSIS OF VHF

Viral haemorrhagic fevers (in order of incidence)
Dengue haemorrhagic fever
Haemorrhagic fever with renal syndrome
Yellow fever
Lassa fever
Crimean–Congo haemorrhagic fever
Argentine, Bolivian and Venezuelan
　haemorrhagic fevers
Rift Valley fever
Omsk haemorrhagic fever and Kyasanur
　Forest disease
Ebola and Marburg haemorrhagic fevers

Treatable causes of fever with rash/haemorrhage
Parasites
　Malaria
Bacteria
　Meningococcal
　Typhoid
　Septicaemic plague
　Shigellosis
　Any severe sepsis with DIC
Rickettsia
　Tick and epidemic typhus
　Rocky Mountain spotted fever
Spirochetes
　Leptospirosis
　Borreliosis

Causes of fulminant hepatic failure
Hepatitis viruses A–E
Paracetamol and other drugs
Reye's syndrome
Alcohol

Arboviral causes of fever with rash
Alphaviruses
　Chikungunya
　O'nyong nyong
　Sindbis
Bunyaviruses
　Oropouche
Phleboviruses
　Sandfly fever
Coltiviruses
　Colorado tick fever

Non-arboviral causes of fever with rash
Entero viruses
　Coxsackie viruses
　Echoviruses
　Enteroviruses 68–71
Paramyxoviruses
　Measles
Herpesviruses
　Herpes zoster virus
　Human herpes virus 6 and 7
Orthomyxoviruses
　Influenza A and B
Rubiviruses
　Rubella

Miscellaneous
Drug reactions
Toxins
Acute surgical emergencies (upper
gastrointestinal bleeding)

DIC, disseminated intravascular coagulation.

Table 41.2 Differential diagnosis of viral haemorrhagic fevers.

out in biosafety level 4 facilities (high levels of staff protection only available in specialist centres).

Management

Encourage oral fluid intake with oral rehydration solution, and a straw if the patient cannot sit up. For patients with suspected Lassa fever, CCHF, Rift Valley fever (RVF) and haemorrhagic fever with renal syndrome (HFRS), ribavirin should be started as soon as possible (30 mg/kg loading dose, then 16 mg/kg q.d.s. for 4 days, then 8 mg/kg t.d.s. for 6 days). Hypovolaemic shock should be treated with crystalloids and colloids, and ionotropes may be needed. Pulmonary oedema and effusions are common because of the increased capillary permeability. Blood transfusions are not required in most patients, but fresh frozen plasma may be needed.

Nosocomial spread is limited by isolating the patient, strict barrier nursing (with goggles and mask), proper decontamination and disposal of clinical waste and sharps and prompt disposal of bodies by specialized burial teams. Laboratory staff must be warned about possible hazardous specimens. The risks of respiratory spread are probably negligible. Negative pressure isolation units are used in the West, but not in the African countries where most cases occur. Here implementation of the measures outlined above has dramatically reduced nosocomial transmission. 'High risk' contacts who were exposed to blood, secretions or body fluids (usually before the diagnosis is suspected) should have their temperature checked twice daily for 3 weeks. Casual contacts at low risk should be told to report if they have fever.

Lassa fever

Lassa fever is the directly transmissible VHF most likely to be seen in returning travellers, because of its wide distribution and long incubation period (5 days to 3 weeks). Lassa virus (genus *Arenavirus*, family Arenaviridae) is found across West Africa and is transmitted naturally between *Mastomys* rodents via their urine and faeces. Humans are infected by contact with these secretions (probably via inhalation of virus). Secondary human cases may occur by nosocomial spread. It is estimated that there are 100 000 cases and 5000 deaths annually, plus many inapparent infections.

Clinically, Lassa fever usually presents as a non-specific febrile illness, followed by conjunctival injection, sore throat with a pharyngeal exudate, retrosternal chest pain, vomiting and diarrhoea. Some patients progress to facial and laryngeal oedema, a mild bleeding diathesis and shock. Sensorineural deafness is a late complication in 30% of patients. The disease should be treated with ribavirin. Control measures include rodent control.

South American haemorrhagic fevers

Related arenaviruses with epidemiological and clinical similarities to Lassa are found in South America: Junin, Machupo and Guanarito viruses cause Argentine, Bolivian and Venezuelan haemorrhagic fever, respectively. Whitewater Arroya virus is a recently identified rare cause of VHF in southern USA.

Ebola and Marburg haemorrhagic fevers

These are caused by Ebola and Marburg viruses (genus *Filovirus*, family Filoviridae), which are presumed to be zoonotic, but the natural reservoir remains unknown despite intensive investigations. Marburg virus first caused disease in 1967 in laboratory workers in Marburg, Germany who were handling tissue from African green monkeys imported from Uganda. Occasional cases followed in Africa, then a large outbreak in the Democratic Republic of Congo in 1999. The first outbreak of Ebola occurred in 1976 in southern Sudan and the Democratic Republic of Congo (formerly Zaire). Subsequent outbreaks occurred in 1979, 1995 (Congo), 2000 (Uganda) and 2001–2003 (Gabon and Congo). Four biotypes have been identified: Ebola Zaire, Ebola Sudan, Ebola Cote d'Ivoire and Ebola Reston (which originated in the Philippines).

Naturally acquired human index cases of Ebola and Marburg always occur in rural areas (sometimes bat infested, and sometimes following contact with diseased primates). Secondary cases are infected by contact with blood or other fluids from primary cases and hence are mostly carers. The route of virus entry is uncertain, but is possibly via small cuts in the skin, or conjunctivae. Reuse of unsterile needles and lack of barrier nursing were important factors in early nosocomial outbreaks.

Clinically, the incubation period is 4–10 days. Patients present with a febrile illness with myalgia, abdominal pain, sore throat, herpetic lesions in the mouth and pharynx, conjunctival injection, diarrhoea and a maculopapular rash. There is sometimes bleeding from the gastrointestinal tract, nose or injection sites. Petechiae, shock and neurological manifestations can occur. The case fatality rate is 30% (Marburg) and 60–90% (Ebola). Supportive treatment only can be given. Convalescent serum from survivors may help and barrier nursing is essential.

Haemorrhagic fever with renal syndrome

This is caused by four viruses (all members of the genus *Hantavirus*, family Bunyaviridae), which are transmitted naturally between various rural rodents in their excreta.
• Hantaan virus causes epidemic HFRS in the Far East.
• Seoul virus causes a milder syndrome in the same geographical area.
• Dobrova virus causes severe HFRS in the Balkans (Europe).
• Puumula virus causes a milder variant across Scandinavia and northern Europe, with renal predominance (also called nephropathia epidemica).

Hantavirus pulmonary syndrome (a related condition with non-cardiogenic pulmonary oedema and shock) occurs in the Americas and is caused by Sin Nombre and other 'new world' hantaviruses. Humans are infected with hantaviruses by contact with rodent excreta. There is no evidence of human–human spread. Classically, HFRS patients have five phases: febrile, hypotensive phase (with haemorrhage), oliguric, diuretic and then convalescent. The illness should be treated with ribavirin. Prevention and control measures include minimizing human exposure to rodent excreta; e.g. by rodent-proofing homes. Formalin-inactivated vaccines are used in Asia.

Crimean–Congo haemorrhagic fever

This virus is transmitted naturally between animals (small mammals and livestock) by ixodid ticks, especially of the *Hyalomma* genus. It is endemic throughout Africa, Asia, the former USSR, eastern Europe and the Middle East. There have been recent outbreaks in Afghanistan, Pakistan, the Russian Federation and South Africa. Humans are infected after being bitten by, or crushing, an infected tick; or by contact with blood from infected livestock or patients (hence barrier nursing is necessary). Unlike other VHFs, major haemorrhage is more important than vascular leakage in the pathophysiology and clinical presentation of CCHF. It is treated with ribavirin and prevention is by minimizing tick exposure.

Rift Valley fever

Rift Valley fever (RVF) is endemic in the African Rift Valley and much of sub-Saharan Africa, Egypt, Saudi Arabia and Yemen. It is transmitted naturally between livestock by many mosquito species, especially *Aedes* and *Culex*. Epidemics are associated with increases in mosquito population following heavy rains, or irrigation projects. Humans are infected by mosquitoes, and by contact with animal products. Direct transmission between humans has not been documented, but barrier nursing is advisable. RVF also causes disease in sheep and cattle (abortions). Clinically, it presents as a mild febrile illness in humans; 5% have haemorrhagic manifestations, meningoencephalitis or retinitis. Ribavirin treatment is probably effective, and control measures include livestock vaccination, personal protection of workers in the livestock industry and mosquito control.

Dengue haemorrhagic fever and yellow fever

These are discussed in Chapter 42.

Further reading

Centers for Disease Control and Prevention and World Health Organization. *Infection Control for Viral Haemorrhagic Fevers in the African Health Care Set-*

ting. Atlanta: Centers for Disease Control and Prevention, 1998. [Very helpful manual for use in the field.]

Solomon T. Viral haemorrhagic fevers. In: Cook G, Zumla A, eds. *Manson's Tropical Diseases*, 21st edn. London: Saunders, 2000: 773–93. [Detailed discussion of the ecology, pathophysiology and clinical aspects of viral haemorrhagic fevers.]

CHAPTER 42

Dengue and Yellow Fever

Dengue, 262   Yellow fever, 265   Further reading, 266

While the most important viral haemorrhagic fevers numerically (dengue and yellow fever) are transmitted exclusively by arthropods, other arboviral haemorrhagic fevers (Crimean–Congo and Rift Valley fevers) can also be transmitted directly by body fluids. A third group of haemorrhagic fever viruses (Lassa, Ebola, Marburg) are only transmitted directly, and are not transmitted by arthropods at all. The directly transmissible viral haemorrhagic fevers are discussed in Chapter 41.

Dengue

Dengue virus is numerically the most important arbovirus infecting humans, with an estimated 100 million cases per year and 2.5 billion people at risk. There are four serotypes of dengue virus, transmitted by *Aedes* mosquitoes, and it is unusual among arboviruses in that humans are the natural hosts. Dengue fever ('breakbone fever') has been around for many hundreds of years; dengue haemorrhagic fever (DHF) emerged as an apparently new disease in South East Asia in the 1950s.

Epidemiology

Dengue has spread dramatically since the end of World War II, in what has been described as a global pandemic. Virtually every country between the tropics of Capricorn and Cancer is now affected (Fig. 42.1).

Factors implicated in the spread of dengue viruses include poor control of its principal vec-

tor (*Aedes aegypti*) as well as reinfestation of this insect into Central and South America (it was largely eradicated in the 1960s). Other factors include intercontinental transport of car tyres containing *Aedes albopictus* eggs, overcrowding of refugee and urban populations and increasing human travel. In hyperendemic areas of Asia, disease is seen mainly in children.

Aedes mosquitoes are 'peri-domestic': they breed in collections of fresh water around the house (e.g. water storage jars). They feed on humans (anthrophilic), mainly by day, and feed repeatedly on different hosts (enhancing their role as vectors).

Clinical features

Dengue virus may cause a non-specific febrile illness or asymptomatic infection, especially in young children. However, there are two main clinical dengue syndromes: dengue fever (DF) and dengue haemorrhagic fever (DHF).

Dengue fever

This is a classical fever–arthralgia–rash syndrome (Chapter 40), with retro-orbital pain, photophobia, lymphadenopathy and, in about 50% of patients, a rash. This is usually maculopapular, but may be mottling or flushing. In addition, there may be petechiae and other bleeding manifestations including gum, nose or gastrointestinal haemorrhage, but these do not define it as DHF according to WHO criteria—see below (Table 42.1). About one-third of patients have a positive tourniquet test (a blood pressure cuff inflated to half way between

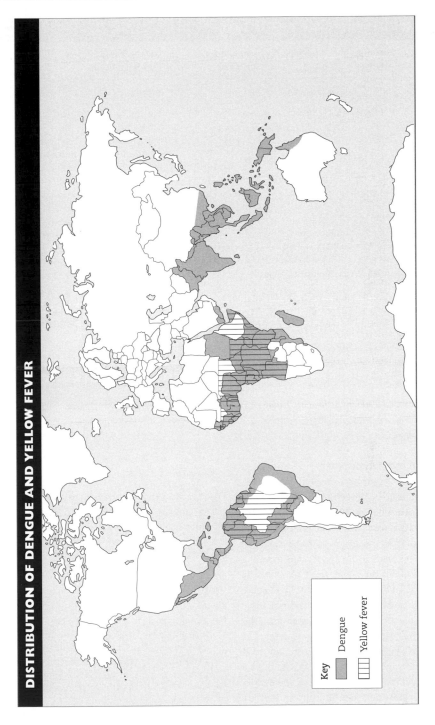

DISTRIBUTION OF DENGUE AND YELLOW FEVER

Key

Dengue

Yellow fever

Fig. 42.1 Map showing the approximate distribution of dengue and yellow fever viruses.

DISTINGUISHING DF FROM DHF

	Plasma leakage*	Platelets (/μL)	Circulatory collapse	Haemorrhagic manifestations
DF	No	Variable	Absent	Variable
DHF I	Present	< 100 000	Absent	Positive tourniquet test (or easy bruising)
DHF II	Present	< 100 000	Absent	Spontaneous bleeding† with or without positive tourniquet test
DHF III	Present	< 100 000	PP < 20 mmHg‡	Spontaneous bleeding and/or positive tourniquet test
DHF IV	Present	< 100 000	Pulse and BP undetectable	Spontaneous bleeding and/or positive tourniquet test

BP, blood pressure; DF, dengue fever; DHF, dengue haemorrhagic fever; PP, pulse pressure.
* Identified by haematocrit 20% above normal, or clinical signs of plasma leakage.
† Skin petechiae, mucosal or gastrointestinal bleeding.
‡ Pulse pressure less than 20 mmHg, or hypotension for age.

Table 42.1 World Health Organization criteria for distinguishing dengue fever (DF) and dengue haemorrhagic fever (DHF) grades I–IV. DHF grades III and IV are collectively known as dengue shock syndrome (DSS).

systolic and diastolic pressure for 5 min produces 20 or more petechiae in a 2.5-cm square on the forearm).

Dengue hemorrhagic fever

Initially, patients have a non-specific febrile illness, which may include a petechial rash. Then on the third to seventh day of illness, as the fever subsides, there is a massive increase in vascular permeability (this is the major pathophysiological process). This leads to plasma leakage from the blood vessels into the tissue, causing an elevated haematocrit, oedema and effusions. In addition, there is thrombocytopenia and haemorrhagic manifestations. If a positive tourniquet test is the only such manifestation, then this is defined as DHF grade I (Table 42.1). If there is spontaneous bleeding this is grade II. In grade III, the plasma leakage is sufficient to cause shock (defined in children as a pulse pressure < 20 mmHg). In grade IV, the blood pressure is unrecordable. Collectively, grades III and IV are known as dengue shock syndrome (DSS). Patients with DHF are restless or lethargic, and often have tender hepatomegaly or abdominal pain.

DIFFERENTIAL DIAGNOSIS OF DENGUE

Fever with arthralgia or rash
- *Arboviruses*: Chikungunya, O'nyong nyong, sindbis, West Nile, Ross River, Oropouche, sandfly fevers, Colorado tick fever.
- *Other viruses*: rubella, measles, herpes, enteroviruses.
- *Bacteria*: meningococcus, typhoid.
- *Spirochaetes*: leptospirosis, Lyme disease, relapsing fevers.
- *Rickettsiae*: tick and endemic typhus, Rocky Mountain spotted fever.
- *Parasites*: malaria.

Fever with haemorrhage
- *Arboviruses*: yellow fever, Crimean–Congo haemorrhagic fever, Rift Valley fever, Omsk haemorrhagic fever.
- *Other viruses*: hantaviruses, fulminant hepatitis (A–E); Lassa; South American haemorrhagic fevers, Ebola, Marburg.
- Any severe sepsis with disseminated intravascular coagulation (DIC).
- Drug reactions.

Box 42.1 Differential diagnosis of dengue.

Investigations

Leucopenia and thrombocytopenia are common. In the first few days of illness dengue virus can be isolated from serum or detected by polymerase chain reaction (PCR). After the fever subsides, IgM and then IgG antibodies can be detected by ELISA. New enzyme immunoassay kits allow rapid diagnosis in the field. A lateral chest X-ray may show a pleural effusion in DHF.

Management

Dengue fever

Most cases are self-limiting. Oral fluids should be encouraged, and paracetamol given. Patients may have a maculopapular recovery rash and prolonged lethargy and depression after recovery are common.

Dengue haemorrhagic fever

For grades I and II DHF, oral fluids should be encouraged, vital signs closely monitored, as well as haematocrit and platelet count, which may warn of deterioration to grades III and IV. For grades III and IV (dengue shock syndrome), central venous pressure (CVP) should be monitored if possible. Intravenous crystalloid (10–20 mL/kg/h)should be given, followed by intravenous colloid if shock persists. Patients should be watched carefully for fluid overload, and infusions reduced accordingly.

Other severe manifestations of dengue infection

These include hepatitis, or fulminant hepatic failure (Reye-like syndrome) as well as neurological complications (metabolic encephalopathy, cerebral oedema or, occasionally, viral encephalitis).

Pathogenesis of dengue haemorrhagic fever

Current evidence suggests two mechanisms may be important:
1 Antibody-dependent enhancement—antibodies against one dengue virus serotype (from a previous infection) enhance the entry of a second dengue virus into macrophages, leading to a more severe infection.

2 Viral strain differences—e.g. increased virulence of South-East Asian strains of dengue-2 virus.

Prevention

Prevention is by control of Aedes mosquitoes. Methods include treating stored water with larvicides (e.g. temephos), educating people to remove collections of water around the house (e.g. in rubbish), and spraying with insecticide during epidemics.

Future developments include tetravalent vaccines (effective against all four dengue serotypes), which are in development, for example, live attenuated vaccines and recombinant copy DNA infectious clone vaccines.

Yellow fever

Epidemiology

Yellow fever virus is naturally transmitted between primates by various mosquitoes in jungle cycles in Central America and Africa (Fig. 42.1). Aedes aegypti transmits the virus to humans in urban cycles. The disease has re-emerged in South America since the 1970s, when the Aedes eradication programme was relaxed. There are an estimated 200 000 cases, with 30 000 deaths annually.

Clinical features

The illness is biphasic and often mild. Severe disease is characterized by jaundice, fulminant hepatic failure and gastrointestinal bleeding. Faget's sign is the failure of the heart rate to increase with a rising temperature and is indicative of cardiac damage. Elevated liver function tests, leucopenia, thrombocytopenia and clotting abnormalities may occur. Liver histology reveals Councilman bodies, which also occur in Crimean–Congo haemorrhagic fever and Rift Valley fever.

Control

Yellow fever control consists of use of the highly effective 17D live attenuated vaccine. Vector control is as for dengue fever.

Further reading

Gibbons RV, Vaughan DW. Dengue—an escalating problem. *Br Med J* 2002; **324**: 1563–6.

Solomon T, Mallewa MJ. Dengue and other emerging flaviviruses. *J Infect* 2001; **42**: 104–15. [Includes important tick-borne viruses.]

World Health Organization. *Dengue Haemorrhagic Fever: Diagnosis, Treatment and Control.* Geneva: World Health Organization, 1997. [Very useful manual on dengue haemorrhagic fever.]

CHAPTER 43

Relapsing Fevers

Epidemiology, 267
Pathology, 267
Clinical features, 268

Differential diagnosis, 268
Diagnosis, 268
Treatment, 268

Prevention and control, 268
Further reading, 269

Relapsing fevers are caused by various species of *Borrelia*. They fall into two main categories: epidemic or louse-borne relapsing fever (LBRF) caused by *Borrelia recurrentis*; and endemic or tick-borne relapsing fever (TBRF) caused by numerous other species of *Borrelia*, depending on the geographical location. Untreated, these infections are characterized by a series of febrile episodes, often associated with systemic symptoms, separated by periods of relative well-being. TBRF is usually clinically milder and may be associated with up to 11 relapses, whereas LBRF is more severe but seldom gives rise to more than three relapses.

Epidemiology

LBRF, in common with many louse-borne infections, tends to occur in epidemics in situations of poor hygiene and overcrowding, such as in prisons and among refugee or displaced populations. The disease is most common in the highland regions of Ethiopia and Burundi and, to a lesser extent, in other highland areas of Africa, India and the Andes. Humans are the reservoir host. The louse, most commonly the human body louse (*Pediculus humanus*) but also occasionally the head louse (*P. capitis*) and, possibly, the crab louse (*Phthirus pubis*), becomes infected following a blood meal and remains infected for life. The louse provokes itching and is crushed when the host scratches, releasing *Borrelia*

which enter the new host via abrasions and mucous membranes. Blood-borne and congenital infections may also occur.

TBRF occurs in geographically widespread endemic foci: central, eastern and southern Africa (*B. duttonii*); north-western Africa and the Iberian peninsula (*B. hispanica*); central Asia and parts of the Middle East, India and China (*B. persica*); and various regions of the Americas (*B. hermsii*, *B. turicatae*, *B. venezuelensis*). Animal reservoirs include wild rodents, lizards, toads and owls. Transmission to humans occurs following the bite of an infected argasid (soft) tick of the genus *Ornithodorus* via tick saliva or coxal fluid. Soft ticks favour cool, relatively humid environments such as caves or the mud walls or thatch of huts. They exhibit 'transovarial transmission': vertical transmission of *Borrelia* from one tick generation to the next without further exposure to a reservoir host. Human congenital infections may also occur.

Pathology

Borreliae multiply in blood by simple fission and are taken up by the reticuloendothelial system. They have a predisposition for the liver (sometimes resulting in intrahepatic biliary obstruction), spleen and the central nervous system (CNS). Widespread vascular endothelial damage and platelet sequestration in the bone marrow occur. Myocardial and pulmonary damage are also common. Clinical severity tends to correlate with the level of spirochaetaemia. Relapses result from antigenic variation.

Clinical features

The incubation period is usually 4–8 days (range 2–15). Typically, there is a sudden onset of high fever accompanied by headache, confusion, meningism, myalgia, arthralgia, nausea, vomiting and, sometimes, dysphagia.

Dyspnoea and cough may be severe and, if productive, sputum may contain *Borrelia*. Hepatomegaly is common and is associated with jaundice in 50% of patients with LBRF and in less than 10% of those with TBRF. Splenomegaly is common and may be associated with an increased risk of rupture. Petechiae, erythematous rashes, epistaxis, conjunctival injection and haemorrhages are more common in LBRF. Complications include pneumonia, nephritis, parotitis, arthritis, cranial and peripheral neuropathies, meningoencephalitis, meningitis, acute ophthalmitis and iritis. Myocarditis may give rise to sudden and fatal arrhythmias. Most complications are more common and more severe in LBRF. Case fatality rates may reach 70% in epidemics of LBRF. In contrast, with the exception of children and pregnant women, case fatality rates rarely exceed 10% in untreated cases of TBRF.

Differential diagnosis

The differential diagnosis is wide and includes malaria, typhus, typhoid, meningococcal septicaemia/meningitis, dengue, hepatitis, leptospirosis, yellow fever and other viral haemorrhagic fevers.

Diagnosis

Borrelia are large spirochaetes measuring 10–30 ×0.2–0.5 μm. They are visible in Giemsa or Field stained blood films, and may be a surprise finding in a patient with suspected malaria. The spirochaetes are also visible unstained using darkfield or phase-contrast microscopy. They may be concentrated above the buffy coat following centrifugation of anticoagulated whole blood. The acridine orange-coated quantitative buffy coat (QBC) technique is also useful. Infected blood or cerebrospinal fluid (CSF) inoculated into mice or rats yields Borreliae in the peripheral blood after 2–3 days. Serology is unreliable. Examination of the vector may also be useful.

Treatment

A single dose of antibiotic is effective in about 95% of cases of LBRF and in up to 80% of those with TBRF. However, the usual practice is to give a 10-day course to minimize the likelihood of relapses. Effective antibiotics include tetracycline, doxycycline, penicillin, erythromycin, chloramphenicol and ciprofloxacin. The choice will depend on drug availability, age, allergies, whether the patient is pregnant and one's confidence in the diagnosis.

A potentially fatal Jarisch–Herxheimer reaction occurs in 80–90% of patients treated for LBRF and in 30–40% of those treated for TBRF. This usually follows within a few hours of the first dose of antibiotic and is characterized by intense rigors, restlessness and anxiety. The temperature rises sharply accompanied by an initial rise in pulse rate and blood pressure. This is followed by marked vasodilation and sweating, sometimes resulting in collapse and shock. Patients must be closely monitored for this complication and may require intravenous fluids to maintain blood pressure. If available, meptazinol, an opioid antagonist, should be given to reduce the severity of the reaction. Steroids are of no benefit.

Prevention and control

Prevention of LBRF is largely a matter of improving hygiene, reducing crowding and delousing. Postexposure antibiotic prophylaxis with tetracycline or doxycycline may be recommended in high-risk situations. TBRF is best prevented by avoiding tick habitats.

Further reading

Parola P, Raoult D. Ticks and tickborne bacterial diseases in humans: an emerging infectious threat. *Clin Infect Dis* 2001; **32**: 897–928. [This article reviews and illustrates various aspects of the biology of ticks and the tickborne bacterial diseases — rickettsioses, ehrlichioses, Lyme disease, relapsing fever borrelioses, tularaemia and Q fever — particularly those regarded as emerging diseases.]

Raoult D, Ndihokubwayo JB, Tissot-Dupont H *et al.* Outbreak of epidemic typhus associated with trench fever in Burundi. *Lancet* 1998; **352**: 353–8. [This epidemic highlights the appalling conditions in central African refugee camps and the failure of public health programmes to serve their inhabitants.]

CHAPTER 44

Rickettsial Infections

Louse-borne typhus, 270   African tick typhus, 271   Further reading, 271
Scrub typhus, 270

There are many species and subspecies of *Rickettsiae* that can infect humans. They may also infect rodents, and are transmitted to humans by the bites, body fluids or faeces of a variety of arthropods. The illness is very variable in intensity, but is characterized by fever and rash. There is therefore often a wide differential diagnosis. In this chapter only the three main types of typhus seen worldwide are considered: louse-borne typhus, scrub typhus and African tick typhus.

Louse-borne typhus

This is caused by *Rickettsia prowazekii*, which is transmitted to humans from the infected faeces of the human body louse, *Pediculus humanus*, usually by being scratched into the skin. Louse-borne typhus may be epidemic, and occurs particularly in malnourished migrant populations with poor hygiene (e.g. in refugee camps). The disease can occur in wide geographical areas; indeed it was common in Europe in the 19th century and was a frequent cause of death in concentration camps in World War II.

Clinical features

The disease incubates for about 12 days following which there is high fever, myalgia, headache and prostration. The conjunctivae may be suffused and delirium is common. A rash appears on about the third or fourth day—it is central and macular, although the lesions may later become petechial or purpuric. Pneumonia and/or meningoencephalitis frequently occur later, as can sometimes myocarditis. Untreated, the disease has a high mortality. Diagnosis is usually made clinically, especially in epidemic situations. The Weil–Felix serological test can still be useful. Modern, more specific serological and polymerase chain reaction (PCR) techniques are rarely available in tropical environments.

Treatment

As well as full supportive medical and nursing care, the disease usually responds well and rapidly to either tetracycline or chloramphenicol as below.

• 500 mg tetracycline four times daily (adult dose) orally or intravenously for 1 week.
• 500 mg chloramphenicol four times daily (adult dose) orally or intravenously for 1 week.
An alternative is 200 mg/day doxycycline for 7 days. Preventive measures are important in epidemics; as well as delousing procedures, 200 mg doxycycline as a single dose to all those at risk may be useful.

Scrub typhus

This is also known as mite typhus or Tsutsugamushi fever. It is caused by *Orientia tsutsugamushi* (previously known as *Rickettsia orientalis* or *Rickettsia tsutsugamushi*). It is a zoonosis of rodents, and humans are infected by the bites of infected larval mites. Scrub typhus occurs in wide parts of South East Asia, Oceania and northern parts of Australia.

Clinical features

The incubation period is 5–10 days, and a small eschar may be noted at the site of the mite bite. There is an abrupt fever, as well as headache, myalgia and prostration, as in louse-borne typhus. The rash is also similar. Lymphadenopathy may be generalized or local (related to the eschar). Hepatosplenomegaly may also occur, as may pneumonia and myocarditis. Delirium is frequently marked, although neuropsychiatric features are not as prominent as in louse-borne typhus and the overall mortality is lower. Diagnosis is usually clinically based. The Weil–Felix test is insensitive in this form of typhus.

Treatment

Tetracycline and chloramphenicol are effective, in regimens as in other forms of typhus (see above). However, the simplest and most optimal treatment is 200 mg doxycycline orally once daily for 3–7 days.

Resistance to both tetracycline and chloramphenicol has been reported in northern Thailand, and here rifampicin or ciprofloxacin may have to be used. Preventive measures include avoidance of mite-infested areas, impregnation of clothing with permethrin and prophylactic doxycycline (200 mg weekly while in high-risk areas).

African tick typhus

There are various forms of tick typhus (e.g. Rocky Mountain spotted fever, Siberian tick typhus, Queensland tick typhus, etc.). African tick typhus occurs in wide areas of Africa, but particularly central and southern parts. The causative organism is usually *Rickettsia conorii*, although in Zimbabwe it is often *Rickettsia africae* which has a reservoir in cattle, domestic cattle and even in the hippopotamus and rhinoceros. A variety of tick species are involved as both vectors and reservoirs. The infection is usually caught by travellers and campers in veld areas or grasslands.

Clinical features

The illness mimics a mild attack of scrub typhus. There is usually a noticeable eschar with local lymphadenopathy, and a mild fever with toxaemic symptoms. A central maculopapular rash later spreads to the limbs. The disease is brief and complications are rare. There is almost no mortality. Diagnosis is usually clinically based.

Treatment

Mild cases may not require treatment. If necessary, tetracycline or chloramphenicol can be used as above, or 200 mg doxycycline for 3–7 days. Preventive measures include tick-avoidance strategies such as appropriate clothing and insect repellents.

Further reading

Watt G, Parola P. Scrub typhus and tropical rickettsioses. *Curr Opin Infect Dis* 2003; **16**: 429–36. [Comprehensive and well-annotated but brief review.]

CHAPTER 45

Leptospirosis

Microbiology, 272
Reservoir of infection, 272
Epidemiology, 272

Transmission, 272
Pathology, 272
Clinical syndromes, 273

Diagnosis, 273
Treatment and prophylaxis 274
Further reading 274

Microbiology

Leptospirosis is a potentially severe infection acquired from animals. Leptospires are tightly coiled thread-like bacteria 5–20 μm in length, each end terminating in a 'hook' resembling an interrogation mark. The genus *Leptospira interrogans* is subdivided into 23 serogroups and over 200 serovars. A recent genomic classification, which is incompatible with the classical system of serogroups, has revealed that serovars are genetically heterogeneous.

Reservoir of infection

The organisms are ubiquitous in wildlife and many domestic animals. Rodents, particularly the common rat, are the main hosts. In reservoir species, leptospires persist indefinitely in the convoluted tubules of the kidney and are shed in the urine in large numbers. There are several well-known serogroup–host associations (however, note that there is much overlap):
- *Leptospira icterohaemorrhagiae*: rats.
- *Leptospira canicola*: dogs.
- *Leptospira hardjo*: cattle.
- *Leptospira pomona*: pigs.

Epidemiology

Although leptospirosis occurs worldwide, it is particularly prevalent in the tropics because warm moist conditions favour the organisms' survival. Disease incidence peaks in the summer and autumn in temperate climates, and in the rainy season in the tropics. Much infection is subclinical. The disease is relatively common in rural and farming communities where there is contact with cattle, pigs and other livestock: it is also a hazard for abattoir and sewerage workers.

Transmission

Leptospires survive well in moist conditions; they live for many days in wet soil and can persist in surface water for several weeks. Water and soil contaminated with infected animal urine are the most common sources of infection; heavy rains and flooding provide ideal conditions for disease transmission, as do flooded rice fields.

Leptospires gain entry to human hosts via skin, mucous membranes and conjunctiva. Although they may be able to penetrate intact skin, their entry is facilitated by cuts and abrasions; prolonged immersion in water will also favour invasion.

Pathology

The incubation period averages 1–2 weeks. After a relatively brief bacteraemia, leptospires are distributed throughout the body: the main organs affected are the kidneys, liver, heart, skeletal muscle, meninges and brain. The mechanism of tissue damage is uncertain as it is associated with the lysis of organisms rather than with their multiplication.

Clinical syndromes

Bacteraemic leptospirosis

Any serogroup may be responsible. Illness is non-specific; high fever is accompanied by weakness, muscle pain and sometimes diarrhoea and vomiting. Abdominal muscle tenderness may be prominent in children. Conjunctival congestion is characteristic but often absent. Illness resolves after about a week or else merges into one of the other clinical syndromes.

Aseptic meningitis

This is very difficult to distinguish from viral meningitis, and is a common presentation in children. The conjunctivae may be congested but there are no other differentiating signs. Clues to the diagnosis include a neutrophil leucocytosis, abnormal liver function tests, and the occasional presence of albumin and casts in the urine. The cerebrospinal fluid (CSF) shows a moderately elevated protein level and a normal glucose content: the cellular response is initially mixed, later lymphocytic.

Icteric leptospirosis (Weil's disease)

Severe icteric illness occurs in less than 10% of symptomatic infections. Classically associated with *Leptospira icterohaemorrhagiae* infection, Weil's disease is a dramatic life-threatening illness. Clinical features include the following.
- Conjunctival hyperaemia.
- Transient macular skin rash.
- Skin purpura and bruising.
- Muscle pain and tenderness.
- Diarrhoea and vomiting.
- Hepatomegaly and jaundice.
- Renal failure.
- Epistaxis, haematemesis and melaena.

- Bleeding into pleural, pericardial or subarachnoid spaces.
- Thrombocytopenia (50%).

There is usually little evidence of hepatic failure or encephalopathy despite deep jaundice. Renal failure is primarily caused by impaired renal perfusion and acute tubular necrosis. Weil's disease may also be accompanied by myocarditis, encephalitis and aseptic meningitis. Uveitis and iritis may appear months after apparent clinical recovery.

Pulmonary syndrome

This syndrome has been recognized in the Far East and in Central America. It is characterized by haemoptysis, patchy lung infiltrates on chest radiography and renal failure. Bilateral lung consolidation and the adult respiratory distress syndrome develop in fatal cases.

Diagnosis

In the tropics, dengue, malaria, typhoid fever, scrub typhus and hantavirus infection are important differential diagnoses.

Results of standard laboratory tests are non-specific. There is often a neutrophil leucocytosis in the peripheral blood. Severe infections are commonly accompanied by elevated blood creatine phosphokinase levels from damaged skeletal muscle, and by thrombocytopenia. The urine may contain protein, casts, neutrophils and erythrocytes. In jaundiced patients, liver function tests are mildly hepatitic in pattern and the prothrombin time may be a little prolonged.

Specific diagnostic methods are as follows.
- *Blood cultures*. Most likely to be positive in the first 10 days. Special media and prolonged incubation necessary.
- *Urine culture* (second week onwards). Urinary excretion of organisms may continue for several weeks.
- *Serological methods*. The microscopic agglutination test (MAT) uses whole organisms and is serogroup specific. Detection of a broad-based IgM response by ELISA is replacing the older

complement fixation test, and is more sensitive than MAT early in the illness.
• *Polymerase chain reaction.* Positive in blood early in the illness, and in urine from the second week and for many months.

Treatment and prophylaxis

In severe illness, general care is critically important, with special attention to fluid balance and replacement of blood loss by transfusion. Renal damage is essentially reversible and peritoneal or haemodialysis may be life-saving.

A week's treatment with either doxycycline or intravenous benzylpenicillin has been shown to shorten illness in controlled clinical trials. However, treatment may not prevent the development of renal impairment. Antibiotics may provoke a mild Jarisch–Herxheimer reaction. Uveitis is treated with antibiotics and local corticosteroids.

Controlled trials in military personnel have shown that infection with *Leptospira icterohaemorrhagiae* can be prevented by taking doxycycline 200 mg weekly.

Further reading

Levett PN. Leptospirosis. *Clin Microbiol Rev* 2001; 14: 296–326. [A wide-ranging review containing detailed information on molecular biology and diagnostic techniques.]

CHAPTER 46

Melioidosis

Epidemiology, 275
Pathogenesis, 275
Clinical features, 275

Diagnosis, 275
Treatment, 275

Prognosis, 276
Further reading, 276

Epidemiology

Melioidosis is caused by the Gram-negative bacillus, *Burkholderia pseudomallei*. In endemic areas, the organism can be easily found in the soil and surface water such as in rice paddies, but only certain strains are pathogenic to humans. Melioidosis was initially recognized as a serious problem during the Vietnam war and now causes clinical disease in a relatively geographical constrained area of South-East Asia. In Thailand, the most affected country, 3000–5000 new cases are diagnosed annually. Clinical cases are also regularly reported from Vietnam, Malaysia, Singapore and northern Australia, although sporadic cases occur over a far greater geographical area including India, China and the Caribbean.

Pathogenesis

Infection is acquired primarily by inoculation of contaminated soil or water, but may also be acquired by inhalation. Most infection is asymptomatic; organisms may remain latent within the macrophages and can cause disease many years after infection. Localized abscesses may develop at the site of inoculation which can lead to bacteraemia and dissemination of the organism. Up to 70% of patients have predisposing diseases. Diabetes mellitus is the most common, but chronic renal impairment, steroid therapy and malignancy are also important. There is no clear association with HIV infection.

Clinical features

Many individuals are found to have positive serology without having had obvious clinical symptoms. Acute presentations can be with localized or septicaemic disease. The most common form of localized disease is pneumonia, but abscesses may also be found in the skin and soft tissue or organs such as the spleen and liver. Localized disease may lead to subsequent bacteraemia. Septicaemic disease is associated with a poor prognosis: an obvious focus of disease cannot always be found. If patients survive the initial stages of septicaemic disease, dissemination can occur to cause abscesses in a number of different sites.

Diagnosis

Definitive diagnosis of melioidosis is by culture of the organism from blood or pus. Serological tests can detect rising titres of IgG or a raised specific IgM in acute infections.

Treatment

Melioidosis is both difficult and expensive to

treat. *Burkholderia pseudomallei* is intrinsically resistant to a large number of antibiotics. The best proven treatment for severe disease is parenteral ceftazidime which is sometimes combined with co-trimoxazole (trimethoprim/sulfamethoxazole). Imipenem and amoxicillin clavulanate are suitable alternatives although the latter is slightly less effective than ceftazidime. A minimum of 10 days' intravenous treatment is required. However, several weeks of intravenous therapy may be needed to produce clinical improvement in patients with visceral abscesses. The response of symptoms to treatment is slow: fever may often persist for over a week and does not imply failure of antibiotic therapy.

Oral maintenance therapy is required following completion of parenteral therapy to prevent relapse: relapse rates may reach 25% in severe disease. The most common regimen used is a combination of chloramphenicol, doxycycline and co-trimoxazole but, although this is cheap and effective, compliance is poor. Amoxicillin clavulanate may also be suitable, although it may be less effective than the four-drug combination and is more expensive. Oral maintenance therapy using the regimens above is very important in preventing relapses of melioidosis and 20 weeks' therapy are advocated to reduce the relapse rate to less than 10%. Oral regimens can also be used for the treatment of severe disease. Aggressive supportive therapy is required for individuals with septicaemic disease and abscesses should be surgically drained when feasible.

Prognosis

There is a very high mortality rate (up to 50%) in septicaemic melioidosis, even with adequate treatment. Long-term follow up is necessary to detect relapse.

Further reading

White NJ. Melioidosis. *Lancet* 2003; **361**: 1715–22. [A comprehensive review.]

CHAPTER 47

Tropical Ulcer

Clinical features, 277
Microbiology, 277
Epidemiology, 277

Treatment, 277
Prevention, 278

Complications, 278
Further reading, 278

Tropical ulcer is a term used to describe ulcers of the ankle and lower leg occurring in the tropics and subtropics that are not typical of other leg ulcers of more definitive aetiology (e.g. Buruli ulcers, diabetes, leprosy).

Clinical features

The vast majority of tropical ulcers occur below the knee, usually around the ankle. They are often initiated by minor trauma, and subjects with poor nutrition are at increased risk (Fig. 47.1, see colour plate facing p. 146). Once developed, the ulcer may become chronic and stable, but also it can run a destructive course with deep tissue invasion, osteitis and risk of amputation. Unlike Buruli ulcer (Chapter 48), tropical ulcers are typically painful.

Microbiology

There is no single agreed causative organism for tropical ulcers, although early lesions may be colonized or infected by *Bacillus fusiformis*, anaerobes and spirochaetes. Later, tropical ulcers may become infected with a wide variety of organisms, notably staphylococci and/or streptococci.

Epidemiology

Tropical ulcer is seen throughout the tropics and subtropics. No figures are available for its prevalence, but it is not seen as frequently as previously. Tropical ulcer has been described as a disease of the 'poor and hungry', and it may be that slowly improving socioeconomic conditions and nutrition account for its decline. Urbanization of populations is another factor, as tropical ulcer is usually a rural problem. More widespread use of shoes and socks also provides protection from initiating trauma. Despite this, susceptible individuals still develop tropical ulcers. Sometimes 'outbreaks' can occur; one was recorded in Tanzania in sugar cane workers (cutting the crop in bare feet). Tropical ulcer can also occur in visitors to the tropics—the disease was very common amongst Allied prisoners of war working on the Thai–Burma railway in the early 1940s. The men often suffered very severe ulcers which frequently required amputation.

Treatment

Antibiotics should be given in adequate dosages. For early ulcers, penicillin is usually sufficient, although later broad-spectrum antibiotics are likely to be needed. Improved nutrition and vitamin supplementation is helpful. The important principle of dressings is that they must be nonadherent (e.g. saline soaks, petroleum jelly impregnated gauze), otherwise they will stick to the ulcer surface and when removed they will disrupt granulation tissue. For sloughy ulcers, honey, sugar paste or paw paw (papaya) are useful inexpensive dressings. Large infected ulcers may require curettage and débridement under anaesthetic. Skin grafting can occasionally be helpful. In extreme cases, amputation may be inevitable.

Prevention

Trauma avoidance is important—in particular wearing adequate footwear. General good health and nutrition also reduce ulcer risk. Adequate and prompt treatment of ankle and leg skin breaks is also important.

Complications

• *Deep tissue invasion*—often with bone involvement, and potentially leading to amputation.

• *Chronic ulceration*—particularly if poorly treated, tropical ulcers may become chronic. In former Far East prisoners of war of World War II, they have been recorded for over 50 years since original development of the ulcer.
• *Recurrent ulceration*—in the same site may occur when a 'paper-thin' scar forms over the ulcer.
• *Squamous cell carcinoma*—may occasionally develop, usually in very chronic cases, and at the edge of the ulcer.
• *Tetanus*—by entry of tetanus bacilli through the ulcer.

Further reading

Parry E. Tropical ulcer and the rural health team. *Africa Health* 1996; **18**; 20–21. [A useful review of tropical ulcer and its individual and community effects; as well as good discussion of preventive strategies.]

CHAPTER 48

Buruli Ulcer

Microbiology, 279
Background and
  epidemiology, 279
Clinical features, 279

Differential diagnosis, 280
Investigations, 280
Management, 280

Prevention and public health
  aspects, 281
Further reading, 281

Buruli ulcer is a highly destructive ulcerating condition caused by *Mycobacterium ulcerans*. Any part of the body may be affected, particularly areas exposed to minor trauma such as the limbs. *Mycobacterium ulcerans* ranks third among mycobacterial infections affecting immunocompetent humans.

Microbiology

Mycobacterium ulcerans, a slowly growing acid- and alcohol-fast organism, belongs to a large group of environmental mycobacteria. Three different genetic strains have been identified but their relationship to virulence remains uncertain. Ulceration and necrosis are thought to be caused by toxins. There is some evidence that intercurrent helminthic infections may also predispose to ulceration.

Background and epidemiology

Buruli ulcer has been reported from several parts of Africa, notably the Buruli region of Uganda, Ghana, Papua New Guinea, the Americas, South East Asia and China. Buruli ulcer has also been described among koala bears and Australian golfers. Predominantly a disease of children, infection is thought to occur following a penetrating injury — usually minor — resulting in inoculation of the organism, which is found naturally in soil or stagnant water. It has been postulated that transmission may also follow the bite of an infected water bug. Person–person transmission is very rare.

Clinical features

A non-ulcerative lesion usually precedes ulceration. Four non-ulcerative presentations are recognized.

1 *Papule* — painless, sometimes itchy, non-tender palpable intradermal lesion (seen in Australia but rare in Africa).

2 *Nodule* — painless palpable firm lesion, 1–2 cm in diameter, situated in the subcutaneous tissue and usually attached to the skin (uncommon in Australia).

3 *Plaque* — painless well-demarcated elevated dry indurated lesion more than 2 cm.

4 *Oedematous* — diffuse extensive non-pitting swelling, ill-defined margin, firm, usually painful, with or without colour change over the affected skin.

In due course, the overlying skin breaks down and an ulcer forms with a necrotic centre, often spreading very rapidly in all directions. The following features are clinically very characteristic.
• The ulcer is usually painless.
• The skin at the edge of the ulcer is deeply undermined.
• Satellite ulcers often communicate with the original ulcer by a subcutaneous tunnel, so the

DIFFERENTIAL DIAGNOSIS OF BURULI ULCER

Papule	Nodule	Plaque	Oedema	Ulcer
Insect bites	Cyst	Leprosy	Cellulitis	Tropical ulcer
Pimple	Lipoma	Cellulitis	Elephantiasis	Tuberculosis
Herpes	Onchocercoma	Mycosis	Actinomycosis	Neurogenic ulcer
Psoriasis	Boil	Psoriasis	Osteomyelitis	Yaws
Pityriasis	Lymphadenitis	Haematoma	Onchocercoma	Syphilis
Granuloma	Mycosis	Insect bites	Necrotizing	Sickle cell disease
annulare	Leishmaniasis	Leishmaniasis	fasciitis	Cutaneous diphtheria
Leishmaniasis				Guinea worm
				Venous ulcer
				Squamous cell carcinoma
				Necrotizing fasciitis
				Leishmaniasis

Note: Infection caused by other mycobacterial organisms can be mistaken for any of the above.

Table 48.1 Differential diagnosis of Buruli ulcer. (Modified from *Diagnosis of* Mycobacterium ulcerans *disease* (Buruli ulcer). WHO/CDS/CPE/GBUI/2001.4)

skin between adjacent ulcers is often unattached to the underlying tissues. The extent of the damage is always much greater than it looks from the surface.

Regional adenitis and systemic symptoms are unusual and, if present, are suggestive of primary or secondary bacterial infection. Complications such as tetanus and primary or secondary osteomyelitis may occur. Eventually, after months or years, healing may result in scarring, ankylosis and contractures. Currently, 25% of those affected develop long-term complications that may include amputation or loss of sight. HIV infection does not appear to be associated with an altered clinical course.

Differential diagnosis

Differential diagnosis is shown in Table 48.1.

Investigations

The slough from the ulcer usually contains numerous acid-fast bacilli on Ziehl–Neelsen stain, but may be negative. Culture is time consuming, expensive and too frequently gives rise to false-positive results to make it worthwhile. Polymerase chain reaction (PCR) has been used as an epidemiological tool and is now increasingly used in diagnosis.

Management

Small preulcerative papules, nodules and plaques can be surgically excised. Necrotic ulcers should also be excised with care to remove all affected tissue by extending the margin into healthy tissue. Excision is followed by primary closure or split-skin grafting. Surgery and physiotherapy may be required for patients with contractures. Medical treatment with antimycobacterial agents is disappointing, but may be worth trying in an attempt to prevent local spread to bone, or for patients with disseminated disease. Rifampicin may promote healing in early preulcerative lesions. Continuous local heating to 40°C (e.g. by circulating water jackets) may also promote healing, even without excision.

Prevention and public health aspects

Long trousers and other mechanical barriers reduce the likelihood of infection. There is no specific vaccine available at present, although bacille Calmette–Guérin (BCG) offers some protection. A prospective vaccine candidate is the environmental mycobacterium *M. vaccae*.

The Global Buruli Ulcer Initiative, launched by the World Health Organization (WHO) in 1998 is an important initiative targeting this neglected disease. The following control strategies are being promoted.
• Health education and staff training in the communities most affected.
• Development of educational materials adapted to the needs of the countries.
• Community-based surveillance system to increase early detection and referral for treatment in collaboration with diseases such as leprosy and Guinea worm.

• Assessment of local health services and resources currently available for the diagnosis and treatment of Buruli ulcer in endemic areas.
• Strengthening of the capacity of health systems in endemic areas by upgrading surgical facilities and improving laboratories.
• Rehabilitation of those already deformed by the disease.

Further reading

van der Werf TS, van der Graaf WT, Tappero JW, Asiedu K. *Mycobacterium ulcerans* infection. *Lancet* 1999; **354**: 1013–18. [Highly recommended, this concise review includes some good clinical photographs and an excellent figure illustrating pathogenesis.]

Extensive information is also available from the WHO at http://www.who.int/gtb-buruli/ [Click on 'publications' and look for the following excellent on-line manuals: *Diagnosis of* Mycobacterium ulcerans *disease*. Portaels F, Johnson P, Meyers WM, eds. 2001; *Management of* Mycobacterium ulcerans *disease*. Buntine J, Crofts K, eds. 2001.]

CHAPTER 49

Myiasis

Tumbu fly, 282   Chiggers, 282   Body cavity myiasis, 283
Bot fly, 282

The term myiasis refers to a variety of conditions characterized by insect larvae invading the subcutaneous tissues or body cavities. There are only three common syndromes: the Tumbu fly, the Bot fly and Chiggers.

Tumbu fly

This is also sometimes known as the 'Putsi fly' in central and southern Africa. It is caused by the larvae of *Cordylobia anthropophaga*, which mostly inhabits sub-Saharan Africa. The fly lays its eggs on clothing (often on a washing line) and these hatch with body warmth when the clothes are worn. The larvae invade the skin and develop over the next 2 weeks, causing a 'blind boil'. The lesion is painful and often 'prickles' as a result of larval movement. The small dark 'head' of the boil is actually the respiratory spiracles of the larva. Multiple lesions are often present.

Treatment is to partly suffocate the larva by putting petroleum jelly or other oil or grease over the spiracles. The larva will become activated, and will partly extrude from the lesion when it can be grasped with forceps and removed intact. Care must be taken as maceration of the larva causes a severe inflammatory reaction. Prevention is by hot-ironing all clean clothes after drying.

Bot fly

This is *Dermatobia hominis*, and is found in Cen-

tral and South America. The Bot fly deposits eggs directly on the skin, rather than via clothes as does the Tumbu fly. The lesion that develops is similar; however, removal is more difficult. Occasionally, mechanical extraction of the larva can be done, but its shape often makes this difficult and incision under local anaesthetic is often needed. After infiltration of lidocaine, a cruciate incision should be made over the lesion, taking care not to incise the larva itself. Following this, extraction with forceps is usually easy. The lesions of both the Bot and Tumbu flies are usually microbiologically sterile, but sometimes secondary infection can occur, and antibiotics may be required.

Chiggers

Chiggers (or 'jiggers') are caused by *Tunga penetrans*, a flea that is widely distributed around the tropics — including much of Central and South America, Africa and the Asian Subcontinent. The gravid jigger flea invades exposed human skin — almost always the feet, and usually the interdigital clefts or the base of the toes. The flea encapsulates itself and produces eggs about 10 days later. A papular — and often later pustular — lesion develops which is painful and itchy. Excoriation helps to expel the eggs. Secondary infection and even ulceration can occur and multiple 'jigger' lesions may be present.

The flea should be carefully removed with a sterile needle, following which the lesion usually heals. Late ulcerative lesions will require antibiotics. The main aspect of prevention is good foot care and wearing shoes.

Body cavity myiasis

A variety of syndromes of myiasis exist in which various larvae invade body cavities — including wounds, urethra, vagina, anus, eye and ear. Nasal myiasis is the most common, caused usu-ally (but not always) by the Old World screw fly (*Chrysomia bezziana*). Cold-like symptoms develop, followed by nasal obstruction and epistaxis. The fly maggots can usually be seen with a nasal speculum. Application of 15% chloroform in vegetable oil to the nasal cavity causes the larva to appear, when it can be re-moved with forceps. In occasional advanced cases, invasions of the nasal sinuses and even the brain can occur.

CHAPTER 50

Cutaneous Larva Migrans

Parasitology, 284   Treatment, 284   Further reading, 285
Clinical features, 284   Prevention, 284

Cutaneous larva migrans is an intensely itchy and slowly moving linear rash under the skin of the foot and ankle. It represents the subcutaneous meanderings of invading dog hookworms, and is one of the most common exotic diseases imported to western countries (usually after tropical beach holidays).

Parasitology

The disease is caused by the larvae of animal hookworms—most commonly the dog hookworm *Ancylostoma braziliense*. Eggs are shed in the faeces of canine hosts to the soil or sand. Humans walking barefoot, or lying on the soil or sand, can become infected by larval invasion through intact skin. Humans are an incidental host, and infection represents a cul-de-sac of the life cycle. The larvae therefore travel aimlessly under the skin, causing the typical clinical eruption, until they eventually die.

Clinical features

A typical cutaneous larva migrans rash is shown in Fig. 50.1 (see colour plate facing p. 146). The rash is a very itchy serpiginous red track, which is often excoriated. The larva advances by only a few millimetres a day, so the rash is relatively static. This is in contrast to the very rapidly moving linear rash of larva currens caused by *Strongyloides stercoralis* (Chapter 52). Although the foot and ankle are by far the most common sites for cutaneous larva migrans, it can occur on other parts of the body in contact with the ground. 'Hookworm folliculitis' is an uncommon form of the disease characterized by pustular folliculitis of the buttocks.

Treatment

There is no constitutional disturbance and the rash will heal spontaneously within a few weeks. However, it is aesthetically unpleasant and the severe itch can be debilitating. Also, some larvae can survive for several months. Older treatments included local freezing of the head of the larval track with an ethyl chloride spray, or occlusive application of 10% tiabendazole in emulsifying ointment. Neither were highly effective, and oral treatments are better. Current options are as follow:
- *Albendazole*—a single dose of 400 mg is usually completely effective, and if available this is the drug of choice.
- *Ivermectin*—is also highly effective in a single dose (12 mg for adults).
- *Tiabendazole*—is less effective than albendazole and ivermectin, and also more prone to side-effects (e.g. dizziness, nausea and vomiting). The dose is 25 mg/kg twice a day for 3 days.

Prevention

Contamination of soil and sand by dog faeces is the cause of the disease. Beaches are a particular hazard, so banning dogs from beaches is an effective option (widely practised in Australia, but difficult to enforce in most developing countries). Programmes to de-worm dogs regularly will also be effective, provided the uptake is high. On an individual basis, wearing shoes or sandals on beaches helps, as well as avoiding lying on dry sand (preferably using sand washed by the tide). It can be seen that both these practices signifi-cantly detract from the pleasures of a tropical beach, and many may prefer to risk infection from these annoying but benign parasites!

Further reading

Caumes F. Treatment of cutaneous larva migrans. *Clin Infect Dis* 2000; **30**: 811–14.

CHAPTER 51

Scabies and Lice

Scabies, 286   Lice, 286   Further reading, 287

Scabies

Scabies is a common skin condition globally, but it is seen particularly frequently and severely in tropical countries. It is caused by infestation with the mite *Sarcoptes scabiei*. Infection occurs by direct skin contact, and there is often then a 4–6-week period before clinical symptoms occur. The mite burrows beneath the skin, causing an inflammatory reaction and an intensely itchy generalized rash ensues. The classical diagnostic lesion is the interdigital burrow, from which the mite can be sometimes extracted with a sterile needle. In practice, however, the diagnosis is frequently made clinically and empirical treatment given.

The classical eruption is not always seen in tropical countries. Secondary infection and/or allergic hypersensitivity to the mite can alter the rash significantly. Thus, infected papules, widespread vesicles and papular urticaria may occur. In all types of rash, excoriation often alters its appearance. The important clinical principle is to *always* think of scabies when presented with an intensely itchy generalized rash in the tropics (especially in a child). A history of nocturnal itch in other family members is good supportive evidence.

'Norwegian scabies' or 'crusted scabies' occurs sometimes in immunologically compromised patients (e.g. HIV or lepromatous leprosy), and represents a massive proliferation of infecting mites (perhaps analogous to the hyperinfection syndrome of *Strongyloides sterco-*ralis). The skin becomes scaly or 'crusted' and is frequently not as itchy as classical scabies. Crusted scabies also occurs sporadically for no obvious reason, and is relatively common in Australian aborigines.

Scabies treatment is usually topical and is applied from the neck down to all parts of the skin, with particular attention to crevices and the genitalia. For severe cases, a second treatment 5–7 days later should be given. All household contacts must be treated at the same time. The preparations available are as follow:
- *Benzyl benzoate 25%*—old-fashioned but still effective and cheap; however, best avoided in children under 4 years of age.
- *Sulphur 6% ointment*—a better alternative for young children (< 4 years).
- *Permethrin 5% cream*—more effective than benzyl benzoate and the treatment of choice, if available.
- *Ivermectin 200 µg/kg*—a single oral dose is effective in especially difficult cases.

The itch may continue for some weeks after effective treatment, and can be controlled with calamine lotion. Preventive measures for scabies include general principles of hygiene, and also the use of Tetmosol (5% tetraethylthiuram monosulphide) soap or rubbing oils.

Lice

Lice infestations of humans include *Pediculus humanus* (the body louse), *Phthirus pubis* (the pubic or crab louse) and *Pediculus capitis* (the head louse). Transmission is by close contact (usually head–head for head lice).

Pediculus humanus can transmit louse-borne relapsing fever, louse-borne typhus and trench fever, but only the local effects are considered here. Body lice cause generalized itch and often a maculopapular rash which may become secondarily infected. Pubic and head lice cause local itch and sometimes excoriation, but their importance is frequently more aesthetic than medically important.

A variety of insecticidal preparations are available for treatment. These include lotions, dusting powders or shampoos of malathion, permethrin and dichlorodiphenyl-trichloroethane (DDT). Head lice can some-times be managed physically with a 'lice comb' (a fine-toothed comb that removes the eggs or 'nits' from the shaft of the hair). This process can be aided by the use of hair conditioner.

Body lice will recur if clothing is not treated by heating to about 70°C for 30 min. It should also be remembered that there are wide geo-graphical variations in the susceptibility of lice to drug treatment, and local information on resistance patterns should be sought.

Further reading

Gibs S. Basic dermatologic treatment for tropical district hospitals. *Tropical Doctor* 1997; **27**: 142–5. [A useful practical guide to the management of common skin conditions in the tropics.]

Mneinking TL, Taplin D, Hermida JL *et al*. The treatment of scabies with ivermectin. *New Engl J Med* 1995; **333**: 26–30.

CHAPTER 52

Strongyloides stercoralis

Usual life cycle, 288

Autoinfection cycle, 288

Clinical features, 289

Hyperinfection syndrome, 289

Diagnosis, 290

Treatment, 290

Epidemiology and control, 291

Further reading, 291

Strongyloides stercoralis is a highly advanced nematode worm that inhabits the small bowel of human hosts. It occurs in widespread areas of the tropics and subtropics and has also been reported in more temperate climates (e.g. southern parts of North America, southern Europe and even the UK).

Most infections cause minor symptoms or none at all. However, because of its 'autoinfective' life cycle, strongyloidiasis can become permanently established in human hosts without the need for reinfection. In this situation, a more chronic clinical syndrome may occur. Of particular importance in such cases is the potential for fatal 'hyperinfection' if host immunity is reduced. Strongyloidiasis has been recorded in patients with HIV infection, but the association appears weak or absent, and the condition is not generally regarded as a classical HIV-associated infection.

A related worm *Strongyloides fülleborni* has been reported to infect children, and to be associated with a condition known as 'swollen belly syndrome' in young children in Papua New Guinea.

Usual life cycle

Adults live in the small intestine of humans only. The females, 2 mm long and very slender, live in the mucosa. They lay eggs that soon release microscopic larvae which usually escape at the non-infective (rhabditiform) stage in the faeces. Adult male worms are rapidly expelled and reproduction is probably usually parthenogenetic.

In the hospitable environment of warm moist soil, the larvae develop into free-living male and female worms within a week. The free-living females produce another generation of rhabditiform larvae, which develop into infective filariform larvae under certain environmental conditions. Humans are infected by penetration of the intact skin. Larvae may persist in the soil for many weeks, and the free-living cycle may be repeated many times. *Stercoralis stercoralis* is the only common soil-transmitted helminth infecting humans in which the worms can multiply in the free-living stage. After penetrating the skin, the larvae are carried to the lungs, migrate through the alveoli to reach the bronchial tree, and are swallowed to reach their normal habitat. From initial infection to maturity probably takes less than 4 weeks.

Autoinfection cycle

The rhabditiform larvae, after their release into the bowel lumen, sometimes change into the infective filariform stage. They may then reinfect the same host by either penetrating the perianal skin or the bowel wall. They then migrate through the tissues and the lungs and re-establish themselves in the intestine as new adult worms. This is how infection can persist for more than 40 years, even in the absence of external reinfection, such as in about 1 in 5 of ex-prisoners of war of the Japanese who worked on the infamous Thai–Burma railway during World War II.

Clinical features

Many infections are asymptomatic. However, both acute and chronic stages of infection can have symptoms that are quite distinct from each other. Untreated acute infections may resolve spontaneously, or become chronic because of the autoinfective cycle. Immunologically, acute strongyloidiasis is characterized by a marked IgE and blood eosinophil response, but these are less constant in the chronic form of the disease, presumably because of the host becoming immunologically tolerant.

Acute infection
1 An itchy eruption at the site of larval penetration (patients seldom recollect this).
2 Cough and wheeze because of larvae in the lungs (also uncommon).
3 Abdominal pain and diarrhoea. Pain is usually vague and ill-defined. Diarrhoea can be marked. Occasionally, steatorrhoea and even bloody diarrhoea occurs.
4 Weight loss (usually associated with diarrhoea).

Chronic strongyloidiasis
1 Larva currens ('creeping eruption'). This is a characteristic, virtually pathognomonic skin eruption (Fig. 52.1, see colour plate facing p. 146). It is caused by the migration of larvae through the skin during autoinfection. The eruption is typically:
  • a serpiginous wheal (a raised line) surrounded by a flare;
  • evanescent (comes and goes in a few hours);
  • very itchy;
  • confined to the trunk between the neck and the knees; and

  • tends to appear in crops at irregular and unpredictable intervals.
2 Intestinal symptoms. These are usually vague, taking the form of irregular bouts of looseness of the stools. Diarrhoea is not constant, and the patient may only recognize that his or her bowels were abnormal in retrospect, when the infection has been eliminated. Bloody diarrhoea is not a feature of uncomplicated chronic strongyloidiasis. Very occasionally, a 'sprue-like' syndrome of diarrhoea and weight loss occurs.

Hyperinfection syndrome

Hyperinfection syndrome is a rare complication of Strongyloides stercoralis infection — usually the chronic form of disease. It occurs when host immunity is significantly and usually abruptly reduced, allowing rapid and disseminated migration of filariform larvae into tissues not involved in the normal human life cycle. Conditions reported to be associated with hyperinfection include the following:
• systemic steroid treatment;
• other immunosuppressives treatment;
• leukaemia and lymphoma;
• postirradiation treatment;
• severe malnutrition;
• diabetic ketoacidosis; and
• lepromatous leprosy.
  Of all the causes, corticosteroid treatment is the most common. A well-reported scenario for hyperinfection is postrenal transplant (resulting from steroid and other immunosuppressant therapy).
  Features of the hyperinfection syndrome are as follows:
1 severe and often bloody diarrhoea;
2 bowel inflammation with multiple microperforations;
3 bacterial peritonitis and paralytic ileus;
4 Gram-negative septicaemia;
5 pulmonary exudates, haemoptysis, pleural effusions and hypoxia; and
6 encephalitis and bacterial meningitis.
  The diagnosis of hyperinfection is essentially clinical and a high index of suspicion is needed.

Eosinophilia is usually absent, but larvae are generally easy to find in stool and other body fluids (e.g. peritoneal and/or pleural exudates, bronchoalveolar lavage samples). Treatment is with standard antihelmintic drugs (see below) as well as full supportive therapy (e.g. fluids, antibiotics); however, the outcome is poor. For patients in endemic areas who are likely to need steroid treatment (e.g. severe asthmatics), there is a case for regular diagnostic screening for strongyloidiasis, with appropriate eradication if the disease is found.

Diagnosis

Diagnosis of strongyloidiasis is notoriously difficult. The following strategies may be useful:
- direct stool microscopy;
- stool microscopy after concentration;
- stool culture;
- microscopy of duodenal juice;
- duodenal biopsy;
- serological tests (e.g. ELISA);
- clinical suspicion; and
- therapeutic trials.

Except in hyperinfection, larvae in the stool are frequently scant and appear intermittently. This is especially true with chronic *Strongyloides* infections. Even with concentration techniques, there are well-documented cases of patients with up to 12 consecutive negative stool samples although subsequently shown to have strongyloidiasis. Stool culture with charcoal may be helpful, but duodenal aspirates can be especially helpful, although difficult to obtain. Flexible gastroduodenoscopy is an option (as well as an aspirate for microscopy, a duodenal biopsy for histological examination should be taken). However, such techniques are frequently not available in tropical countries, although an option is the Enterotest capsule (sometimes known as the 'hairy string' test). This is a small, weighted capsule containing a hairy nylon thread, the end of which is taped to the cheek while the capsule is swallowed. The capsule dissolves in the stomach, releasing the thread, which is carried through to the duodenum, usu-

ally in 2–3 h. The string is withdrawn and the part which has entered the duodenum is stained yellow with bile. The duodenal juices are squeezed into a Petri dish between the thumb and forefinger of a gloved hand, and the fluid is then examined microscopically. Finally, a variety of serological tests have been developed, the best of which is probably the ELISA method, which is highly sensitive and specific.

Despite all these techniques, diagnosis may remain in doubt, and clinical features should remain part of the diagnostic process. High eosinophilia, unexplained diarrhoea and a typical larva currens rash are all highly suggestive evidence of *Strongyloides* infection in at-risk subjects. In such cases, and even in the absence of supportive parasitology, a therapeutic trial may be worthwhile, and is often rewarding.

Treatment

There are three possible therapeutic agents. Single courses of treatment are reasonable for acute infections, but in severe and/or longstanding strongyloidiasis the course should be repeated 2 weeks later.
- *Albendazole*. This is currently the most widely used drug, as it is well-tolerated and effective. The usual recommended dose is 400 mg/day for 3 days, but there is evidence that 400 mg twice daily is more effective.
- *Ivermectin*. Recent evidence suggests that this drug is more effective than albendazole, and it is likely to become the drug of choice (where available). The dose is 200 μg/kg/day for 3 days, although single doses of 6 mg (again, daily for 3 days) have been effective. A parenteral preparation of ivermectin is available, and a case of *Strongyloides* hyperinfection syndrome has been reported to have been successfully treated with 12 mg of subcutaneous ivermectin given as a single dose.
- *Tiabendazole*. This is a more traditional treatment, but it is less effective than albendazole or ivermectin and more prone to side-effects (nausea, vomiting, dizziness and occasional neuropsychiatric problems). However, it may be the

only drug available in many developing countries. The dose is 25 mg/kg twice daily for 3 days (usually 1.5 g twice daily). It should be given as syrup, or tablets which are chewed before swallowing.

Epidemiology and control

The occurrence of strongyloidiasis in the tropics is variable — with intense infection in some parts and apparent absence in others. The variability is partly climatic — prevalence is increased in wetter and humid areas. The free-living cycle of Strongyloides does better in such conditions than, for example, hookworm. Diagnostic problems may account for the apparent absence or rarity in some areas.

Because infection enters the human host by larval soil transmission through intact skin, encouragement to wear footwear is the mainstay of control strategies. The only case of strongyloidiasis recorded in Britain was in a young woman who was in the habit of walking barefoot in the local park!

The major control method for prevention of the hyperinfection syndrome is to screen people who need, or are likely to need, steroid or immunosuppressive therapy. Asthmatics are the most common group, but others include those with ulcerative colitis, collagen vascular disease, leukaemias, lymphomas, other malignancies and those on transplant waiting lists. Amoebiasis and tuberculosis should also be screened for in such individuals. Like strongyloidiasis, these conditions also may be seriously exacerbated by immunosuppressive therapy.

Further reading

Ashford RW. *Strongyloides fülleborni kellyi:* infection and disease in Papua New Guinea. *Parasitology Today* 1992; **8**: 314–17. [A review of this 'other' strongyloid worm, and the 'swollen belly syndrome'.]

Chiodini PL, Reid AJ, Wiselka MJ, Firmin R, Foweraker J. Parenteral ivermectin in *Strongyloides* hyperinfection. *Lancet* 2000; **335**: 43–4. [An important report on the successful treatment of a case of hyperinfection with subcutaneous ivermectin.]

Grove DI, ed. *Strongyloidiasis: a Major Roundworm Infection.* London: Taylor and Francis, 1989. [Main source textbook on strongyloidiasis.]

Siddiqui AA, Berk SL. Diagnosis of *Strongyloides stercoralis* infection. *Clin Infect Dis* 2001; **33**: 1040–47. [Useful up-to-date review on diagnostic methods and their difficulties.]

Toma H, Sato Y, Shiroma Y, et al. Comparative studies of the efficacy of three antihelminthics on treatment of human *Strongyloides* in Okinawa, Japan. *South East Asia J Trop Med Public Health* 2000; **31**: 147–51. [Efficacy study of the effectiveness of drug treatment for uncomplicated *Strongyloides* infections, including albendazole and ivermectin. A 97% cure rate was found with ivermectin.]

CHAPTER 53

Guinea Worm Infection (Dracunculiasis)

Life cycle, 292

Clinical features, 292

Diagnosis, 292

Treatment, 292

Control, 294

Further reading, 294

Guinea worm infection, a subcutaneous parasitic disease caused by *Dracunculus medinensis*, was a major cause of disability in Asia and Africa but is now confined to certain African countries and is expected to become the second disease after smallpox to be eradicated by public health efforts (Fig. 53.1).

Life cycle

Larvae of guinea worm are drunk in water containing their intermediate host, the freshwater copepod (water flea) *Cyclops*. The larvae then penetrate the gut wall and develop within subcutaneous tissue into adults over about 3 months. Adult female worms grow to about 50–100 cm long and, as they become distended with millions of larvae, they migrate to dependent parts of the body after about 1 year. Here they secrete enzymes that allow them to emerge through the skin and discharge huge numbers of larvae once the skin is immersed in water. The active larvae that emerge swim for 2–3 days and must be ingested by a suitable *Cyclops* within which to develop for a further 2 weeks before they become infective to humans.

Clinical features

Patent human infections are usually highly seasonal and are often most frequent in the height of the dry season when water is scarce. Developing worms do not usually cause symptoms but as guinea worms emerge they cause burning pain that motivates patients to immerse the limb in water, thus encouraging transmission. Sometimes, emerging worms provoke allergic responses including urticaria or even asthma. A blister forms at the point of emergence; this is usually on the foot or lower leg but sometimes the arm, scrotum or indeed any part of the body. After discharging larvae, the worm dies and may gradually extrude or become absorbed. However, the process often takes many weeks and local ulceration with spreading secondary bacterial infection can cause disability for months, especially if there are multiple worms. Abscess formation is common. Worms migrating near a joint sometimes cause arthritis with effusion and, rarely, aberrant migration of a worm to the spinal cord causes paraplegia. The prolonged disability caused by guinea worm disrupts childrens' schooling and agricultural work.

Diagnosis

This is clinical. The white cloud of larvae extruded from a female worm immersed in water is characteristic. Dead calcified worms are sometimes seen on X-rays.

Treatment

There is no specific drug treatment. Courses of albendazole or metronidazole have been recommended as a means of reducing the inflam-

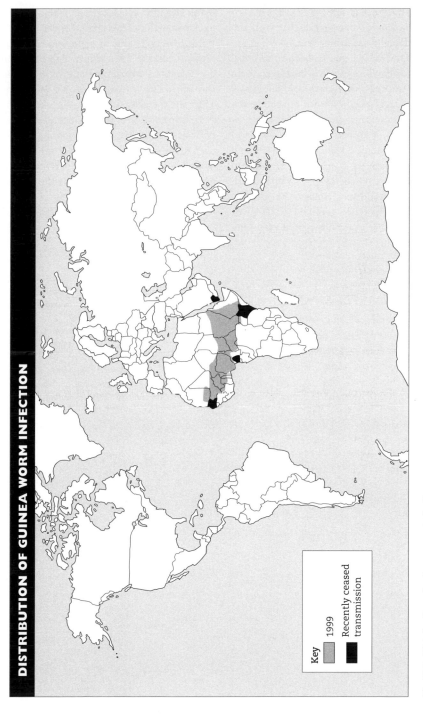

DISTRIBUTION OF GUINEA WORM INFECTION

Key

1999

Recently ceased
transmission

Fig. 53.1 Distribution of guinea worm infection.

matory response, but they are of marginal benefit. If the uterus has emerged, discharge the larvae by immersion in water and take care to dispose of the water hygienically. The traditional method is to tie the end of the emerging worm to a small stick and wind the worm out slowly over many days, taking care not to rupture the worm as this can cause severe allergic responses. Ulceration requires antiseptic dressings. Tetanus immunization should be checked and updated if necessary. Surgical removal of the worm before it emerges, by extraction through small transverse incisions, reduces the risk of infection and disability. However, surgical facilities are usually lacking in the poor villages where the disease occurs.

Control

Provision of a safe drinking water supply is the key to control. Wells must be protected to prevent contamination by people bathing infected limbs. Even straining water through cloth or unglazed pottery will filter out *Cyclops* and prevent infection. Filters made of monofilament nylon cloth for individuals, or within oil drums for a community, have proved useful in eradication projects. Efforts at prevention have been highly successful; from a peak of perhaps 50 million infections annually in the 1950s there were thought to be less than 100000 infections in 1999 and the disease has been eradicated from Asia. Transmission is now concentrated in parts of West Africa, particularly among the wartorn areas of Sudan where over two-thirds of all recorded cases occurred in 1999. WHO is hoping for global disease eradication in the not too distant future.

Further reading

Anonymous. Dracunculiasis eradication: certification of absence of transmission. *Weekly Epidemiol Record* 2000; **75**: 77–9. [A review of the WHO-led guinea worm eradication programme.]

Bloch P, Simonsen PE. Immunology of *Dracunculus medinensis* infections II. Variation in antibody responses in relation to transmission status and patency. *Am J Trop Med Hyg* 1998; **59**: 985–90. [A useful account of the relationship of antibody responses to infection and immunity.]

CHAPTER 54

Histoplasmosis

Classical histoplasmosis, 295   African histoplasmosis, 296   Further reading, 296

There are two main types of histoplasmosis. 'Classical' histoplasmosis is caused by the dimorphic fungus *Histoplasma capsulatum* var *capsulatum*, and occurs in many parts of the world outside Europe. Central and South America are areas of particularly high occurrence, although it also occurs in the USA, the Asian Subcontinent and the Far East. African histoplasmosis is caused by *Histoplasma capsulatum* var *duboisii*, and is confined to Africa — usually Central and West Africa.

Classical histoplasmosis

Clinical features

The fungus is present in bat and bird excreta, and is a particular hazard for cave explorers in endemic areas. Infection is by inhalation, following which a number of clinical syndromes can result.

• *Asymptomatic.* Many infected individuals develop no illness, but may have serological evidence of past exposure.

• *Acute pulmonary histoplasmosis.* This is a febrile bronchitic illness occurring 10–14 days after exposure. There is usually systemic malaise and myalgia, and the chest X-ray shows generalized diffuse pulmonary shadows and sometimes hilar lymphadenopathy. The illness may resolve spontaneously with no treatment.

• *Chronic pulmonary histoplasmosis.* Sometimes the disease can cause asymptomatic single or multiple pulmonary nodules, often found on routine chest radiography. More importantly, and usually in patients with underlying chronic lung damage, focal consolidation and cavitation can occur, often in the lung apices, and this can mimic pulmonary tuberculosis (cough and haemoptysis may occur).

• *Acute disseminated histoplasmosis.* In this form of the disease liver, spleen, bone marrow and lymph glands are infected. Patients are usually ill and febrile, with weight loss, lymphadenopathy and/or hepatosplenomegaly. This form of the disease is often associated with AIDS.

• *Chronic disseminated histoplasmosis.* Sometimes, years after exposure, various organ-specific syndromes resulting from histoplasmosis can present. These include oral ulceration, hypoadrenalism, meningitis and endocarditis.

Diagnosis

This is ideally made by finding the fungus in body secretions or tissues (e.g. sputum, buffy coat layer, lymph node aspirates, bone marrow samples and biopsies of liver or pulmonary nodules). If culture facilities are available, the yeast can be grown from sputum or blood. There are a variety of serological tests available, as well as an intradermal histoplasmin skin test.

Treatment

Treatment should be reserved for severe cases or immunocompromised patients. Ideally, 200–400 mg/day itraconazole should be given. Fluconazole is probably also effective, but there is less experience. Amphotericin B is more difficult and toxic to use, but it is widely available. The dosage is 0.6–1.0 mg/kg/day by slow intravenous infusion. In AIDS patients, relapses are inevitable and, if possible, patients should receive long-term itraconazole or fluconazole as secondary prophylaxis.

African histoplasmosis

The portal of entry and source of the fungus is poorly understood. The most common presentation is with skin nodules or ulcers, enlarged lymph nodes or lytic lesions in bones. Disseminated disease can occur with lung and gastrointestinal involvement. Diagnosis is usually made histologically from biopsy specimens, and treatment is as for classical histoplasmosis. Although African histoplasmosis has been recorded with HIV infection, the association is much less certain than with acute disseminated histoplasmosis caused by *Histoplasma capsulatum* var *capsulatum*.

Further reading

Wheat J, Sarosi G, McKinsey D *et al.* Practical guidelines for the management of patients with histoplasmosis. Infectious Diseases Society of America. *Clin Infect Dis* 2000; **30**: 688–695.

CHAPTER 55

Other Fungal Infections

Mycetoma (Madura foot), 297   Sporotrichosis, 298   Further reading, 299
Chromoblastomycosis, 298

Mycetoma (Madura foot)

Mycetoma is defined as chronic swelling, induration and sinus formation with the discharge of fungal grains, involving the skin, subcutaneous tissue and bone, usually of the foot. The clinical syndrome is caused by a variety of different fungi (Eumycetes) and also by aerobic actinomycete bacteria. Differentiation is important because of the differing response to treatment. *Actinomycetes* are Gram-positive organisms with branching filaments whose width is generally less than 1 mm. Fungal hyphae stain with special fungal stains (periodic acid–Schiff (PAS) or methenamine silver) and are usually more than 2 mm in diameter; chlamydospores may also be seen.

Epidemiology
These are saprophytic organisms introduced through the skin by a thorn prick. They are not contagious, but can also occur in animals. The diseases are widely distributed in tropical areas from 18°N–18°S, commonly among barefoot farmers. Some areas, particularly in the Sudan and India, have a high incidence. The chief agents of mycetoma differ in different areas (e.g. Mexico 80% *Nocardia brasiliensis*, India chiefly *Madurella mycetomatis* and *Streptomyces somaliensis*).

Clinical features
Mycetoma presents with painless (80%) swelling usually involving the foot, less commonly the hands, back or head. After several years, nodules form in the skin and break down to form discharging sinuses from which pus and coloured fungal grains emerge. There are no systemic effects unless secondary infection occurs. The condition progresses slowly and relentlessly but lymphatic spread is late and is more likely in actinomycotic infection. Eumycetomas are better circumscribed with a palpable edge, while actinomyectomas are more diffuse.

Diagnosis
The colour, size and consistency of grains obtained from deep within a sinus together with microscopy after they are crushed in 10% sodium hydroxide gives a provisional diagnosis. Cultures are necessary and are best obtained by deep biopsy and the material also sent for histology. Antibiotic sensitivities should be obtained for actinomycotic infections. Serology (immunodiffusion, counterimmunoelectrophoresis or ELISA tests) can also be used to distinguish maduromycotic from actinomycotic infections and to follow the effects of treatment. Radiological examination may show large erosions of bone, especially in eumycotic infections, while many small cavities and extensive bone sclerosis is more likely to be caused by actinomycotic disease.

Treatment
Drug treatment has generally been disappointing in true fungal infections despite organisms that are sensitive to antifungal drugs *in vitro*.

However, succesful treatment of *Madurella mycetomatis* infections with prolonged use of ketoconazole 200 mg twice daily has been recorded. Many actinomycotic infections respond to treatment with antibiotics. A useful regimen is intramuscular streptomycin 14 mg/kg/day for 1 month and then on alternate days plus co-trimoxazole (trimethoprim/sulfamethoxazole) 960 mg twice daily given for 10 months or longer. Treatment should be continued until repeat biopsies are negative for fungus and, if available, serological titres are falling.

Surgery involves the complete excision of the affected area, taking care not to rupture the 'capsule' that often surrounds the infection. Small nodules are often successfully treated in this way. Often a below-knee amputation is needed, but sometimes the heel can be conserved. In poor farmers, amputation is often best left until the limb has become useless. Recurrences are quite common after surgery and so surgery should be combined with chemotherapy whenever possible.

Chromoblastomycosis

This is warty violaceous, often ulcerated, chronic skin lesions usually involving the leg and causing itching. Further spread is by scratching or via lymphatics. The condition is usually found in Latin America or Africa. Several different fungi are responsible (e.g. *Fonsecaea pedrosoi*). They are saprophytes of wood and transmitted by skin trauma.

Diagnosis
Diagnosis is by finding the chestnut brown thick-walled fungal cells often in a wall-like pattern, or branching hyphae in skin smears or histological sections.

Treatment
Treatment is unsatisfactory. Drug treatment with 5-flucytosine 1 g orally 4 times daily combined with tiabendazole 0.5 g twice daily for about 6 months is often useful and oral itraconazole has sometimes been effective.

Surgical treatment tends to spread infection but cryosurgery is useful for small lesions and long-term use of local heat packs has been successful.

Sporotrichosis

This infection is caused by *Sporothrix schenckii*, a worldwide saprophyte of decaying vegetation, sphagnum moss and soil. It typically affects farmers and florists. Cats and other animals can inoculate the organism by scratching. An epidemic in South African Witwatersrand miners was caused by infected wooden pit props.

Clinical features
Infection manifests as a pustule or nodule typically on a finger or hand, often followed by spread along lymphatics causing nodular ulcerating lesions. Osteoarticular, disseminated and pulmonary forms are uncommon except in the immunosuppressed.

Diagnosis
Diagnosis is by microscopy and culture of material from pus, crusts or biopsies. Yeast forms, hyphae or asteroid bodies (from antigen–antibody complexes on the fungal surface) can be demonstrated, often with a background of polymorph leucocytes. The differential diagnosis of the 'sporotrichoid' lesions along lymphatics includes cutaneous leishmaniasis, nocardiosis, tuberculosis and atypical mycobacterial infection, especially *Mycobacterium marinum* (fish tank granuloma).

Treatment
The cutaneous lymphatic form responds to saturated potassium iodide orally, starting with 1 mL three times daily and rising to 15 mL/day as tolerated. Itraconazole 100 mg/day orally is a preferable alternative when it can be afforded. Treatment should be continued for at least 3 months.

Further reading

Hay RJ, ed. Tropical fungal infections. In: *Ballière's Clinical Tropical Medicine and Communicable Diseases*. London: Baillière Tindall, 1989. [An older but still useful review that includes the fungal infections mentioned above.]

Kauffman CA, Hajjeh R, Chapman SW. Practice guidelines for the management of patients with sporotrichosis. For the Mycoses Study Group, Infectious Diseases Society of America. *Clin Infect Dis* 2000; **30**: 684–7.

Welsh O, Salinas MC, Rodriguez MA. Treatment of eumycetoma and actinomycetoma. *Curr Top Med Mycol* 1995; **6**: 47–71.

CHAPTER 56

Haemoglobinopathies and Red Cell Enzymopathies

Haemoglobinopathies, 300   Enzymopathies, 302   Further reading, 302

Defects in the haemoglobin or enzymes within red cells interfere with their normal function and cause a range of clinical features depending on the type and degree of abnormality.

Haemoglobinopathies

Sickle cell disease

Sickle cell disease is the result of a β-globin chain mutation that alters the structure and function of haemoglobin. When the circulating red cells containing sickle haemoglobin encounter conditions of low oxygen tension they take up a sickle shape (Fig. 56.1, see colour plate facing p. 146). Although normal red cells have a diameter of 7 μm they are very flexible and able to squeeze through small capillaries such as those in the spleen, which are only 3 μm wide. In addition to being prematurely destroyed (haemolysed), the sickle cells get stuck in these small vessels causing microthrombi and ischaemia of the tissues. This is the basis for the anaemia, chronic organ damage and pain crises so characteristic of sickle cell disease. In an individual with sickle cell disease, 80–95% of the haemoglobin will be sickle haemoglobin (HbS) with the remainder being made up of HbF. In carriers of the sickle cell gene (sickle cell trait; HbAS) only about 30% of haemoglobin is HbS, the rest comprising predominantly HbA.

Epidemiology

The distribution of sickle cell disease follows that of malaria transmission. Carriers of the sickle cell gene have up to 10 times better protection against malaria than those with normal haemoglobin. The highest frequencies of the carrier state are generally found in Africa (e.g. up to 30% of all births are HbAS in parts of Nigeria, Ghana, Gabon and Zaire) but are also high in parts of eastern Saudi Arabia and east central India.

Clinical features

At all ages, chronic haemolysis of abnormal red cells means that sickle cell disease is associated with steady state haemoglobin levels of 6–8 g/dL. The response to this anaemia is pronounced bone marrow expansion visible as bossing of the frontal bones in the skull and overgrowth of the maxillae. In young children, clinical features include stunting, bony deformities, pain and swelling of the small bones in the hands and feet (dactylitis), acute sequestration of red cells in the spleen, aplastic crises and strokes. Before the introduction of systematic care for young children with sickle disease, the mortality in under-fives exceeded 95%. Much of this mortality can be prevented by neonatal screening programmes for sickle cell disease which enable the infants to be monitored more closely during the critical first few years of life. The lives of older children and adults are punctuated by acute severe episodes of pain in the bones of the trunk and limbs, and by repeated infections particularly by encapsulated organisms, secondary to splenic atrophy. This makes it difficult for them to achieve adequate schooling and to commit to regular employment. The carrier state, HbAS, is not normally associated with any clinical problems. Conditions that are clinically similar to sickle cell disease can result from a combination of HbS with other haemoglobin variants (e.g. HbSC, HbS thalassaemia and HbSD).

Diagnosis

Often the family history and clinical findings clearly point towards a diagnosis of sickle cell disease and during an acute crisis abundant sickled red cells can be seen on a blood film. The presence of sickle haemoglobin (e.g. HbAS, HbSS, HbSC) can be confirmed by a simple sickle slide or solubility test. Haemoglobin electrophoresis will distinguish between HbAS and HbSS.

Management

Individuals with sickle cell disease are best managed by a multidisciplinary team as they may require a variety of specialist inputs including ophthalmology, nephrology, obstetrics, orthopaedics and physiotherapy. In steady state it is usual practice to give sickle cell patients folate supplements (5 mg/day) because their high rates of haemopoiesis put them at risk of deficiency. They should also receive prophylactic penicillin (250 mg twice a day) and be monitored closely for signs of infection. Severe painful crises are best managed in hospital with intravenous fluids and adequate, often opiate, analgesia. If the crisis was precipitated by an infection this should also be treated. Because of the increased risk of thrombosis in sickle cell disease, blood transfusions should only be given for emergencies such as sequestration or aplastic crises. The replacement of the patient's blood with normal blood to reduce the level of HbS to 30% (exchange transfusion) is only beneficial for specific clinical indications such as respiratory distress syndrome or incipient stroke. The repeated microthrombi eventually cause organ failure and chronic bony deformities (Fig. 56.2, see colour plate facing p. 146).

β Thalassaemia

βThalassaemia is a genetic defect that results in insufficient production of the β-globin chains needed to make HbA. HbA comprises 97% of adult haemoglobin. The amount of β chain produced can vary from none (β0) to almost normal levels (β+) and the degree of anaemia and compensatory bone marrow overactivity determines the clinical picture.

Epidemiology

βThalassaemia is present in all ethnic groups but has the highest incidence in the Mediterranean basin (15–20%) and South East Asia and Africa (5–10%).

Clinical features

Classification of β thalassaemias is based on clinical criteria, not laboratory results.
- *β Thalassaemia trait* — clinically well with normal haemoglobin in steady state.
- *β Thalassaemia intermedia* — symptoms of anaemia (Hb 7–8 g/dl) but not transfusion dependent.
- *βThalassaemia major* — dysmorphic and transfusion dependent (Hb 2–3 g/dL).

Diagnosis

Clinical features and family history should indicate the diagnosis but definitive investigations require measurement of HbA2 levels, which are usually increased, or molecular studies. The blood film shows hypochromic microcytic red cells with more target cells than in iron deficiency. If there is severe anaemia then bone marrow overactivity may be evident by the presence of polychromatic red cells (seen as reticulocytes if special stain is used) or even nucleated red cells.

Management

This depends on the severity of the anaemia. To prevent death in β thalassaemia major in the first few years of life regular transfusions must be given. This will soon lead to iron overload and death in the second decade if these are not accompanied by an iron chelation programme which ideally involves nightly subcutaneous infusions of desferrioxamine. This is expensive treatment and may not be affordable for poorer patients.

Enzymopathies

Glucose-6-phosphate dehydrogenase deficiency

Epidemiology

The enzyme glucose-6-phosphate dehydrogenase (G6PD) is present in red cells and protects them from oxidant damage (e.g. infection, drugs, fava beans). G6PD deficiency is an X-linked genetic defect manifest predominantly in males. Early red cells have higher levels of enzyme than older cells so the degree of haemolysis, and hence clinical severity, is dependent on the quantity and half-life of the enzyme. Like HbS, G6PD deficiency has a protective effect against malaria and has its highest prevalence in the Mediterranean basin (35–40%) and Africa (25%).

Clinical features

G6PD deficiency can cause neonatal anaemia and jaundice with a risk of kernicterus. In older children and adults, oxidant stress (such as drugs — primaquine, sulphonamides, nitrofurans — infections or, in the Mediterranean variety, fava beans) causes sudden haemolysis. The severity is dependent on the levels of G6PD, which are genetically determined. The African type (G6PD A–) tends to be mild and self-limiting whereas the Mediterranean variety can cause life-threatening haemolysis.

Diagnosis

During an acute haemolytic episode caused by G6PD deficiency, the blood film appearances are very characteristic. The haemoglobin in the red cells appears to be pushed into the middle or to one side of the cell ('bite' and 'helmet' cells; Fig. 56.3, see colour plate facing p. 146). Screening tests such as the methaemoglobin reduction test can detect an 80% reduction in G6PD levels and are generally available in district hospitals where G6PD is common. Enzyme assays are the definitive test but need a specialist laboratory. Tests for G6PD deficiency should be carried out in the steady state 6–8 weeks after an acute attack. During the haemolytic episode, the cells that are deficient in G6PD are destroyed leaving only cells with normal levels of enzyme and a test carried out during acute haemolysis will therefore not detect a deficiency.

Management

As many episodes of G6PD haemolysis are self-limiting, particularly in Africa, transfusions are rarely required. In very severe cases, such as those associated with fava bean ingestion, acute renal failure necessitating dialysis may supervene. Once a patient has been identified as G6PD deficient they should be advised to avoid drugs that may precipitate a haemolytic episode (Box 56.1).

COMMON DRUGS THAT CAN CAUSE HAEMOLYSIS IN G6PD-DEFICIENT INDIVIDUALS

Analgesics — acetylsalicylic acid
Antimalarials — primaquine, pamoquine, dapsone
Antimicrobials — sulphonamides, nitrofurantoin
Others — vitamin K analogues, probenecid, PAS

Box 56.1 Common drugs that can cause haemolysis in G6PD-deficient individuals

---

Further reading

Davies SC, Cronin E, Gill M et al. Screening for sickle cell disease and thalassaemia: a systematic review with supplementary research. *Health Technol Assess* 2000; 4: iii–v, 1–99. [A detailed overview of the methods and rationale for establishing newborn screening programmes for haemoglobinopathies and of the evidence for its efficacy.]

Lewis SM, Bain BJ, Bates I. *Dacie and Lewis' Practical Haematology*, 9th edn. London: Churchill Livingstone, 2001. [Details and rationale of standard methods for laboratory diagnosis of haemoglobinopathies and red cell enzymopathies, covering a range of tests from the most basic to the most sophisticated.]

Serjeant GR. The emerging understanding of sickle cell disease. *Brit J Haem* 2001; **112**: 3–18. [Comprehensive review of all aspects of sickle cell disease with extensive list of references for specific topics.]

Weatherall DJ. Provan AB. Red cells I: inherited anaemias. *Lancet* 2000; **355**: 1169–75. [General, easily readable review of haemoglobinopathies and enzymopathies with useful list of references.]

CHAPTER 57

Haematinic Deficiencies

Iron deficiency, 304

Folate deficiency, 305

Cobalamin (vitamin B$_{12}$)
deficiency, 306

Further reading, 306

Iron deficiency

Iron deficiency is the most common cause of anaemia and affects 20–50% of the world's population. In developing countries, about half of all cases of anaemia in women and children are the result of iron deficiency. It is usually caused by excessive loss of red cell iron from the body but may also be caused by insufficient intake. Combinations of excessive loss and reduced intake are also common. Iron deficiency is a particular problem in childhood and pregnancy when physiological requirements are high. The body has very little capacity to regulate either iron absorption or iron loss. The maximum absorptive capacity of the gut for iron is about 3.5 mg/day and iron requirements in pregnancy are approximately 2.0 mg/day. Other factors that commonly coexist with iron deficiency and can contribute to anaemia include hookworm infestation, HIV infection and folate deficiency. Anaemia is one of the later manifestations of iron deficiency and tissue function can be impaired even before there is a detectable reduction in haemoglobin level. This can lead to subtle changes in behaviour, cognition and psychomotor development in children.

Clinical features

If iron deficiency develops slowly, as in chronic hookworm infestation, physiological compensation mechanisms ensure that symptoms do not become significant until the haemoglobin has reached very low levels. Physical signs specifically associated with iron-deficiency anaemia include spoon-shaped nails (koilonychia) and angular stomatitis. The degree of anaemia can be assessed from the patient's symptoms and signs, which may also indicate the cause of the iron deficiency. Common causes of excessive iron loss include menorrhagia, haemorrhoids, hookworm, bowel carcinoma, hiatus hernia and treatment with aspirin or non-steroidal anti-inflammatory drugs. The best dietary source of iron is red meat so reduced iron intake is commonly associated with a vegetarian diet and can be exacerbated by phytates and tannates in cereals and tea, which inhibit iron absorption.

Investigations

In rural health facilities where specific tests to measure iron status may not be available, a reasonably firm diagnosis can be made from a well-prepared blood film. In iron deficiency the red cells appear hypochromic (over half of the diameter of the cell is pale rather than only one-third as in normal cells) and microcytic (significantly smaller than a small lymphocyte) (Fig. 57.1, see colour plate facing p. 146). The red cells may also appear flattened like a pencil. If an automated blood count is available the mean corpuscular volume (MCV), mean corpuscular haemoglobin concentration (MCHC) and mean corpuscular haemoglobin (MCH) will all be reduced. The 'gold standard' investigation for iron deficiency is demonstration of a lack of iron stores in aspirated bone marrow. If iron studies are available they will demonstrate low serum iron with raised total iron-binding capacity and a low ferritin. Within 4–5 days of starting iron therapy, early red cells should appear in the peripheral blood. These are slightly larger and bluer (polychromatic) than normal red cells. Iron deficiency is often associated with a mildly raised platelet count that resolves as the anaemia improves. Tests to determine the cause of the iron deficiency should be performed and may include stool examination for hookworm ova or blood, and endoscopy or radiography of the gastrointestinal tract.

The diagnosis of iron-deficiency anaemia in young children and pregnancy may be particularly difficult and there is no single diagnostic test that will confirm iron deficiency. In children, ferritin may be unreliable because of the high numbers of intercurrent infections, and raised zinc protoporphyrin levels may be the best single indicator of iron deficiency. In pregnancy, the haemoglobin is affected by the physiological changes in plasma volume and red cell mass and the use of a low MCV may be misleading because of the higher proportion of larger younger red cells. For the purposes of screening in pregnancy, ferritin is probably a better single indicator of storage iron than MCV, serum iron or zinc protoporphyrin.

Management

Treatment of the iron deficiency itself comprises iron sulphate 200 mg three times a day; absorption can be enhanced by ascorbic acid. Iron supplementation should be continued for at least 3 months after a normal haemoglobin is achieved. The cause of the iron deficiency should also be rectified to prevent recurrence of the anaemia. This may involve encouraging inclusion of locally available iron-rich foods in the diet, or explaining the need for farmers to wear shoes when working in their fields to prevent hookworm infestations. Iron supplementation can reverse some developmental delays in anaemic children, even in the absence of an overall increase in haemoglobin.

Folate deficiency

Folate deficiency is usually caused by insufficient intake and less often by malabsorption. It can be exacerbated or precipitated by the excessive physiological demands for folate that occur in pregnancy and childhood, and in chronic haemolytic states. Folate is widely available in liver, yeast, spinach, green leafy vegetables and nuts but it is easily destroyed by boiling. Bone marrow stores only last a few months. Mixed iron and folate deficiencies are not uncommon and are usually caused by poor diet. The clinical and laboratory features that result from combined deficiency are a mixture of those occurring in isolated iron and folate deficiencies.

Clinical features

In addition to general symptoms of anaemia, folate deficiency can cause anorexia, change in bowel habit, glossitis and a mild haemolytic anaemia.

Investigations

Folate deficiency results in enlarged, slightly oval red cells and hypersegmented neutrophils (six or more nuclear lobes) on the peripheral blood film (Fig 57.2, see colour plate facing p. 146). This combination strongly suggests either folate or vitamin B_{12} deficiency and a bone marrow examination will show typical changes of megaloblastic anaemia. On automated blood counts, an MCV over 100 indicates macrocytosis. Serum and red cell folate assays can provide a definitive diagnosis but are not always available in rural laboratories in poorer countries. As folate is required for DNA synthesis, severe chronic deficiency can eventually cause a reduction in white cells and platelets as well as red cells (pancytopenia).

Management

Treatment comprises folic acid 5 mg/day but, as with all anaemias, the underlying cause should be corrected and appropriate advice given to avoid the overcooking of vegetables.

Cobalamin (vitamin B$_{12}$) deficiency

Deficiency of vitamin B$_{12}$ is much less common than folate deficiency as replete body stores can last for about 2 years and absorption mechanisms are efficient. Deficiency is usually the result of impaired absorption because of gastrointestinal disorders, especially those that affect the small bowel and ileum such as Crohn's disease, tuberculosis and tropical sprue. Cobalamin deficiency can occasionally be caused by lack of dietary B$_{12}$. This is particularly common in vegans and strict vegetarians.

Pernicious anaemia is a specific failure of B$_{12}$ absorption because of a lack of intrinsic factor production by gastric parietal cells. It occurs in all races and up to 30% of patients have relatives with the same disorder.

Clinical features

The typical neurological symptoms of cobalamin deficiency—posterolateral column degeneration, peripheral neuropathy and optic atrophy—can occur in the absence of anaemia. In severe cases, profound life-threatening anaemia may develop. As with folate deficiency, a reduction in platelets and white cells can occur. These abnormalities are generally mild so severe infections or bleeding episodes are unusual. Features that may indicate an underlying cause of small bowel dysfunction, such as diarrhoea or abdominal pain, should be sought. Pernicious anaemia may be associated with thyroid disease, vitiligo, Addison's disease and other autoimmune disorders, and clinical evidence of these may be apparent.

Investigations

The peripheral blood film is indistinguishable from that seen in folate deficiency. Assays of vitamin B$_{12}$ levels may be available at central laboratories and provide a definitive diagnosis. It is important to carry out investigations to determine any underlying cause.

Management

Treatment is with hydroxycobalamin injections 1 mg every 3 months after an initial loading dose of 1 mg/day for 6 days. Any underlying condition should be treated appropriately. Unless the cause of the deficiency can be eliminated, treatment will be needed for life.

Further reading

Bain B. Blood cells: A Practical Guide, 2nd edn. Oxford: Blackwell Science, 1995. [Atlas of haematological morphology with clear illustrations and accompanying text. Good guide to the use of the diagnostic laboratory.]

Saloojee H, Pettifor JM. Iron deficiency and impaired child development. Brit Med J 2001; 323: 1377–8. [Good overview of the current information about iron deficiency and child development with up-to-date references.]

Stoltfuz RJ, Kvalsvig JD, Chwaya HM et al. Effects of iron supplementation and antihelminthic treatment on motor and language development of preschool children in Zanzibar: double blind, placebo controlled study. Brit Med J 2001; 323: 1389–93. [Effect of multiple two WHO-recommended interventions to combat anaemia on the childhood development in a country with a high prevalence of severe anaemia.]

Van den Broek NR, Letsky EA, White SA, Shenkin A. Iron status in pregnant women: which measurements are valid? Brit J Haem 1998; 103: 817–24. [A critical review describing the usefulness of different methods available for measuring iron status in the context of anaemia in a developing country.]

CHAPTER 58

Bites and Stings

Snakebite, 307

Scorpion stings, 309

Spider bites, 310

Marine envenoming, 310

Further reading, 310

Snakebite

Clinical features

There are a large number of species of venomous snakes throughout the world. These can be divided into three main categories: vipers, elapids and sea snakes. The pattern of envenoming depends upon the biting species. Only 50–70% of patients bitten by venomous snakes develop signs of envenoming.

The major clinical effects following snakebite can be divided into:

1 *Local effects.* Pain, swelling or blistering of the bitten limb. Necrosis at the site of the wound can sometimes develop.

2 *Systemic effects.*
  • Non-specific symptoms: vomiting, headache, collapse.
  • Painful regional lymph node enlargement, indicating absorption of venom.
  • Specific signs:
    (a) non-clotting blood;
    (b) bleeding from gums, old wounds, sores;
    (c) neurotoxicity: ptosis, bulbar palsy and respiratory paralysis;
    (d) rhabdomyolysis; muscle pains and black urine; and
    (e) shock; hypotension, usually resulting from hypovolaemia.

Vipers most commonly cause local swelling, shock, bleeding and non-clotting blood. Elapids cause neurotoxicity and usually minimal signs at the bite site (with the exception of some cobras which also cause necrosis). Sea snakes cause myotoxicity and subsequent paresis. There are exceptions to this general rule, some vipers cause neurotoxicity and Australian elapids also cause non-clotting blood and haemorrhage.

First aid for snakebites

1 Reassure the patient. Many symptoms following snakebite are caused by anxiety.

2 Immobilize the limb. Moving the limb may increase systemic absorption of venom. Splinting is especially helpful in children.

3 Wipe the site with a clean cloth.

4 Avoid harmful manoeuvres such as cutting, suction or tourniquets.

5 Consider a pressure bandage in regions where snakebite does not cause tissue necrosis, particularly if rapid transport to hospital is not possible. This is especially important for snakes that cause neurotoxicity. A firm crêpe bandage should be applied over the bite site and wound up the limb.

6 Transport the patient to hospital as soon as possible.

7 If the snake has been killed, take it to hospital with the patient.

Diagnosis and initial assessment

Think of envenoming in unusual cases.

1 Carefully examine bitten limb for local signs.

2 Measure the pulse, respiration rate, blood pressure and urine output. The blood pressure must be watched if patients are unwell, are bleeding or have significant swelling; shock is common in viper bites.

THE 20-MIN WHOLE BLOOD CLOTTING TEST

• Place a few millilitres of freshly sampled blood in a new, clean, dry glass tube or bottle.
• Leave undisturbed for 20 min at ambient temperature.
• Tip the vessel once.
• If the blood is still liquid (unclotted) and runs out, the patient has hypofibrinogenaemia (incoagulable blood) as a result of venom-induced consumption coagulopathy.

Box 58.1 The 20-min whole blood clotting test

3 Look for non-clotting blood. This may be the only sign of envenoming in some viper bites. The 20-min whole blood clotting test (WBCT20; Box 58.1) is an extremely easy and useful test. This should be performed on admission and repeated 6 h later.

4 Look carefully for signs of bleeding, which may be subtle (gums/old wounds/sores). Bleeding internally (most often intracranial) may cause clinical signs.

5 Look for early signs of neurotoxicity; ptosis (this may be interpreted as feeling sleepy), limb weakness, or difficulties in talking, swallowing or breathing.

6 Check for muscle tenderness and myoglobin-uria in sea-snake bites.

7 Take blood for:
 • haemoglobin, white cell count and platelet count;
 • prothrombin time, activated partial thromboplastin time and fibrinogen levels if available;
 • serum urea and creatinine; and
 • creatine phosphokinase, reflecting skeletal muscle damage.

8 ECG if available.

Management

General management

All patients should be observed in hospital for 24 h, even if there are no signs of envenoming initially. They should be regularly reviewed; envenoming can develop quite rapidly. Nurse patients on their side with a slight head down tilt to prevent aspiration of blood or secretions. Avoid intramuscular injections and invasive procedures in patients with incoagulable blood. Tetanus prophylaxis should be given, but routine antibiotic prophylaxis is not required unless necrosis is present.

Antivenom

Antivenom is indicated for signs of systemic envenoming. Evidence for its efficacy in severe local envenoming is poor, but it is usually indicated if swelling extends over more than half the bitten limb. Monospecific (monovalent) antivenom can be used for a single species of snake; polyspecific (polyvalent) for a number of different species. The choice and dose of antivenom depends upon manufacturers' recommendations and local experience (see Theakston & Warrell 1991 for a list of available antivenoms; reference details in Further reading). Children require exactly the same dose as adults as the dose is dependent upon amount of venom injected, *not* bodyweight. There is little evidence to support routine prophylaxis against antivenom reactions.

• Antivenom should be diluted in 2–3 volumes of dextrose/saline and infused over an hour or so. The infusion rate should be slow initially and gradually increased.

• Adrenaline (epinephrine) should be drawn up in a syringe ready for use.

• Patients should be observed closely during antivenom administration. Common early signs of an antivenom reaction are urticaria and itching, restlessness, fever, cough or feeling of constriction in the throat.

• Patients with these signs should be treated with adrenaline (0.01 mg/kg) intramuscularly. An antihistamine, e.g. chlorpheniramine (0.2 mg/kg IM or IV) should also be given.

• Unless life-threatening anaphylaxis has occurred, antivenom can cautiously be restarted after this treatment. Routine prophylaxis against antivenom reactions is currently unproven and should not generally be used.

The response to antivenom should be monitored. In the presence of a coagulopathy, restoration of clotting depends upon hepatic resynthesis of clotting factors. The WBCT20 should be repeated 6 h after antivenom; if blood is still non-clotting, further antivenom is indicated. After restoration of coagulation, measurement of the WBCT20 should be repeated every 6 h as a coagulopathy may recur because of late absorption of venom from the bite site.

The response of neurotoxicity to antivenom is less predictable. In species with predominantly postsynaptically acting toxins, antivenom may reverse neurotoxicity; failure to do so is an indication for further doses. However, response to antivenom is poor in species with presynaptically acting toxins.

Other therapy
• Sloughs from necrotic wounds should be excised. Skin grafting may be necessary. Severe swelling may lead to a suspicion of compartment syndromes. Fasciotomy should *not* be performed unless there is definite evidence of raised intracompartmental pressure (>45 mmHg) and any coagulopathy has been corrected.
• Blood products are not necessary to treat a coagulopathy if adequate antivenom has been given.
• Endotracheal intubation should be performed to prevent aspiration if bulbar palsy develops, often obvious when difficulty in swallowing leads to pooling of secretions.
• Paralysis of intercostal muscles and diaphragm requires artificial ventilation. This can be performed by manual bagging and may need to be maintained for days, using relays of relatives if necessary.
• Anticholinesterases may reverse neurotoxicity following envenoming by some species.
• Careful fluid balance should be maintained to treat shock and prevent renal failure.
• Some cobras spit venom into the eyes of their victims. Rapid irrigation with water will prevent severe inflammation. 0.5% adrenaline drops may help to reduce pain and inflammation.

Epidemiology and prevention
Snakebite is mainly a rural and occupational haz-ard: farmers, plantation workers, herdsmen and hunter-gatherers are at greatest risk. Children also are frequently bitten as a result of their inquisitive nature. Most bites occur in the daytime and involve the foot, toe or lower leg as a result of accidentally disturbing a snake. However, some species of snake (e.g. kraits) may bite sleeping victims at night. In some areas of the world, snakebite is one of the most common causes of death and severe morbidity can result from snakebite. Sensible footwear, discouraging handling of potentially venomous animals and keeping the grass short around dwellings can all reduce the chance of snakebite.

Scorpion stings

In some areas of the world, scorpion stings are more common than snakebites and cause significant mortality. The stinging scorpion is often not seen. A number of different species have broadly similar clinical effects. The major feature of envenoming is severe pain around the bite site, which may last for many hours or even days. Systemic envenoming is more common In children and may occur within minutes of a bite. Major clinical features are caused by activation of the autonomic nervous system (Table 58.1). Severe hypertension, myocardial failure and pulmonary oedema are particularly prominent in severe envenoming.

Management
Patients should be taken to hospital immediate-

CLINICAL FEATURES OF SCORPION STINGS	
Tachypnoea	Muscle twitches and
Excessive salivation	spasms
Nausea and vomiting	Hypertension
Lachrymation	Pulmonary oedema
Sweating	Cardiac arrhythmias
Abdominal pain	Hypotension
	Respiratory failure

Table 58.1 Clinical features of scorpion stings.

ly; delay is a frequent cause of death. Control pain with infiltration of lidocaine around the wound or systemic opiates (with care). Scorpion antivenom is available for some species. It should be given intravenously in systemic envenoming, but intramuscular injection has been used with good effect. Prazosin is particularly effective for treating hypertension and cardiac failure. Severe pulmonary oedema requires aggressive treatment with diuretics and vasodilators.

Spider bites

Three genera of spiders cause significant envenoming in the tropics: widow, recluse and banana spiders. Each cause different clinical effects, but fatal envenoming is rare.

Widow spiders
Widow spiders (*Latrodectus* spp.) are found throughout the world. Severe pain at the bite site is common. Rare cases develop systemic envenoming with abdominal and generalized pain and other features resulting from transmitter release from autonomic nerves. Hypertension is characteristic of severe envenoming. Antivenom is available in some regions and is effective for relief of pain and systemic symptoms. Opiates and diazepam are also useful for treatment of pain.

Recluse spiders
Recluse spiders (*Loxosceles* sp.) have a wide distribution and cause bites in which pain develops over a number of hours. A white ischaemic area gradually breaks down to form a black eschar over 7 days or so. Healing can be prolonged and occasionally causes severe scarring. The efficacy of antivenom and other advocated treatment (dapsone, steroids, hyperbaric oxygen) remains uncertain.

Banana spiders
Banana spiders (*Phoneutria* sp.) occur only in South America. They usually cause severe burning pain at the site of the bite, but in severe cases can cause systemic envenoming with tachycardia, hypertension, sweating and priapism. A polyspecific antivenom is available in some regions.

Marine envenoming

Venomous fish
Many different venomous fish can sting patients if they are stood on or touched. Systemic envenoming is rare. Excruciating pain at the site of the sting is the major effect. Regional nerve blocks and local infiltration of lidocaine may be effective, but most marine venoms are heat labile. Immersing the stung part into hot water is extremely effective in relieving pain. Care should be taken to avoid scalding; the envenomed limb may have abnormal sensation. Clinicians should check the water temperature with their own hand. Asking the patient to also immerse the non-bitten limb may help to avoid scalding.

Jellyfish
Venomous jellyfish have a large number of stinging capsules (nematocysts) on their tentacles that inject venom when tentacles contact skin. Pain and wheals are the usual effects but, rarely, systemic envenoming can be life-threatening. Many of the nematocysts remain undischarged on tentacles that adhere to the victim and rubbing the area of the sting causes further discharge and worsens envenoming. In box jellyfish stings, pouring vinegar over the sting prevents the discharge of nematocysts. For most other jellyfish, seawater should be poured over the stings and adherent tentacles gently removed. Ice may be useful for pain relief. Box jellyfish stings can occasionally be rapidly life-threatening. Antivenom is available and can be administered intramuscularly.

Further reading

Isbister GK, Gray MR. A prospective study of 750 definite spider bites, with expert spider identification.

Quart J Med 2002; **95**: 723–31. [An up-to-date and extensive review of venomous spider bites from Australia.]

Meier J, White J. *Handbook of Clinical Toxicology of Animal Venoms and Poisons*. Florida: CRC Press, 1995. [A comprehensive summary of venomous animals and the effets of their bites.]

Theakston RDG, Lalloo DG. Venomous bites and stings. In: Zuckerman J, ed. *Principles and Practice of Travel Medicine*, pp. 321–41. Chichester: John Wiley and Sons, 2001. [A practical approach to venomous bites and stings.]

Theakston RDG, Warrell DA. Antivenoms: a list of hyperimmune sera currently available for the treatment of envenoming by bites and stings. *Toxicon* 1991; **29**: 1419–70.

Warrell DA. Injuries, envenoming, poisoning and allergic reactions caused by animals. In: Warrell DA, Cox TM, Firth JD, Benz EJ, eds, *Oxford Textbook of Medicine*, 4th edn, Section 8.2, pp. 923–46. Oxford: Oxford University Press, 2003.

CHAPTER 59

Non-Communicable Diseases

Introduction, 312   Hypertension, 318   Epilepsy, 324
Diabetes mellitus, 313   Asthma, 320   Further reading, 326

Introduction

Disease spectrum

Non-communicable diseases (NCDs) are of increasing importance all over the world, but particularly in developing countries. NCDs include all chronic disease processes, which may be treatable but are frequently not curable. By definition, they are not caused by infectious agents. Common NCDs are as follow:

- hypertension;
- cardiovascular disease;
- diabetes;
- asthma;
- epilepsy;
- stroke;
- accidents;
- cancer;
- arthritis; and
- mental disease.

The pattern of NCDs encountered in tropical countries is often different from that seen in the western world. Thus, coronary artery disease is the most important form of cardiac disease in developed countries, but in the tropics the most important is usually rheumatic heart disease. Cardiomyopathy and hypertensive heart disease are more frequent. Similarly, although chronic obstructive pulmonary disease (COPD) is seen in the tropics, asthma is usually the most common respiratory problem encountered. Road traffic accidents (RTAs) are a particularly problematic form of trauma in developing countries, as well as the effects of war and civil unrest. Malignancy patterns also vary: hepatoma, Burkitt's lymphoma, Kaposi's sarcoma, nasopharyngeal carcinoma and bladder cancer are particular tropical problems.

Mortality patterns

These are often difficult to determine in tropical countries because of diagnostic difficulties, problems of enumerating deaths outside hospital, and variable and mobile populations. In general, NCDs are globally the major cause of death, although in tropical countries infectious disease remains the major killer, as shown in Table 59.1. However, there is good evidence that the proportion of deaths caused by NCDs in developing countries is steadily increasing. Thus, in the Gambia, NCDs made up about 32% of total deaths in the 1970s but in the 1990s the proportion had risen to 49%. General projections are that, despite the HIV/AIDS epidemic, proportionate NCD mortality will overtake that caused by communicable diseases in most developing countries in the next 10–20 years. Currently, the leading causes of NCD deaths in the tropics are:

- trauma;
- heart failure;
- stroke;
- cancer; and
- diabetes.

WORLD PATTERNS OF MORTALITY

Population of mortality	Communicable disease (%)	Non-communicable disease (%)
Total world	35	65
Poorest countries	60	40
Richest countries	10	90

Table 59.1 World patterns of mortality.

URBANIZATION AND NCDS

	Rural (%)	Urban (%)
Hypertension	13	23
Diabetes	1	6
Childhood asthma	11	26

Table 59.2 Non-communicable diseases and rural–urban migration in East Africa.

COCA-COLONIZATION

- Urban–rural migration
- Adoption of 'western' lifestyles
- Increased food intake
- Reduced dietary quality
- Reduced exercise
- Increased alcohol intake
- Smoking
- Higher salt intake
- Pollution
- Family and social breakdown

Box 59.1 Coca-Colonization

Demographic transition

The reasons for the rising mortality from NCDs in tropical countries are various. Overall mortality caused by infectious disease is falling; and even in poor countries life expectancy is slowly increasing. As many NCDs increase in prevalence with rising age (e.g. diabetes and hypertension), extended life expectancy necessarily increases the rates of such diseases. Increased vehicle use and social unrest are also leading to more traumatic deaths. A further problem is that of 'epidemiological transition'. This term refers broadly to sociocultural population changes that can have profound effects on disease patterns. Population transition is a complex process, but approximately equates to what is often referred to as 'westernization' or sometimes, more light-heartedly, as 'coca-colonization' (Box 59.1).

These processes are best seen at work in the effect on tropical populations migrating to urban environments from the country. Many studies have shown that dramatic increases in NCD prevalence occur following rural–urban migration, as shown by the figures from East Africa in Table 59.2. Obesity is a major effect of rural–urban migration, and is a major risk factor for several important NCDs (in particular, accidents, hypertension and diabetes). In Africa, obesity is an especial problem in women, and in some areas up to 40% of adult urban women are significantly obese, compared with less than 5% in rural environments.

Diabetes mellitus

Epidemiology

Diabetes is an especially important NCD in the tropics as it is common, rapidly increasing, is difficult to manage adequately, and is associated with morbidity and mortality from specific acute and chronic complications. The latter two factors in particular make it different from and more problematic than other NCDs. Not surprisingly, its mortality is high: 20 years ago a study from Zimbabwe showed that 6 years after diagnosis, nearly 50% of diabetic patients had died — predominantly from metabolic problems

DIAGNOSIS OF DIABETES

- Fasting plasma glucose > 7.0 mmol/L
- Random plasma glucose > 11.1 mmol/L
- If no symptoms present, then abnormal tests are needed on two separate occasions

Box 59.2 WHO criteria for diagnosis of diabetes (WHO, 1999)

TYPE 2 DIABETES PREVALENCE

Region	Prevalence (%)
Europe	3
Africa	1–3
India	4–6
Middle East	10–30

Table 59.3 Type 2 diabetes prevalence.

(hypoglycaemia, ketoacidosis and non-ketotic hyperosmolar coma) or infections. There have been relatively few studies since, but what information there is suggests that there have been only slight improvements. However, in some communities renal failure (as a result of diabetic nephropathy) and cardiovascular disease are emerging causes of mortality. Diabetes outcome everywhere is highly dependent on local medical services, patient education and the availability of insulin and other appropriate medication.

Diagnosis

All type 1 diabetes, and much type 2 diabetes, present no diagnostic challenges, as patients present with classical symptoms (thirst, polyuria, weight loss, etc.) and obviously raised blood glucose levels. Some type 2 diabetic patients, however, have borderline values (particularly those who may be found accidentally). The World Health Organization (WHO) has recently reviewed their diagnostic criteria, as shown in Box 59.2.

A glucose tolerance test (GTT) should be rarely needed, but if so a 75-g glucose load should be used, and tests performed at 0 and 2 h only. The basal and 2-h cut-off levels are the same as for the fasting and random levels above. It should be remembered that although most European and North American laboratories have standardized to *plasma* glucose levels, many tropical laboratories still measure *blood* glucose. The values are not the same: a plasma glucose of 7.0 is equivalent to a blood glucose of 6.1 mmol/L; and a plasma glucose of 11.1 mmol/L equivalent to a blood glucose of 10.0 mmol/L.

Prevalence

Fifty years ago diabetes was thought to be rare or even non-existent in tropical countries. It was probably genuinely less common than in western countries, but high mortality prior to presentation at hospital may well have accounted for much of this 'rarity' (this remains a problem). In the last decade, type 2 diabetes in particular has reached epidemic rates all over the world, and the rate of expansion appears to be faster in developing as compared to developed countries. The total world number of people with diabetes will double in the next 10–15 years, and by 2020 diabetes will (in terms of mortality) almost certainly be the most important NCD. Actual prevalence rates of type 2 diabetes vary enormously, but Table 59.3 gives some idea of current geographical trends. Particularly high rates are seen in the elderly, and also in Asian migrants. In the tropics, urbanization and westernization are particular risk factors. These lead to increased body weight and reduced exercise, both of which are potent causes of insulin resistance, which is the hallmark of type 2 diabetes.

Type 1 diabetes appears less common than in western countries, and contributes usually less than 10% of the total diabetic population. The incidence in Africa is about 3–5 per 100 000 per year (compared to about 15 per 100 000 per year in Europe). Enumeration difficulties, as referred to previously, may partly account for this difference. Other possible explanations include genetic factors and prolonged breast-feeding (there is some evidence that delayed introduction of cow's milk protein protects against future type 1 diabetes).

Causes

The causes of diabetes in tropical countries are essentially similar to elsewhere. Type 2 diabetes can be seen as a 'lifestyle' disease, rapidly increasing as a result of excessive eating and a sedentary lifestyle. The tropical townships in particular are highly 'diabetogenic'. One currently popular explanation for this is the 'thrifty genotype' theory. This theory suggests that genetically predisposed individuals may have reduced insulin secretory capacity which is actually beneficial in a 'subsistence' or 'hunter-gatherer' situation, as ingested carbohydrate tends to be stored as fuel rather than rapidly burnt off. This 'thrifty' genotype loses its advantage when food supplies increase and exercise reduces (the urbanization situation). Here the reduced insulin reserves are overwhelmed and type 2 diabetes results. Although still theoretical, possible examples of the 'thrifty genotype' in action can be seen. Thus, in the late 1980s, a group of Africans from a famine area in northern Ethiopia were moved to Israel and within 4 years type 2 diabetes prevalence had risen from 0.1 to 8.9%.

A particular debate over the last 20 years has concerned the possible existence of a separate type of diabetes in tropical countries, distinct from type 1 and type 2 disease. This is closely associated with malnutrition — hence its usual name of malnutrition-related diabetes mellitus (MRDM). The features include the following:

- restricted to the tropics;
- variable geographical occurrence;
- young age;
- male excess;
- past or present malnutrition;
- low body weight.
- resistance to ketosis;
- sometimes, steatorrhoea; and
- sometimes, pancreatic fibrosis/calcification.

There are two possible types: one in which there is definite evidence of generalized pancreatic damage (fibrocalculous pancreatic diabetes; FCPD); and one without such features but with marked evidence of malnutrition (malnutrition-modulated diabetes mellitus; MMDM). The exact cause of these syndromes is unknown,

and even their definite status as truly separate types of diabetes is controversial.

Complications

Acute complications

These are hypoglycaemia, ketoacidosis (DKA) and hyperosmolar non-ketotic coma (HNK). They are seen more frequently in the tropics and carry a higher mortality. DKA mortality is now below 5% in Europe, but ranges from 10 to 30% or more in developing countries. Hypoglycaemia is not infrequently caused by sulphonylurea drugs; the commonly used chlorpropamide is particularly long-acting and may cause severe and prolonged hypoglycaemia.

Chronic complications

These include the classic specific complications of retinopathy, neuropathy and nephropathy, as well as the non-specific large vessel complications caused by atherosclerosis of the lower limb, coronary and cerebral arteries. Other complications include cataracts and erectile dysfunction. Diabetic complications — in particular those caused by small vessel disease — are strongly related to the degree of glycaemic control and the duration of disease. In some poorly resourced areas, the occurrence of diabetic complications may appear low; however, this may be because of inadequate surveillance or patients simply dying prematurely before complications have time to appear.

Infection

Infections are more common and more severe in diabetic patients, particularly those with poor control. Infective complications can be dramatic in the tropics — caused both by high blood glucose levels and delayed presentation. As well as standard infections involving, e.g. the chest, urinary tract and foot, some particularly dramatic and specific diabetic infections can be seen in the tropics. These include severe deep sepsis of the hand and orofacial mucormycosis.

Management

Diet

Diet and exercise are the prime treatments for type 2 diabetes, particularly when associated with obesity. Unfortunately, even with good provision of expert dietetic and patient educational support, lifestyle change rarely controls diabetes adequately. Nevertheless, even in resource-poor situations, simple but firm advice should be offered.

Oral agents

Metformin should be used for obese patients inadequately controlled on diet alone, and sulphonylureas for the non-obese with similarly inadequate control. A body mass index (BMI) of 27.0 can be used for the obese/non-obese cut-off. All sulphonylurea drugs can cause hypoglycaemia, but glibenclamide and chlorpropamide are the most problematic. Oral hypoglycaemic agents should be started in low doses, and then increased as necessary. Combination treatment (metformin plus sulphonylureas) can be used if control is poor on maximal doses of one drug, but beyond this insulin may be needed.

Insulin

All type 1 diabetic patients obviously require insulin for survival, but many type 2 patients ideally require insulin for control. If insulin is in short supply, it may need to be reserved for type 1 patients only. The potential benefits of insulin in type 2 diabetes also need to be weighed against potential hypoglycaemic risks (especially if patient self blood glucose monitoring is not available). When insulin is used, as simple and safe a system as possible should be used. A twice-daily intermediate-acting insulin system (e.g. lente or isophane) may be successful. A common misconception is that patients need a refrigerator to store their insulin. Insulin is relatively stable in all but the hottest conditions, and storage in a cool and shady area of the home is sufficient.

Ketoacidosis treatment

Successful and standard protocols are widely available for the management of DKA with intravenous insulin and fluids. However, such systems often require equipment not available in developing countries. A simple system requiring no special technology is shown in Box 59.3, which assumes that only bedside reagent-strip

TREATMENT OF KETOACIDOSIS IN RESOURCE-POOR SETTINGS

Fluids	Give 500 mL 0.9% saline quickly, then 500 mL hourly for 4–6 h
Potassium	Give none for the first hour, then 20 mmol KCl hourly for 3 h, then 10 mmol hourly for 2 h (diluted in the saline infusion)
Insulin	Give any soluble insulin 20 units i.m. stat, then 10 units i.m. hourly. If no soluble is available, give lente or isophane i.m. similarly
Bicarbonate	Only give if the patient is very ill and not improving. Give 50 mmol NaHCO$_3$ slowly i.v.
Monitoring	Blood glucose hourly by reagent strip
Later	When BG < 15 mmol/L, convert intravenous saline to 5% dextrose. Continue i.m. insulin, and change to a subcutaneous regimen when the patient can eat

Box 59.3 Treatment of ketoacidosis in resource-poor settings. BG, blood glucose

monitoring is available. In all cases of DKA it is also important to consider an infective precipitant, such as pneumonia, urinary infection, skin sepsis (e.g. foot or hand) or malaria.

Organization of care

Although the skilled use of individual drugs and insulin is important, the real challenge of diabetes management in the tropics is of delivery of care in difficult circumstances. Particular problems are:

• late presentation;
• low and irregular food supply;
• lack of insulin and oral agents;
• absence of dietitians and podiatrists;
• lack of monitoring equipment; and
• poor laboratory support.

Insulin shortage is a particular problem. It is an expensive drug in developing countries, and supplies are often poor and erratic. With all these problems, the aims of treatment may need to be compromised, with relief of symptoms and avoidance of hypoglycaemia the main aims. Diabetes is an NCD ideally suited to nurse-led and community-delivered care. The major factors in setting up a district diabetic service in a tropical country are shown in Box 59.4. Obviously, the list will need to be adapted to the geographical situation, resources available and particular clinical problems present.

Nurse-led care of type 2 diabetic patients in primary health care units is especially appropriate, and has been shown to be highly successful. Figure 59.1 shows an algorithm that has been used successfully. It assumes metformin and glibenclamide are the available drugs, but obviously must be adapted to local conditions. If metformin is not available, the sulphonylureas only must be used. Glibenclamide is commonly available, but any currently used sulphonylurea can be substituted in the protocol.

ELEMENTS OF A TROPICAL DIABETES SERVICE

- Organization and delegation
- Central hospital referral clinic
- Decentralized peripheral clinic care
- Nurse-led protocols for type 2 diabetes
- Patient and staff education system
- Medical protocols for DKA treatment
- Simple dietetic and foot care
- Sensible use of drugs and insulin
- Hypertension management
- Complication surveillance
- Gestational diabetic care

Box 59.4 Elements of a tropical diabetes service

Fig. 59.1 Nurse-led algorithm for treating patients with type 2 diabetes. Note:
1 The treatment target must be set locally depending on resources. Without laboratory support it should be 'absence from hyperglycaemic symptoms, and drug-induced hypoglycaemia'. Laboratory targets may be a fasting blood glucose < 8.0 mmol/L, or an HbA_{1c} < 8.0%.
2 Ideally, patients should see a doctor at diagnosis for complication screening.
3 Obesity can be defined as a body mass index (BMI) > 27.0.
4 The drugs are given in stepwise increments increasing each month as follows: metformin 500 mg/day and then 500 mg b.d., then 500 mg t.d.s., then 1 g b.d., then 1 g t.d.s. For glibenclamide, give 2.5 mg/day, then 5 mg/day, then 5 mg b.d., then t.d.s.
5 When using combination treatment, use the second drug according to the dose regimen above.
6 At each visit reinforce lifestyle advice, check for diabetic symptoms and possible drug side-effects (hypoglycaemia with glibenclamide, and dyspepsia or diarrhoea with metformin).
7 Weigh patient and check urine at clinic visits. Always check blood pressure and, if constantly > 140/90 mmHg, treat vigorously.

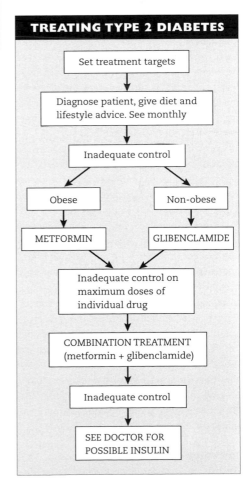

TREATING TYPE 2 DIABETES

Set treatment targets
↓
Diagnose patient, give diet and lifestyle advice. See monthly
↓
Inadequate control
↓
Obese / Non-obese
↓
METFORMIN / GLIBENCLAMIDE
↓
Inadequate control on maximum doses of individual drug
↓
COMBINATION TREATMENT (metformin + glibenclamide)
↓
Inadequate control
↓
SEE DOCTOR FOR POSSIBLE INSULIN

Hypertension

Epidemiology

Hypertension is globally by far the most common NCD, and in the tropics it makes up approximately 40% of the total NCD burden. The major importance of hypertension is that it is a potent risk factor for stroke and, because of the frequency of raised blood pressure (BP), it leads to a very high population stroke risk. Effective treatment of hypertension lowers risk by at least 40%, and such treatment can involve relatively simple and inexpensive drugs. Sadly, however, much hypertension remains undiagnosed, and of those cases diagnosed many are inadequately controlled.

Diagnosis

Population BP levels are continuously distributed, and there is therefore no clear dividing line between hypertensive and non-hypertensive levels. Diagnostic cut-off levels are therefore decided on the basis of risk, and these levels have been progressively decreasing. Until recently, WHO defined hypertension as a BP level of over 160/95 mmHg, but this has been revised to over 140/90 mmHg. There is growing evidence that in certain high-risk groups (notably those with diabetes) the cut-off level should be lower—probably 130/85 mmHg.

Accurate measurement is important in diagnosing hypertension. A good quality mercury sphygmomanometer, with an adequate-sized and well-fitting cuff, remains the 'gold standard' system. However, the use of such equipment is declining, as a number of countries are restricting their use because of concerns over mercury toxicity. The alternative electronic machines are expensive and difficult to maintain. Aneroid models are a reasonable compromise, although they can read slightly lower than other systems. Overall, the major problem in accurate BP measurement is failure to use a large-sized cuff in an obese patient.

Prevalence

As may be expected, prevalence rates vary widely geographically. In some poor rural areas of the tropics, where obesity is rare, rates of well below 5% are found. Conversely, in many urban areas of sub-Saharan Africa, where obesity is very common, hypertension may be found in 40–50% of the population or more. Most tropical doctors will find themselves working in areas with hypertension prevalence rates of 10–20% at least. It should always be remembered that the prevalence of *known* hypertension (those on antihypertensive medication) always considerably underestimates the total hypertension prevalence—often by about one-half. In view of this, and the essentially asymptomatic nature of the disease, opportunist hypertension screening should be undertaken whenever possible (e.g. at inpatient or outpatient hospital attendances, regardless of the reason for the consultation).

Causes and risk factors

About 95% of cases of hypertension are 'essential' (with no definable underlying cause). Secondary hypertension is always rare. Causes include Cushing's syndrome, Conn's syndrome, phaeochromacytoma and a variety of renal disorders. A number of causes have been suggested for essential hypertension, including genetic factors, fetal malnutrition (the 'fetal origins' hypothesis), salt retention and subtle abnormalities of the renin–aldosterone system. It is more useful to think in terms of risk factors for essential hypertension, rather than actual causes. In the tropics, the main risk factors are as follows:
- obesity (especially central);
- high salt intake;
- urbanization;
- excess alcohol intake; and
- reduced activity.

It can be seen that these make up a 'package' of adverse lifestyle factors similar to those predisposing to urban diabetes. Reduced activity and central obesity leads to insulin resistance, and this is strongly associated with hypertension, as well as diabetes. Increased salt intake is probably a major feature of tropical urbanization, and some individuals may be particularly prone to the potential hypertensive effect of increased dietary salt.

Complications

The long-term result of uncontrolled hypertension can be a variety of complications resulting from end-organ damage. These are as follows:

- stroke;
- left ventricular hypertrophy;
- renal failure;
- coronary artery disease; and
- hypertensive retinopathy.

Of these, stroke is by far the most important, with hypertension leading to an excess risk of up to 10 times that in non-hypertensive subjects. Stroke in hypertensive patients can be caused by cerebral thrombosis as well as cerebral haemorrhage. Left ventricular hypertrophy is also a serious complication, which may lead to hypertensive heart failure. Hypertensive renal disease is one of the most common causes of chronic renal failure.

A rare acute complication is the 'hypertensive crisis' that occurs in severe and accelerated disease. Patients usually have a diastolic BP in excess of 140 mmHg, encephalopathy and hypertensive retinopathy (usually with papilloedema).

Management
Principles of treatment

Treatment of hypertension should follow a logical pathway, bearing in mind some important basic principles of management.

- *Non-drug treatment* is important. This includes weight reduction, reduced salt intake, reduced alcohol, avoidance of smoking, and increased exercise.
- *Compliance* is a major barrier to good BP control. Try to use once daily drug systems and be sensitive to possible side-effects.
- *Side-effects* can be minimized by using low doses of more than one drug, rather than very high doses of a single drug.
- *Organization* of hypertension care is very important, as the number of patients involved is likely to be high.

Drug treatment

Patient education and non-drug treatments should always be offered first, but most patients will require some form of long-term drug treatment. This should follow a logical cascade of treatment, adding second or third drugs if control is not reached (Box 59.5).

Thiazides and beta-blockers have a firm evidence base for reducing mortality. They are cheap and generally widely available. Side-effects are not usually troublesome unless high doses are used. Glucose intolerance, dyslipidaemia and hyperuricaemia do not occur with low-dose thiazides (and higher doses do not give added antihypertensive effect). Beta-blockers must *not* be used in asthma, and should be used with extreme caution in peripheral vascular disease and heart failure. Unfortunately, beta-blockers are relatively ineffective in black people, and are best avoided (step 4 in Box 59.4 should be omitted in such cases). The same is true of angiotensin-converting enzyme (ACE) inhibitors *alone*, but not when used in combination with thiazides.

Third-line drugs depend on local availability and costs. It should be remembered that modern ACE inhibitors and calcium-channel blockers are very expensive, and apart from specific

---

HYPERTENSION DRUG TREATMENT

1 *Confirm diagnosis.* At least two BP levels > 140/90 mmHg
2 *Non-drug treatment*, e.g. weight reduction, reduced salt intake
3 *Thiazides.* Low dose, e.g. bendrofluazide 2.5 mg/day or hydrochlorthiazide 25 mg/day
4 *Beta-blockers,** e.g. atenolol 50–100 mg/day, or propranolol 40–80 mg b.d. or t.d.s.
5 *Third-line drug*:
- calcium blocker
- ACE inhibitor*
- alpha-blocker
- others, e.g. methyldopa, reserpine, hydralazine

* Ineffective in black people (see text)

Box 59.5 Hypertension drug treatment

indications (see below), have no special benefit over cheaper traditional treatments. Methyldopa, reserpine and hydralazine remain good, cheap and effective treatments. Side-effects are also rarely encountered.

Special situations

There are specific clinical situations where the standard progression of treatment (first-line, second-line, etc.) should not be followed, as specific drugs may be indicated or contraindicated. Beta-blockers should be avoided in asthma, but they are ideal first-line drugs in the presence of angina. Similarly, ACE inhibitors (if available) are ideally suited to patients with heart failure, or diabetes with renal complications (nephropathy or microalbuminaemia).

Hypertensive crisis

Patients should be bed-rested and observed closely, with hourly BP measurements. A simple, but often surprisingly effective treatment is simply to give methyldopa orally 500 mg 4-hourly. Otherwise, hydralazine 10 mg i.m. every 2–4 h is usually effective.

Organization of care

As with type 2 diabetes, routine hypertension management is ideally suited to primary health clinic care by suitably trained nurses. In most areas, there are large numbers of hypertensive patients, and such a system will allow medical staff to concentrate on more difficult cases. Again, as with diabetes, the protocol needs to be adapted to local needs and drug supplies. Figure 59.2 assumes that a thiazide drug and methyldopa are the main available, cheap and effective medications in use.

Initial patient education is important, and should carry two messages. First, advice should be given on non-drug aspects of treatment. Secondly, the nature and importance of hypertension should be carefully explained. In most cases hypertension is asymptomatic, and patients will not comply with long-term treatment unless they understand the benefits.

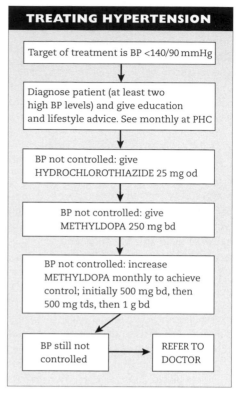

TREATING HYPERTENSION

Target of treatment is BP <140/90 mmHg

↓

Diagnose patient (at least two high BP levels) and give education and lifestyle advice. See monthly at PHC

↓

BP not controlled: give HYDROCHLOROTHIAZIDE 25 mg od

↓

BP not controlled: give METHYLDOPA 250 mg bd

↓

BP not controlled: increase METHYLDOPA monthly to achieve control; initially 500 mg bd, then 500 mg tds, then 1 g bd

↓

BP still not controlled → REFER TO DOCTOR

Fig. 59.2 Nurse-led algorithm for treating patients with hypertension. Note:
1 Any low-dose thiazide can be used.
2 Make sure blood pressure is measured carefully and correctly.
3 At each visit, reinforce lifestyle advice and check compliance.
PHC, primary health clinic.

Asthma

Epidemiology

Asthma is a common chronic condition that particularly affects younger age groups and, if poorly controlled, is likely to interfere with school or work attendance. A significant amount of asthma is undiagnosed and misdiagnosed. In the tropics, asthma is increasing and becoming a major problem in urban environments. Although the disease is usually relatively easy to treat, effective drugs are expensive and often poorly available in tropical countries. Even in resource-rich western countries, asthma has a small but significant mortality. In tropical countries, the death rate is usually unknown, but is certainly significantly higher.

Diagnosis

The hallmark of asthma is reversible bronchospasm, and the key feature in the history is wheeze. Rhonchi may be present on chest auscultation, but their absence does not exclude the diagnosis, which is essentially made on the history. The wheeze is often worse at night, and may be provoked by allergic trigger factors (animals, seasons, etc.), or by exercise. Sometimes, particularly in children, a dry (and often nocturnal) cough may be the only feature. In adults, breathlessness without wheeze may indicate asthma. If available, patient records of peak flow (PF) measurements are useful in making the diagnosis. PF readings will usually be variable, often low, but in particular show classical 'early morning dipping'. In doubtful cases spirometry with and without bronchodilation is useful diagnostically: classically, the FEV_1 is reduced (to below 80% of the predicted value), as is the FEV_1:FVC ratio (to below 75%). Without this equipment a simple clinic PF measurement, repeated 20 min after two puffs of a salbutamol inhaler, can give similar information.

Frequently in developing countries, the diagnosis is made on history, perhaps confirmed by a trial of bronchodilator treatment. The diagnosis of asthma is summarized in Box 59.6.

Some care must be taken with the differential diagnosis. In children, simple wheezy bronchitis may be mistaken for asthma. In adults, chronic obstructive pulmonary disease (COPD) is the main alternative diagnosis. Tropical pulmonary eosinophilia should also be considered (see Chapter 30).

DIAGNOSIS OF ASTHMA

- History
- Patient's peak flow (PF) charts
- Spirometry ± bronchodilation
- PF with bronchodilation
- Response to treatment

Box 59.6 Diagnosis of asthma

Prevalence

Studies of prevalence of asthma in the tropics vary widely. This is partly a result of methodological problems; e.g. studies of simple 'reported asthma' will clearly give lower levels than those using objective tests such as spirometry. Other factors affecting prevalence results include patient age (rates will be higher in younger age groups) and whether the environment is rural or urban. Rates are generally low in rural areas (usually below 5%), but in towns can rise to 10–15% or more. In western countries there is good evidence that asthma prevalence is increasing, and the same is probably true in the developing world.

Causes

General

There are several factors related to the causation of asthma, although the overall explanation for individual disease (and indeed increasing population trends) is often uncertain.

- *Allergens*—allergy is very important in many cases. The most important allergen is the house dust mite, but in tropical countries cockroaches and bed bugs can also be important.
- *Genetic factors*—there is familial clustering of atopy and allergy in general, and asthma in particular.
- *'Hygiene hypothesis'*—this hypothesis is based on the observation that children who have relatively little infection or vaccination exposure early in life have a greater risk of later asthma.
- *Obesity*—the reason for this association is uncertain but asthma is certainly more common in the obese.

In addition to the above, there are a number of environmental precipitants or triggers to asthma. These include some drugs (e.g. non-steroidal anti-inflammatory drugs, beta-blockers), active or passive smoking, occupational sensitizers, exercise, respiratory infections and household pollution. Atmospheric pollution in general is not thought nowadays to be a true cause of asthma. Asthma rates are steadily rising in Europe while the air gets cleaner, and some of the highest asthma rates in the world are observed in New Zealand, where pollution rates are very low.

Urban asthma

There are several potential causes of the marked rise in asthma prevalence seen in tropical urban, compared with rural, environments (Box 59.7). The major reason is probably increased exposure to house dust mites, and this is supported from studies in Ethiopia. Rural dwellings tend to be spacious and well ventilated, and have rudimentary furniture. In towns, houses are smaller, and have doors and windows that reduce ventilation. There may also be curtains, carpets, easy chairs, etc. Overall, the environment is ideal for house dust mites! A further interesting factor may be the reduced childhood intestinal parasite load found in urban areas. There is evidence that increased parasite exposure in childhood reduces the risk of later asthma. Obesity increases with urbanization. Finally, the urban environment has more 'triggers' to asthma for those with a predisposition to the disease (e.g. smoking, household fires, occupational irritants).

Complications

The main complication of asthma is sudden and severe decompensation to status asthmaticus (or acute severe asthma). This is usually brought on by infection and is heralded by increasing need for reliever aerosols, reduced effect of such treatment, and severe nocturnal breathlessness. Patients at presentation usually have tachypnoea, distress and marked airways obstruction. There is often tachycardia and sometimes pulsus paradoxus. Blood gas levels are initially well-preserved, but a rising pCO_2 is a dangerous sign. Cyanosis and a 'silent chest' are similarly serious signs suggesting the likely impending need for ventilatory assistance. Status asthmaticus is a serious complication of asthma with a small but significant mortality.

Management

Routine treatment

The most effective treatment is based on beta-2 agonist and steroid aerosols. Beta-2 agonists can be considered to be 'relievers'. The most commonly used is salbutamol (Ventolin) although there are several others (e.g. terbutaline). Steroid inhalers can be considered to be 'suppressors', and again there are several, although the most common is beclometasone (Becotide). Reliever treatment gives rapid improvement and can be given either as needed or, if necessary, on a regular basis. Suppressor aerosols are given regularly (two puffs, twice daily), regardless of current symptoms. As with oral hypoglycaemic agents, and antihypertensive drugs, asthma can be managed in a simple drug cascade system, adding in further treatment if initial management fails to control the disease (Box 59.8).

Tropical adaptations to treatment

Asthma inhalers are expensive and often difficult to obtain in developing countries. Sometimes old-fashioned non-selective beta-agonist inhalers (e.g. isoprenaline) may have to be used. They are effective, but potentially cardiotoxic in

---

CAUSES OF URBAN ASTHMA

- Increased allergen exposure
- Reduced parasite load
- Obesity
- Increase in trigger factors

Box 59.7 Causes of urban asthma

ASTHMA MANAGEMENT

- Ventolin inhaler as necessary
- Ventolin inhaler regularly (two puffs q.d.s.)
- Becotide inhaler regularly (two puffs b.d.)
- High-dose steroid inhaler (e.g. Becloforte)
- Oral theophylline preparations

Note: Ventolin and Becotide are used as examples of beta-2 agonist and steroid metered aerosols—there are other equally effective preparations.

Box 59.8 Asthma mangement

excessive doses, and patients must be given careful advice on maximal allowable daily doses. Oral bronchodilators may have to be the mainstay of treatment. These are difficult drugs to use, with a narrow therapeutic window and relatively weak bronchodilator activity. Cromoglycate (Intal) is now rarely used in modern asthma practice, but if available (and steroid inhalers are not) it is worth using as a suppressor aerosol.

Treatment of exacerbations

Infective exacerbations should be treated with antibiotics and a brief course of steroids (e.g. prednisone or prednisolone 40 mg/day for 5–10 days). Assuming the patient is on a beta-2 agonist inhaler (e.g. salbutamol), this should be given in a high dose (e.g. 4–8 puffs q.d.s.) via a spacer inhaler (which can, if necessary, be home made — see below).

Status asthmaticus treatment

If asthmatic exacerbations are treated promptly and adequately, this is an avoidable complication. Nevertheless, when it does occur, it must be treated vigorously. The principles of management are shown in Box 59.9.

If nebulized salbutamol is not available, high-dose inhalers with a spacer device should be used, but intravenous aminophylline is likely to be needed. If nothing else is available, adrenaline (epinephrine) can be life-saving, used in a subcutaneous dose of 0.1 mL of 1/1000 solution per

10 kg body weight (give 0.75 mL if the patient is too ill to be weighed).

Organization of care

Patient education

Patient education is especially important if asthma treatment is to be successful, and all newly diagnosed patients should be given simple advice and training as follows:
- understanding asthma;
- avoiding allergic triggers;
- importance of regular treatment;
- correct use of inhalers;
- concept of reliever and suppressor treatment; and
- recognizing deterioration and seeking help.

Particularly in younger asthmatic patients, attention to allergic triggers can be very successful. This includes avoiding domestic pets, good home ventilation, and regular beating and cleaning of bedclothes, mattresses and settees. Inhaler technique is vitally important — many patients find these difficult to use, and without training they will be ineffective and a waste of resources. Spacer devices are very effective, and these can be homemade. An empty plastic milk or fruit juice container is ideal — a hole is cut at the bottom to fit the mouth piece of the inhaler, the required dose is delivered into the container, and then rebreathed through the top. These systems have been shown to be as effective as expensive proprietary spacers.

Doctor education

Doctor education is also important. Protocols of routine and emergency management should be widely displayed and distributed. Many patients develop acute severe asthma because of inadequate preceding medical treatment. In particular, steroids are widely underused — a course of prednisolone 40 mg/day for 5–7 days given on an outpatient basis frequently resolves significantly deteriorating asthma. There are no dangers with such short courses and, in particular, adrenal suppression does not occur. In areas where *Entamoeba histolytica* and/or *Strongyloides stercoralis* are found, asthmatic patients should be regularly screened for these infections, as

---

TREATING STATUS ASTHMATICUS

- High concentration oxygen
- High dose steroid:
  prednisolone 40 mg/day, *plus* initially
  hydrocortisone 100 mg i.v. q.d.s.
- Broad-spectrum antibiotics
- Bronchodilators:
  nebulized salbutamol
  add intravenous aminophylline if necessary
- Ventilation in extreme life-threatening cases

Box 59.9 Treating status asthmaticus

steroids may cause serious deterioration of amoebic dysentery, or the hyperinfection syndrome of strongyloidiasis.

Nurse-led treatment protocols

As with other NCDs, basic asthma treatment is well-suited to protocol-based stepwise care, delivered by nurses (Fig. 59.3). Although PF measurements can be useful to detect improvement and deterioration, the basic aim of treatment should be freedom from significant asthmatic symptoms.

Epilepsy

Epidemiology

Although epilepsy is much less common than hypertension or asthma, it is an important NCD for several reasons. Diagnosis is often difficult, and there are many with undiagnosed epilepsy.

Those with known disease are frequently poorly controlled, and their continuing seizures carry a significant social stigma and life burden (driving and working may be difficult). The condition also carries an excess mortality of 2–5 times that of non-epileptic people. Treatment is often made difficult by a lack of modern drugs, and many doctors do not understand the basic principles of anticonvulsant therapy. There are also cultural problems affecting the person with epilepsy in the tropics. It is not always seen as a disease at all in some cultures (it may be considered a form of bewitching or possession by demons). In other societies it may be thought to be contagious.

Diagnosis

The diagnosis rests on a history of at least two typical attacks, preferably witnessed, and with the history obtained from the witness. A good description of typical grand mal (tonic–clonic)

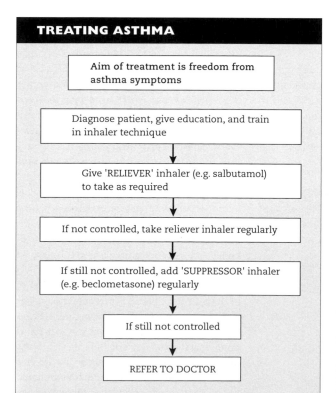

TREATING ASTHMA

Aim of treatment is freedom from asthma symptoms

Diagnose patient, give education, and train in inhaler technique

Give 'RELIEVER' inhaler (e.g. salbutamol) to take as required

If not controlled, take reliever inhaler regularly

If still not controlled, add 'SUPPRESSOR' inhaler (e.g. beclometasone) regularly

If still not controlled

REFER TO DOCTOR

Fig. 59.3 Nurse-led algorithm for treating patients with asthma. Note:
1 At each clinic visit record symptoms, measure peak flow and check inhaler technique.
2 Beyond currently taken reliever and suppressor aerosols, medical staff may have to consider oral theophylline or aminophylline preparations, or even sometimes low-dose maintenance steroids.

CLASSIFICATION OF EPILEPSY

- Grand mal (tonic–clonic)
- Absence attacks (petit mal)
- Simple partial seizures (patient conscious), e.g. focal motor or temporal lobe
- Complex partial seizures (as above, but with loss of consciousness)
- Myoclonic epilepsy (rare)

Note: Partial seizures may sometimes progress to grand mal attacks.

Box 59.10 Classification of epilepsy

seizures should lead to a firm diagnosis, but it must be remembered that there are other less obvious types of seizure: absence attacks (or petit mal — usually in childhood); and a variety of partial seizures, with or without impairment of consciousness (e.g. focal motor epilepsy, temporal lobe epilepsy). A simple and useful modern classification of seizures is as shown in Box 59.10.

When attacks are atypical, and especially where they are not witnessed, diagnosis may be difficult. Electroencephalography (EEG) is rarely available in tropical countries, and is anyway often unreliable. A trial of anticonvulsant therapy is sometimes a reasonable option, with the patient keeping a close record of attacks.

Prevalence

Rates of reported epilepsy in developing countries greatly underestimate the extent of the problem. Their variability also reflects diagnostic difficulties, which may alter geographically, and also local causative factors (e.g. prevalence of neurocysticercosis). Overall prevalence rates are probably around 2%. Reported studies have demonstrated prevalences of 0.5% in Ethiopia, 1.0% in Uganda, 3.0% in Tanzania and 5.0% in India.

Causes

Most epilepsy in developed countries is idiopathic, but in developing countries proportionally more cases have a definable cause. Tropical causes of epilepsy include the following:

- birth hypoxia/injury;
- past head injury;
- past meningitis/encephalitis (e.g. cerebral malaria, bacterial meningitis, sleeping sickness);
- HIV infection (usually caused by an associated opportunist infection or tumour, such as toxoplasmosis, cryptococcosis, tuberculoma or lymphoma);
- brain tumours, cysts (e.g. hydatid);
- neurocysticercosis (*Taenia solium*);
- past stroke; and
- idiopathic.

Complications

Burns and other trauma occurring during seizures are the most common complications of epilepsy. Sudden death may occur during fits. Status epilepticus is a serious complication. It normally occurs in pre-existing epilepsy and can be provoked by infection or alcohol excess. It carries significant mortality and must be treated vigorously.

Management

Drug treatment

Infections leading to epilepsy should always be treated. It has been shown that albendazole or praziquantel treatment for neurocysticercosis reduces seizure frequency. However, long-term anticonvulsant treatment is generally necessary. Valproate and carbamazepine are the commonly used anticonvulsants in western countries, but these are rarely available in the tropics. Phenobarbital and phenytoin are usually available, and if used properly can be very effective. Side-effects can sometimes be problematic. Phenobarbital can cause drowsiness, and sometimes behaviour change in children. Phenytoin may lead to gum hypertrophy, acne and hirsutism; as well as sedation and, in excessive doses, vertigo and incoordination. Like other anticonvulsants, phenytoin and phenobarbital are potentially difficult drugs to use as they have a 'therapeutic window' of blood levels. Below this window they are ineffective, but above it they are toxic. In the absence of drug level estimations in the blood, the drugs should be started at a low dose and gradually titrated up until seizure control is obtained. Experience from India suggests that adult doses of phenobarbital should start at 30 mg/day, and slowly titrate up to 90 mg/day (in 30 mg increments). For phenytoin the range is 100–300 mg (in 50 mg increments). Both drugs should be given once daily at night. Effective seizure control can often be obtained using either of these drugs in this way.

Status epilepticus

If available, intravenous or rectal diazepam is ideal, but if not intravenous phenytoin can be used (this can also be given if diazepam has failed). The dose schedules are as follows:

• *Diazepam* — 2 mg slow i.v. per minute (up to 20 mg)

• *Phenytoin* — 150–250 mg slow i.v. over 20 min. Can give a further 100–150 mg i.v. similarly if necessary 30–60 min later.

Old-fashioned paraldehyde (5 mL deep i.m. in each buttock) can still be effective (remember to use a glass syringe!). In severe refractory cases, general anaesthesia with muscle relaxation and anaesthesia may be needed.

Organization of care

Education

Successful epilepsy management programmes in the tropics involve education of patients, carers and the community. As with other NCDs, patients need to understand the nature of their disease and the importance of regular treatment, particularly as most fits in established epileptic patients result from poor compliance. Other precipitants are infections, alcohol excess and some drugs which lower seizure threshold. Patients may recognize their own triggers (e.g. excessive fatigue, psychological stress and, in women, prior to menstruation). Carers should if possible be involved to provide support and sometimes to supervise medication. Attempts should be made at a community level to increase understanding of epilepsy, and to prevent distrust of or discrimination against the epileptic patient.

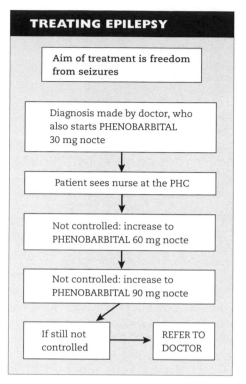

TREATING EPILEPSY

Aim of treatment is freedom from seizures

Diagnosis made by doctor, who also starts PHENOBARBITAL 30 mg nocte

Patient sees nurse at the PHC

Not controlled: increase to PHENOBARBITAL 60 mg nocte

Not controlled: increase to PHENOBARBITAL 90 mg nocte

If still not controlled → REFER TO DOCTOR

Fig. 59.4 Nurse-led algorithm for treating patients with epilepsy. Note:
1 At each visit check compliance, disease understanding and any side-effects of treatment.
2 Patients are asked to keep a diary of fits. Interval between follow-up depends on frequency of fits.
PHC, primary health clinic.

Drug treatment

Although there is less experience with nurse-led epilepsy care than with other NCDs, once the diagnosis is made and treatment initiated by medical staff, there is no reason why nursing staff cannot supervise the dose increases of phenobarbital or phenytoin mentioned above. Figure 59.4 illustrates a suggested protocol using phenobarbital, although an identical one could be produced with phenytoin.

Further reading

Coleman R, Gill G, Wilkinson D. Non-communicable disease management in resource-poor settings: a primary care model from rural South Africa. *Bull World Health Organ* 1998; **76**: 633–40. [Describes the successful use of nurse-led protocol-based care of chronic diseases.]

Cullinan P. Asthma in African cities. *Thorax* 1998; **53**: 909–10. [Useful review of the problem of 'urban asthma'.]

Gill GV, Mbanya J-C, Alberti KGMM, eds. *Diabetes in Africa*. FSG Communications, 1997. [Standard textbook describing diabetes and delivery of care in Africa.]

Guidelines Subcommittee of the World Health Organization, International Society of Hypertension Liaison Committee, 1999. *J Hypertens* 1999; **17**: 151–83. [Most recent WHO document on hypertension and its diagnosis.]

Harpham T. Urbanization and health in transition. *Lancet* 1997; **349**: SM11–14. [Useful review of urbanization and its health effects.]

Kok P. Epilepsy in practice. *Tropical Doctor* 1998; **28**: 68–72. [Thoughtful and practical article describing epilepsy care in poor communities.]

World Health Organization. Definition, diagnosis and classification of diabetes and its complications. Geneva: WHO, 1999. [Up-to-date information on diabetes diagnosis and classification.]

CHAPTER 60

Refugee Health

Humanitarian emergencies in resource-poor countries, 329

Responding to humanitarian emergencies, 329

Health issues among asylum seekers in developed countries, 333

Further reading, 334

The term 'refugee' is usually applied to a person seeking shelter, protection, retreat or sanctuary. The common usage of this term merges important distinctions that are made in international humanitarian law. The following definitions are widely recognized internationally.

Refugee. A person who, owing to a well-founded fear of being persecuted for reasons of race, religion, nationality, membership of a particular social group or political opinion, is outside their country of nationality and is unable, or owing to such fear, is unwilling to avail of the protection of that country. (*United Nations Convention 1951*)

Asylum seeker. A person who is outside their country of nationality and has applied for refugee status.

Internally displaced person (IDP). A person forced to flee from their home but who remains within their national boundaries. IDPs may seek refuge for similar reasons to refugees and may face similar problems but, because they remain within their country's borders, they do not enjoy the same legal status for protection and for assistance. Prior to 1998, no international agency had been given the responsibility of ensuring adequate care and protection for IDPs. Recently, 'Guiding Principles' have been developed by the United Nations (UN) to define standards for the treatment of IDPs.

Unaccompanied child. A person who is under the age of 18, unless, under the law applicable to the child, majority is attained earlier, and who is separated from both parents and is not being cared for by an adult who by law or custom has responsibility to do so. (*UNHCR Geneva 1997, Guidelines on policies and procedures in dealing with unaccompanied children seeking asylum*). Alternative terms are 'unaccompanied minor' or 'separated child'.

For the sake of brevity, the word 'refugee' refers to all the above groups throughout this chapter, unless it is necessary to make a specific distinction.

The UN High Commission for Refugees (UNHCR), is the lead UN agency with responsibility for refugees. At the start of the year 2002, the number of people of concern to UNHCR was 19.9 million. They included 12 million refugees (61%), 940 800 asylum seekers (5%), 462 700 returned refugees (3%), 5.3 million IDPs (25%), 241 000 returned IDPs (1%) and 1 million others of concern (5%).

An estimated 7.7 million people under the care of UNHCR are children below the age of 18. The percentage of children compared with the overall refugee population ranges from 57% in central Africa to 20% in central and eastern Europe. Refugees above 60 years of age constitute more than 15% of the refugee population in eastern Europe and the Balkans, whereas in Africa they generally represent less than 5% of

EMERGENCY SEVERITY INDICATORS

	CMR (per 10 000/day)	UFM (per 10 000/day)
Major catastrophe	>5	>10
Emergency: out of control	>2	>4
Emergency/relief programme: situation serious	>1	>2
Emergency/relief programme: under control	<1	<2
Normal rate for stable developing country	0.5	1.0

Table 60.1 Crude mortality rate (CMR) and under-fives mortality (UFM) used to indicate the severity of an emergency.

the refugee population. In most regions, women and girls of all ages constitute between 45 and 55% of the refugee population.

In 2001, when all persons of concern to UNHCR are included, Asia hosted 44.6%, Africa 21.1%, Europe 24.6%, North America 5.5%, Latin America and the Caribbean 3.8% and Oceania 0.4%. Compared to the size of the national population, the main refugee-hosting countries were Armenia (70 refugees per 1000 inhabitants), followed by Congo (40 per 1000), Yugoslavia (38 per 1000), Djibouti (37 per 1000) and Zambia (27 per 1000).

Humanitarian emergencies in resource-poor countries

Most refugees are displaced because of complex humanitarian emergencies (CHEs) or natural disasters. CHEs are primarily internal wars in which conflicting groups compete for limited resources. CHEs are characterized by administrative, economic, political and social disruption, usually accompanied by high levels of violence. Major violations of the Geneva Conventions and Universal Declaration of Human Rights are common and cultural, religious and ethnic groups may be at risk of extinction. CHEs frequently result in catastrophic public health emergencies as coping capacities are exceeded

by need, placing vulnerable populations at greatest risk of epidemic diseases and malnutrition. CHEs may smoulder on as 'chronic emergencies' for many years, with fluctuating levels of violence, displacement and disease.

Natural disasters (floods, earthquakes, famine) may also become complex humanitarian emergencies, particularly when they affect vulnerable populations in weakened or disrupted states.

Responding to humanitarian emergencies

Emergency phase

The initial phase of an emergency is characterized by need overwhelming available resources and resulting in increased mortality rates. Crude mortality rate (CMR) is generally used as an indicator of the severity of an emergency. Under-fives mortality (UFM) may also be used (Table 60.1). The emergency phase of an intervention can be regarded as over when the CMR is < 1 and basic needs have been met.

Effective action in response to a humanitarian emergency depends on individuals, governmental and non-governmental organizations (NGOs), working in a coordinated and complementary manner. Intersectoral collaboration is the key to success in emergency interventions. Médecins sans frontières (MSF) emphasize the following 10 priorities in an emergency response.

1 Initial assessment

The initial assessment is usually conducted in two phases.

Phase I

- Objective to enable rapid decision regarding the need for and scale of an intervention.
- Completed within 3 days.
- Focus:
  (a) geopolitical context, including the background to the displacement;
  (b) demographic description of the population and map of site;
  (c) characteristics of the environment in which refugees have settled;
  (d) availability of water, food and shelter;
  (e) major health problems, epidemic diseases and mortality rates;
  (f) human and material resources required; and
  (g) operating partners (e.g. local and national authorities, local and international organizations).

Phase II

- Conducted simultaneously with the implementation of relief actions.
- Allows more detailed programme planning and wider dissemination of information.
- Similar range of issues to Phase I, but with greater depth and emphasis.

2 Measles immunization

Measles can be a devastating illness in refugee populations, particularly if there is also a high level of malnutrition and vitamin A deficiency. Herd immunity is only achieved when over 90% of those susceptible have been immunized, a figure rarely achieved in any country in the world. Therefore, all children aged between 6 months and 15 years should be immunized against measles and given vitamin A.

3 Water and sanitation

Access to adequate supplies of water is critical in reducing the likelihood of diarrhoeal diseases. Quantity is more important than quality. A minimum of 5 L per person per day is required for essential needs such as drinking and cooking in the acute stages of an emergency. This should be increased to at least 15–20 L per person per day as soon as possible to reduce the risk of water-borne and water-washed diseases. Having sourced a sufficient quantity of water, attention can be given to assessing quality by determining the level of faecal coliforms using a field testing kit such as the Del Agua/Oxfam kit. Ideally, there should be fewer than 10 faecal coliforms/100 mL. Chlorination is the most effective way of treating water in emergencies.

During the initial stages of an emergency, it may be necessary to identify defecation areas or fields for excreta disposal. However, construction of shallow trenches or collective latrines is preferable. The target should be one latrine or trench per 50–100 people. As the situation stabilizes, the aim is for one latrine per 20 people or, ideally, one latrine per family.

4 Food and nutrition

Existing malnutrition in a population is exacerbated by circumstances giving rise to humanitarian emergencies. Establishing and maintaining food security (access by all people at all times to enough food for an active healthy life) can present an enormous logistical challenge. The minimum mean population requirement is 2100 kcal per person per day. Attention must be given to ensuring that the basic food ration includes an appropriate mix of vitamins and other micronutrients. The prevalence of malnutrition among children aged less than 5 years is usually an indication of the level of malnutrition in the entire population. Therefore, the initial health assessment should include a nutritional survey of a sample of these children by measuring the weight-for-height (W/H) index and identifying children with bilateral pedal oedema.

Prevalence rates can be calculated for:
- global malnutrition (% children with moderate or severe malnutrition);
- severe malnutrition (% children with W/H <-3 Z scores and/or oedema);
- moderate malnutrition (% children with W/H <-2 Z scores but ≥-3 Z scores);
where Z scores refer to the number of standard

deviations above (+) or below (−) the median in the reference population.

Mid-upper arm circumference (MUAC) is less accurate but is sometimes used for rapid screening in children aged 6–59 months if a W/H survey is impossible. MUAC < 125 mm indicates global acute malnutrition; MUAC < 110 mm indicates severe malnutrition. Bilateral pedal oedema is used as an indicator of severe malnutrition. The results of the survey can then be used to guide the level of intervention.

In addition to ensuring that general food distributions provide adequate food rations for all, there is often need for selective feeding programmes (SFPs). Blanket SFPs provide food supplements to vulnerable groups (e.g. pregnant women). Targeted SFPs provide food supplements and medical follow-up for the moderately malnourished. Therapeutic feeding programmes are used in the management of severely malnourished children and include an initial 'intensive care' phase for careful resuscitation, management of medical problems and initiation of nutritional treatment. Once complications have been brought under control and feeding has been established, the child can be transferred to a day-care unit for ongoing nutritional and medical management and follow-up.

5 Shelter and site planning

Overcrowding and lack of hygiene and sanitation are major factors in the spread of communicable diseases among large populations of refugees in camps. In some situations, it may be possible and preferable for refugees to be integrated among the host population.

In planning a refugee camp site, attention must be given to security, access, protection from environmental health risks and provision of essential services (reception, administration, storage, distribution, water, sanitation, cemetery, health facilities, nutrition centre(s), places for social, commercial, educational and religious activities). Careful consideration should also be given to the social, economic, health and environmental impact that refugees may have on the host population and services.

RECOMMENDED STANDARDS IN SITE PLANNING OF REFUGEE CAMP SITES

Area available per person	30 m²
Shelter space per person	3.5 m²
Distance between two shelters	2 m minimum
Number of people per water point	250
Number of people per latrine	20
Distance to water point	150 m maximum
Distance to latrine	30 m
Distance between water point and latrine	100 m
Firebreaks	75 m every 300 m

Box 60.1 Recommended standards in site planning of refugee camp sites

Small camps, accommodating up to about 10 000 persons, are preferable to larger camps and should be organized according to social and cultural norms, shelter and accommodation being provided on the basis of family groupings. Recommended standards in site planning are shown in Box 60.1.

6 Health care in the emergency phase

Whenever possible, health care should be planned and implemented in consultation with national and local health authorities. Responding to the essential health care needs of refugees in camps usually requires the establishment of specific services for this population, focusing on basic curative care. Having established a central health facility, priority should be given to providing a network of peripheral health centres and health posts and developing outreach activities. There should be standardized systems for data collection and surveillance, clear management protocols appropriate to different levels of health facility, and clear guidelines for referral. Contingency plans should be developed so that appropriate action can be taken as circumstances change, e.g. in response to an epidemic, or in managing a sudden influx of new arrivals.

7 Control of communicable diseases and epidemics

Diarrhoeal diseases, especially cholera and *Shigella dysenteriae* type 1, acute respiratory infections, measles, meningitis, and in many regions malaria, are among the leading infectious causes of excess mortality in refugee populations. There may also be a risk of epidemic louse-borne typhus or relapsing fever, and a variety of other infections depending on the region or circumstances.

The emphasis should be on prevention, epidemic surveillance using agreed case definitions, contingency planning and epidemic response appropriate to the specific disease and circumstances of the outbreak. The number of clinical cases may overwhelm existing facilities and specific treatment centres may have to be set up and staffed.

Sexually transmitted infections (STIs) including HIV can be a serious problem in refugee populations as a result of sexual violence or social disruption. Their long-term impact may be as devastating as any of the diseases mentioned above. Therefore, implementation of measures to prevent and treat STIs should be an early consideration following the initial emergency response.

The risk of reactivation and transmission of tuberculosis is increased in refugee populations. However, WHO recommend that a TB control programme should not be initiated until the following criteria have been fulfilled: data indicate that TB is an important health problem; the emergency phase is over; basic needs of water, adequate food, shelter and sanitation are available; essential clinical services and basic drugs are available; security in, and stability of, the camp envisaged for at least 6 months; sufficient funding is available for at least 12 months; and laboratory services for sputum smear microscopy are available.

Detailed information on the control and management of specific epidemic diseases is available in the texts and on websites listed at the end of this chapter.

8 Public health surveillance

Effective programme planning, implementation, monitoring and evaluation depends on a system for the collection of data on demography, morbidity, mortality, basic needs and programme activities. Simple standardized indicators should be used. A minimum data set in the initial emergency phase should include CMR, cause-specific mortality and morbidity using simple case definitions, malnutrition rate among under-fives, and indicators of access to water and sanitation facilities.

9 Human resources and training

Recruitment, training and management of staff are key to the success of an intervention. It is important to determine appropriate staffing levels for a given task, prepare specific job descriptions, and establish lines of management and communication. Consideration should be given to contracts, legal status and salaries. Local services can be severely disrupted when their staff are lured by superior salaries and conditions offered by refugee agencies.

10 Coordination

A reliable coordination mechanism and leadership should be established early in the initial phase of an emergency. Coordination requires regular meetings involving representatives from key ministries in the host government, the host community, the refugee population and the agencies involved in the response. Common objectives should be agreed and tasks determined and formally allocated. Technical guidelines and standardized policies should be introduced from the outset, and procedures agreed for the reporting and dissemination of data.

Post-emergency phase

The post-emergency phase commences when basic needs have been met and excess mortality has been controlled (CMR < 1 in 10 000/day). Continued attention is given to the 'top 10' priorities indicated in the emergency phase, while adapting health programmes to address additional issues that may not have been receiving particular attention, e.g. reproductive and mental health. The post-emergency phase may end with repatriation and resettlement or, if a population faces long-term displacement, there may be a gradual process of adaptation of services according to need, or integration into mainstream services for the host population.

Health issues among asylum seekers in developed countries

According to UNHCR, 614 100 asylum applications were submitted in 30 mostly industrialized countries in 2001. The European Union received 384 530 applications. The UK received a total of 71 365 asylum applicants mainly from Afghanistan, Iraq, Somalia, Sri Lanka and Turkey. Asylum was granted in 9% of cases and 17% were granted exceptional leave to remain on humanitarian grounds.

There is a common perception in industrialized countries that asylum seekers are a threat to the health of the host population. Although asylum seekers may be at risk of certain infectious diseases because of possible exposure either in their country of origin or during their migration in search of asylum, some risk is also posed by members of the host population who travel abroad on holiday or business in numbers far exceeding those of asylum seekers entering industrialized countries. Over 7 million UK residents visited countries other than western Europe and North America in 2001. That is 100 times more than the number of people seeking asylum in the UK. It is in everyone's best interest to keep a sense of proportion and encourage asylum seekers to regard health as an opportunity rather than as a threat.

Health assessment of asylum seekers

In considering the health needs of asylum seekers it is important to consider the person and their predicament. Health care providers should be aware of issues related to language, communication, culture, religion, gender, family, community and social circumstances. Social, economic, physical, psychological and emotional needs may greatly outweigh medical needs.

The 'health assessment' approach is preferable to 'screening', as the latter is often narrowly focused and may have negative connotations from the viewpoint of the asylum seeker. The key elements in a programme for health care of asylum seekers and refugees include the following:
- information and encouragement;
- early initial health assessment;
- specialist follow-up according to need;
- continuing access to specialist services;
- integration into mainstream services;
- communication and continuity of care; and
- close liaison with other services.

The history and examination should include assessment of:
- vaccination status;
- nutritional status;
- mental health status;
- infectious diseases;
- non-infectious diseases (e.g. haematological—sickle cell, thalassaemia, G6PD; cardiac—rheumatic heart disease, endomyocardial fibrosis);
- obstetric and gynaecological problems;
- vision, hearing and dental problems;
- presence and level of disability;
- evidence of torture;
- evidence of substance misuse; and
- risk of exposure to toxins (e.g. lead).

The choice of laboratory investigations will depend on the findings of the health assessment, and whether specific 'screening' programmes are to be followed. Assessment of children should also include: whether the child is accompanied or unaccompanied; growth and development (including language, hearing, vision); and the possibility of congenital diseases routinely screened for in neonates in the UK (e.g. hypothyroidism, phenylketonuria). Another important issue to consider in children is that of child protection. Certain cultural practices, such as female genital mutilation, may be traditional in the child's country of origin but illegal in the country of asylum.

Infectious diseases

Infectious diseases that are likely to be more prevalent among refugees and asylum seekers include tuberculosis, hepatitis B and C, HIV, STIs, gastrointestinal infections (bacterial/ parasitic), malaria, typhoid and other tropical diseases, infestations (lice, scabies) and multi-drug-resistant bacterial infections. In most cases, these infections are likely to be asymptomatic. Therefore, screening programmes may be recommended for specific infections, particularly those of greatest public health interest (e.g. TB, hepatitis B and C, HIV, STIs). Issues concerning ethics, consent, confidentiality, cost effectiveness and continuity of care deserve consideration.

Decisions regarding routine screening for intestinal helminths should be made on the basis of risk assessment. Empirical treatment with albendazole has been shown to be more cost effective than routine screening of immigrants in the USA.

Many latent imported infections may cause clinical disease following immunosuppression, including TB, leprosy, amoebiasis, strongyloidiasis, visceral leishmaniasis, histoplasmosis, malaria, filariasis and American trypanosomiasis. Some imported infections, if untreated, may persist for years, e.g. strongyloidiasis (> 60 years), schistosomiasis (> 30 years), melioidosis (> 25 years), hydatid disease (> 20 years) and trichinellosis, cysticercosis, and onchocerciasis (all > 15 years).

Torture and other traumatic experiences

It is estimated that 10–30% of asylum seekers have survived torture, sexual violence or other seriously traumatizing experiences. Many will experience ongoing psychological problems including nightmares, hallucinations, flashbacks, panic attacks, sexual problems, phobias, difficulty trusting people or forming relationships, de-

pressive illness and anxiety. The medical documentation and reporting of such experiences may be a critical factor in the asylum determination process.

Reactions to trauma and loss can include poor concentration, memory impairment, daydreaming, intrusive thoughts and images, irritability, confusion, tiredness, lethargy, sleep difficulties and loss of motivation. Sufferers can become withdrawn and isolated and may self-harm. It is often very difficult for a person to articulate these experiences. Somatization is common and unexplained backache, headache, stomach ache or other body pains should prompt further careful enquiry. Children may, in addition, experience interrupted or uneven emotional development or failure to thrive.

Further reading

Papers

Burkle FM. Lessons learnt and future expectations of complex emergencies. *Br Med J* 1999; **319**: 422–6.

Coker R, Lambregts van Weezenbeek K. Mandatory screening and treatment of immigrants for latent tuberculosis in the USA: just restraint? *Lancet Infect Dis* 2001; **1**: 270–6.

Goma Epidemiology Group. Public health impact of Rwandan refugee crisis: what happened in Goma, Zaire, in July 1994. *Lancet* 1995; **345**: 339–44.

Lifson AR, Thai D, O'Fallon A, Mills WA, Hang K. Prevalence of tuberculosis, hepatitis B virus, and intestinal parasitic infections among refugees to Minnesota. *Public Health Rep* 2002; **117**: 69–77.

Muennig P, Pallin A, Sell RL, Chan M-S. The cost effectiveness of strategies for the treatment of intestinal parasites in immigrants. *N Engl J Med* 1999; **340**: 773–9.

Books

Médecins Sans Frontières. *Refugee Health: An Approach to Emergency Situations*. London: Macmillan, 1997. [This is essential reading for anyone involved in humanitarian emergencies in developing countries. Refugee Health and other key MSF books are also available on CD-ROM, and on-line at: http://www.msf.org/source/refbooks/E_MSFdoc-menu.htm]

Perrin P. *War and Public Health*. Geneva: ICRC, 1996.

[Clear and systematic text with the added bonus of chapters on international humanitarian law and ethics. Also available on-line at WHO/PAHO Health Library for Disasters.]

Web sites

Female Genital Mutilation. British Medical Association (BMA) guidelines (revised April 2001): http://www.bma.org.uk/public/ethics.nsf/

Medical Foundation for the Care of Victims of Torture is a registered charity that provides care and rehabilitation to survivors of torture and other forms of organized violence:
http://www.torturecare.org.uk/

Physicians for Human Rights website provides access to reports and training materials, including the Istanbul protocol (manual on the effective investigation and documentation of torture and other cruel, inhuman or degrading treatment or punishment): http://www.phrusa.org

Sphere Project. Humanitarian charter and minimum standards in disaster response:
http://www.sphereproject.org/

WHO Emergency Humanitarian Action website has excellent links including access to a huge range of electronic handbooks, manuals and emergency bibliography, many of which are also available on CD ROM (the WHO/PAHO 'Health Library for Disasters' is particularly good):
http://www.who.int/disasters/

Index

Note: page numbers in *italics* refer to figures, those in **bold** refer to tables and boxes

abdomen, fever 27
abdominal pain 3, 42, *45*
abortion, malaria 61
abscess
  amoebic 155
  guinea worm 292
  melioidosis 276
  paratyphoid 251
  pyogenic 156
  spinal epidural 230
  typhoid 247
  see also liver abscess, amoebic
accidents 313
achlorhydria 167
acidaemia, malaria 63
actinomycosis 155
activity, reduced 318
acupuncture, hepatitis B
      transmission 182
acute inflammatory demyelinating
      polyneuropathy (AIDP)
      227, 228
acute motor axonal neuropathy
      (AMAN) 227, 228
aciclovir 225
Addison's disease 34, 306
adrenaline 323
adult respiratory distress syndrome
      (ARDS) 273
aflatoxins 187, 192
African tick typhus 271
AIDS 105
  economic impact 102–3
  malabsorption 4
  mortality *85*
  surveillance 99
  see also HIV infection
AIDS-defining infections 105
AIDS-related complex 104–5
AIDS–dementia complex 230
airway in coma 19
albendazole 170, 325
  alveolar hydatid disease 195
  ascariasis 175, 176

cestode infections 173
  cutaneous larva migrans 284
  cysticercosis 172
  filariasis 118, 119
  giardiasis 168
  hookworm 177
  hydatid cyst disease 194
  strongyloidiasis 290
  treatment of immigrants
      334
  trichuriasis 178
  tropical pulmonary eosinophilia
      209–10
  visceral larva migrans 178
alcohol abuse 187
  hypertension 318–19
allergens
  asthma 321
  avoidance 323
alpha-fetoprotein 189
aminophylline 323
aminosidine 77
amodiaquine 65
amoebiasis 153–9
  amoebic dysentery 324
  amoeboma 155, 158
  clinical features 154–6
  cyst eradication 157
  epidemiology 153
  intestinal 154–5, 157
  investigations 156–7
  management 157–8
  pathogenesis 154
  prevention 158
  see also liver abscess, amoebic
amoxicillin 201
  pneumonia 203
  typhoid 249, 250
amoxicillin clavulanate 276
amphotericin B
  cryptococcal meningitis 220
  histoplasmosis 295
  liposomal 77
  visceral leishmaniasis 77

amputation
  Madura foot 298
  tropical ulcers 277, 278
anaemia 36–9
  aplastic 38
  breathing difficulties 12
  clinical diagnosis 36
  haemolytic in typhoid 247,
      248
  hookworm 176, 177
  iron deficiency 304–5
  laboratory investigations
      36–7
  macrocytic 37–8
  malaria 57
  management 38–9
  megaloblastic 305
  microcytic 37
  normocytic 38
  pernicious 306
  severe in *P. falciparum* malaria 59,
      60
anaesthesia in leprosy 143, 144
Ancylostoma braziliense (dog
      hookworm) 284
Ancylostoma duodenale (hookworm)
      174, 176–7
angina 320
angiotensin converting enzyme
      (ACE) inhibitors 319, 320
antibioma 155
antibiotics
  cholera 166
  dysentery 161
  malaria 64
  tuberculosis 91
  see also named drugs
antimonials, pentavalent
  cutaneous leishmaniasis 82
  visceral leishmaniasis 76–7
antipyretics 26
antiretroviral drugs 102–3
  generic 109–10
  HIV therapy 109–10

antivenom
  snake bites 308–9
  spider bites 310
appendicitis 3
  ascariasis 175
  fever 27
appendix, mass 155
arboviruses 221, 222–3, 224–5,
    252–3
  clinical syndromes 252–3
  fever–arthritis–rash 252–3
  hosts 252
  vectors 252
arsenic, chronic poisoning 34
artemisinin drugs
  malaria therapy 66–7
  schistosomiasis 138
arthritis 161
  guinea worm 292
  typhoid 247
Ascaris lumbricoides (ascariasis)
    174–6
  clinical features 175
  investigations 175
  life cycle 174–5
  management 175
  prevention 175–6
  tropical pulmonary eosinophilia
      differential diagnosis
      209
aspirin 26, 304
asthma 15, 320–4
  beta-blocker contraindication
      320
  care 323–4
  causes 321–2
  complications 322
  diagnosis 321
  education 323–4
  epidemiology 320–1
  exacerbations 323
  guinea worm 292
  intestinal parasite load in children
      322
  management 322–3
  nurse-led treatment 324
  pneumonia differential diagnosis
      199
  prevalence 321
  treatment use 323
  urban 322
asylum seekers 328
  health assessment 333–4
  health issues 333–4
  infectious diseases 334
atovaquone–proguanil
  malaria prophylaxis 69
  malaria therapy 66

autoimmune disorders 306
azithromycin 248, 249

bacille Calmette–Guérin (BCG)
    vaccine 94
  Buruli ulcer 281
  hookworm prevalence 177
  leprosy 149
Bacillus fusiformis 277
bacterial endocarditis 35
  fever 27
barbering 182
bed nets, impregnated
  filariasis 119
  leishmaniasis 78
  malaria 71
bedbugs 182–3
bedsores 225
benzidazole 128
benzyl benzoate 286
benzylpenicillin 274
beta-2 agonists 322, 323
beta-blockers 319, 320
  asthma trigger 321
bilharzia see schistosomiasis
biliary cirrhosis 192
biliary colic 3
  ascariasis 175
  liver fluke 191
biliary tract, Giardia lamblia
    colonization 167
birth
  HIV transmission 101
  hypoxia/injury 325
  leprosy reactions 148–9
birth weight, low in malaria 61, 68
bithionol 191
blackflies 112
blackwater fever 61
Blantyre coma scale 22
blindness
  leprosy 145
  onchocerciasis 114
blood
  cultures in respiratory disease
      14
  HIV transmission 101–2
  prescribing 38
blood clotting test 308
blood pressure 318
  see also hypertension
blood transfusion 38, 39
  HIV transmission 101–2
  malaria 63
Borrelia (relapsing fever) 267
  pathology 267–8
Borrelia recurrentis (louse-borne
    relapsing fever) 267

bot fly 282
botulism 228
  breathing difficulties 12
bowel cancer 304
bowel tumours 5
brain
  cysticercosis 172
  midline herniation syndromes
      21, 22
  P.falciparum malaria 60
  tumours 325
brainstem
  damage 20–1
  signs 23
breast feeding
  HIV transmission 101
  leprosy 149
breathing
  assessment 22
  coma 19
breathlessness see dyspnoea
bronchiectasis 11
bronchitis 321
bronchodilators, oral 323
bronchospasm 11, 321
Brucella abortus 240
Brucella canis 240
Brucella melitensis 240
Brucella suis 240
brucellosis 240–4
  clinical features 240–1
  control 243–4
  diagnosis 241–2
  fever 27
  follow-up 243
  pregnancy 243
  public health 243–4
  treatment 242–3
Brugia malayi (filariasis) 118, 209
bullae 35
Burkholderia pseudomallei
    (melioidosis) 275, 276
Burkitt's lymphoma 30
  malaria endemic areas 62
burns
  drug eruptions 35
  epilepsy 325
Buruli ulcer 279–81
  clinical features 279–80
  differential diagnosis 280
  investigations 280
  management 280–1
  prevention 281
  public health 281

calcium channel blockers 319–20
Campylobacter 6
Campylobacter jejuni 227

candidiasis
  mouth 27, 104, 108
  oesophageal 3, 108
carbamazepine 325
card agglutination test 123
card indirect agglutination test
      123–4
CD4 cells 103
  counts 104
  leprosy 143
CD8 cells 143
ceftazidime 276
central nervous system (CNS)
  HIV infection 109
  infections 23–5
  schistosomiasis 135, 137, 138
  space occupying lesions 19
  trypanosomiasis 122, 123, 125
cephalosporins, third-generation
  Haemophilus influenzae meningitis
      216
  meningococcal disease 215
  pneumococcal disease 216
  typhoid 248, 249
cerebrospinal fluid (CSF)
  brucellosis 241
  cryptococcosis 219
  investigations 25
  Japanese encephalitis 224
  pyogenic meningitis 214, 215
  trypanosomiasis 123
cervical erosion 101
cestodes 171–3
  management of infections 173
Chagas' disease 125, 127–8
  acute 127
  chronic 127–8
  control 128
  diagnosis 128
  epidemiology 128
  megaoesophagus 3
  reactivation 128
  transmission 128
  treatment 128
chancre, trypanosomal 120–1, 123
chancroid 42, 46
  HIV infection risk 101
chemicals, ingestion of corrosive 3
chest
  fever 27
  pain 12
chest X-ray
  pneumonia 200–1
  respiratory disease 14
Cheyne–Stokes breathing 22
chiggers 282–3
Chikungunya 252
child protection 334

children
  anaemia 39
    iron deficiency 305
  ascariasis 174, 176
  breathlessness 14–15
  brucellosis 243
  in care of UNHCR 328
  Chagas' disease 127
  cholera 163
  fever 26, 28
  hepatitis B vaccination 185
  HIV infection 29–30, 101,
      109
    prevalence 100
    transmission reduction 102
  intestinal parasite load 322
  leprosy 149
  malaria 39
    cerebral 60
  pneumonia 196, 200, 203
    risk 198
  pyogenic meningitis 211
  pyrexia of unknown origin 31
  refugees 333–4
  respiratory illness 13
  Reye's syndrome 26
  schistosomiasis 132
  shigellosis 161
  snake bite 309
  swollen belly syndrome 288
  tetanus vaccination 238
  toxocariasis 179
  typhoid 248
    vaccine 250
  unaccompanied 328
  see also infants
Chinese paralytic syndrome 227,
      228
chlamydia 42
  HIV infection risk 101
chloramphenicol
  African tick typhus 271
  Haemophilus influenzae meningitis
      216
  louse-borne typhus 270
  meningococcal disease 215,
      216
  pneumococcal disease 216
  resistance 271
  scrub typhus 271
  typhoid 248, 249, 250
chloroquine
  amoebiasis 157
  malaria prophylaxis 69
  malaria therapy 64–5
chlorpropamide 315
cholangiocarcinoma, oriental liver
      fluke 192

cholangitis
  ascariasis 175
  oriental liver fluke 192
cholecystitis 3
cholera 5, 6–7, 163–6
  antibiotics 166
  control 166
  dehydration 163–4
  diagnosis 164
  diarrhoea 163, 164, 165
  epidemiology 166
  maintenance hydration 165
  oral rehydration solution 165–6
  pandemic 166
  rehydration 164–5, 166
  toxin 163
    ascariasis effect 176
  treatment 164–5
  vaccine 166
chromoblastomycosis 298
chronic obstructive pulmonary
      disease 11
  asthma differential diagnosis 321
Chrysomia bezziana (Old World
      screw fly) 283
chyluria 119
cinchonism 65
ciprofloxacin 170
  pneumonia 203
  typhoid carriers 250
circulation in coma 19
circumcision
  hepatitis B transmission 182
  HIV risk 101
clarithromycin 147
clofazimine 146–7, 149
  pregnancy 149
Clonorchis sinensis (liver fluke) 191
Clostridium difficile 6
Clostridium tetani 237
co-artem malaria therapy 67
co-trimoxazole 170, 201, 203, 204
  brucellosis 243
  HIV prophylaxis 96–7, 110
  Madura foot 298
  melioidosis 276
  typhoid 249, 250
cobalamin see vitamin B_{12} deficiency
coca-colonization 313
colitis
  amoebic 155, 158
  post-dysenteric 161
colon
  polyps 137
  pseudopolyposis 135
  strictures in amoebiasis 154
colonoscopy in amoebiasis 156–7
Colorado tick fever 253

coma
  cerebral malaria 60
  hyperosmolar non-ketotic 315
  P. falciparum malaria 58–9, 60
  rapid assessment 19–23
coma scale 20, 22
commercial sex workers 102
complex humanitarian emergencies
    329
condoms 102, 187
Conn's syndrome 318
consciousness, altered 58–9
contractures in viral encephalitis
    225
convulsions in malaria 63
Cordylobia anthropophaga (Tumbu
    fly) 282
cosmetic surgery 148
cough 11–12, 15
  brucellosis 241
  HIV infection 106–7
  immunosuppression 16
  relapsing fevers 268
  tuberculosis 89
Councilman bodies 265
Crimean–Congo haemorrhagic
    fever 254, 255, 260
  Councilman bodies 265
  management 258
  rash 35
Crohn's disease 306
cromoglycate 323
croup 12
cryptococcosis 30, 219–20
  HIV infection 109
Cryptococcus neoformans 219
Cryptosporidium (cryptosporidiosis)
    108, 169, 170
Cushing's syndrome 318
Cyclops (copepod) 292, 294
Cyclospora cayetanensis 169, 170
cysticercosis 171–2
  epilepsy 325
  nodules 34
cytomegalovirus (CMV) 27
  retinitis 103

dairy products, brucellosis 240, 243
dapsone 146
  pregnancy 149
  DDS syndrome 146
DDT
  louse treatment 287
  malaria 71
  sandfly eradication 79
dehydration
  cholera 163–4
  diarrhoea 6

see also oral rehydration solution;
    rehydration, in cholera
delirium, scrub typhus 271
demographic transition 313
dengue fever 253, 255, 262, 263,
    264–5
  clinical features 262, 264
  differential diagnosis 264
  epidemiology 262
  fever in HIV infection 29
  investigations 265
  leptospirosis differential diagnosis
    273
  management 265
  prevention 265
  rash 35
  vaccines 265
dengue haemorrhagic fever 35,
    262, 264, 265
dengue shock syndrome 264
Dermatobia hominis (bot fly) 282
desferrioxamine 302
dexamethasone 208, 217
diabetes mellitus 313–17
  care organization 316–17
  causes 314
  complications 315
  diagnosis 314
  diet 315–16
  epidemiology 313–14
  fibrocalculous pancreatic 315
  leprosy differential diagnosis 146
  malnutrition-modulated 315
  malnutrition-related 315
  management 315–16
  nurse-led care 317
  pneumonia risk 198
  prevalence 314
  type 1 314
  type 2 314
diabetic ketoacidosis 12, 315, 316
  management 316
  pneumonia differential diagnosis
    199
diarrhoea 4–8, 9, 10
  antibiotic-associated 5
  Campylobacter jejuni 227
  cholera 163, 164, 165
  clinical syndromes 5–6, 7
  dehydration 6
  examination 5
  giardiasis 167–8
  history 4–5
  HIV infection 8, 30, 108
  investigations 6–7
  management 7–8, 9, 10
  mortality 85
  nosocomial 5

pathogens 5, 8
pneumonia 199
protozoal infections 169
diazepam 326
diencephalon damage 20–1
diet in diabetes mellitus 315–16
diethylcarbamazine citrate 118, 119
  tropical pulmonary eosinophilia
    209
  visceral larva migrans 178
diloxanide furoate 157
diphtheria 19
  neuropathy 228
Diphyllobothrium latum
    (diphyllobothriasis) 172
Dipylidium caninum (dipylidiasis) 173
direct agglutination test (DAT) 76
direct observation of therapy
    (DOT) 95–6
  DOTS-Plus pilot schemes 96
disseminated intravascular
    coagulation (DIC) 35
  malaria 64
    P. falciparum 61
  splenectomy 51
  typhoid 248
Dobrova virus 260
dogs
  control 236
  hydatid disease 194–5
  leishmaniasis 78
  rabies 232, 236
dot immunoassay tests, HIV testing
    99
doxycycline
  African tick typhus 271
  brucellosis 242, 243
  leptospirosis 274
  louse-borne typhus 270
  malaria prophylaxis 69
Dracunculus medinensis (guinea
    worm; dracunculiasis) 292
drug users, injecting
  hepatitis B 182
  hepatitis D 185
  HIV prevalence 100
  HIV transmission 101
Duffy blood group antigen 62
Durban's sign 155
dysentery
  amoebic 324
  bacillary 160–2
  trichuris 177
dysphagia 3
  HIV infection 30
dyspnoea 11, 12, 14–15
  malaria 60
  relapsing fevers 268

Eastern equine encephalitis *223,* 225
Ebola fever 35, 255, 259–60
Echinococcus granulosus (cystic hydatid disease) 193–4
Echinococcus multilocularis (alveolar hydatid disease) 194–5
eczema 32
education
  asthma 323–4
  epilepsy 326
  HIV infection 102
  hypertension 320
eflornithine 124, 125
elephantiasis, filariasis 116, 119
ELISA, HIV testing 99
empyema 27
encephalitis 19, 221–5, 226
  complications 225
  epilepsy 325
  management 225
  tick-borne *223,* 224–5
  viral 265
encephalopathy 17, 20
  metabolic 265
  reactive arsenical 124
Entamoeba dispar (amoebiasis) 153, 158
Entamoeba histolytica (amoebiasis) 6, 153, 158
  life cycle 153–4
  liver invasion 155
  virulence factor 158
enteric fever 5
  see also typhoid
Enterobius vermicularis (pinworm; threadworm) 174
Enterocytozoon bieneusi 169, 170
Enterotest capsule 290
enterovirus 71 226
envenoming
  marine 310
  scorpion stings 309
  snake bite 35, 307, 308
  spider bites 310
enzymopathies 302
  see also glucose-6-phosphate dehydrogenase (G6PD) deficiency
epidemics, control 332
epilepsy 324–6
  care 326
  causes 325
  complications 325
  cysticercosis 171–2
  diagnosis 324–5
  education 326
  epidemiology 324

management 325–6
prevalence 325
see also seizures
epinephrine see adrenaline
epistaxis in brucellosis 241
Epstein–Barr virus (EBV) 27
  malaria endemic areas 62
erythema nodosum 34
erythema nodosum leprosum (ENL) reactions 143, 145, 146, 148
  pregnancy 149
Escherichia coli O157:H7 161
espundia 81
ethambutol 91
eye disease
  leprosy 145
  onchocerciasis 114
eyes
  movements 22
  venom spitting 309

fasciola excretory–secretory (FES) antigen 191
Fasciola gigantica (liver fluke) 190
Fasciola hepatica (liver fluke) 190
Fasciolopsis buski (intestinal fluke) 192
feeding programmes, selective/therapeutic 331
fever 26–31
  African tick typhus 271
  brucellosis 240–1
  chronic 29
  clinical problems 30–1
  examination 26–7
  history 26
  HIV infection 27, 29–30, 106–8
  investigations 27–8
  louse-borne typhus 270
  pathogenesis 26
  respiratory illness 14–15
  scrub typhus 271
  symptomatic treatment 26
  treatment 28–9
filariasis 116, *117,* 118–19
  bancroftian 116, *117,* 118
  brugian 118
  early disease 116
  lymphatic 116, *117,* 118
    clinical effects 116, 118
    diagnosis 118
    obstruction 116, 118
    treatment 118–19
  obstructive disease management 119
  tropical pulmonary eosinophilia 209

finger clubbing in tuberculosis 89
fish, venomous 310
fish tank granuloma 298
fisherman's itch 132
flaviviruses 221, 226–7
fluconazole
  cryptococcal meningitis 220
  histoplasmosis 295, 296
  HIV prophylaxis 110
flucytosine 220, 298
fluoroquinolones 248, 249
folate 37
  deficiency 305, 306
  supplements in sickle cell disease 301
folic acid supplements 305
Fonsecaea pedrosoi (chromoblastomycosis) 298
food, humanitarian emergencies 330–1
food contamination
  paratyphoid 245, 250–1
  typhoid 245
food poisoning 5
foreign bodies, ingestion 3
formol gel test (FGT) 75
fungal infections 34
  nails 106
  skin 32
furuncle 34

gastritis 3
  see also stomach
gastrointestinal infections in refugees/asylum seekers 334
gastrointestinal tract 3–10
  see also named regions
Geneva Convention 329
genital herpes simplex 101
genital mutilation, female 334
genital ulcers 42, 43, 46, *46*
Giardia lamblia 4, 167
giardiasis 167–9
  clinical features 167–8
  diarrhoea 167–8
  differential diagnosis 168
  epidemiology 167
  investigations 168
  management 168
  prevention 169
  public health 169
gibbus 229
glibenclamide 317
Global Programme on AIDS (GPA) 98

glucose, blood levels
  malaria 63
  see also hypoglycaemia
glucose-6-phosphate
      dehydrogenase (G6PD)
      deficiency 61, 62, 68, 302
    dapsone contraindication 146
    drugs causing 302
    haemolysis 302
    typhoid 247, 248
glucose tolerance test 314
glucose–electrolyte solutions 165
gonorrhoea 42
  HIV infection risk 101
ground itch 176
growth retardation, whipworm 177
Guillain–Barré syndrome 19, 227,
      228
guinea worm 292, 293, 294
  clinical features 292
  control 294
  diagnosis 292
  life cycle 292
  treatment 292, 294
gut perforation 246

haematemesis 3
haemoglobin
  genotype 62
  measurement 36–7
  sickle 300
haemoglobinopathies 300–2
  massive tropical splenomegaly 49
  see also sickle cell disease;
      thalassaemias
haemoglobinuria, malaria 61
haemolytic uraemic syndrome 161
Haemophilus ducreyi 101
Haemophilus influenzae type b
  acute respiratory infection in
      children 15
  epidemiology 212–13
  lower respiratory infection in
      HIV 106
  pyogenic meningitis 211, 216
  vaccination 204, 213
haemoptysis 11
haemorrhagic fever with renal
      syndrome 255, 258, 260
haemorrhoids 304
halofantrine 66
halzoun 190
Hansen's disease 141
  see also leprosy
Hantaan fever 255, 260
hantavirus 273
  pulmonary syndrome 260
HBeAg 182, 183

HbF 300
HbS 300, 301
HBsAg 181, 183
  endemicity 184
head injury 325
headache 17
Heaf test 85
health care workers
  HIV prophylaxis 110
  HIV transmission 102
heart 27
  failure, hypertensive 319
Helicobacter pylori 3
helminths, soil-transmitted 174–9
  incidence of infection 174
hemiparesis 22
hemispheric signs 23
Henoch–Schönlein purpura 35
hepatitis
  amoebic liver abscess differential
      diagnosis 156
  dengue 265
  fever 27
  viral 180–9
hepatitis A 180–1
  hepatitis E differential diagnosis
      187
hepatitis B 180, 181–3, 184, 185
  asylum seekers 334
  carriers 182
  haematemesis 3
  hepatitis D association 185–6
  prevalence 183
  refugees 334
  risk with blood transfusion 39
  sexual transmission 182
  vaccine 185
hepatitis C 180, 186–7
  refugees/asylum seekers 334
hepatitis D 180, 185–6
hepatitis E 180, 187, 188
hepatocellular carcinoma 156, 181,
      186, 187, 189
  fever 27
hepatoma 187
  see also hepatocellular carcinoma
hepatomegaly
  fever 27
  leishmaniasis 74
  liver fluke 190
  oriental liver fluke 191
hepatosplenomegaly, scrub typhus
      271
herpes simplex 101
herpes zoster 30, 105
hiatus hernia 304
highly active antiretroviral therapy
      (HAART) 75

Histoplasma capsulatum var
      capsulatum (classical
      histoplasmosis) 295
Histoplasma capsulatum var duboisii
      (African histoplasmosis)
      295
histoplasmosis 295–6
  African 295, 296
  classical 295–6
  disseminated 295, 296
  HIV infection 296
  pulmonary 295
HIV-1 98–9
HIV-2 98–9
HIV infection 98–110
  antiretroviral therapy 109–10
  asylum seekers 334
  brucellosis differential diagnosis
      241
  candidiasis 27
  Chagas' disease reactivation 128
  children 29–30
  clinical problems 105–9
  control strategies 102–3
  cough and fever 106–7
  cryptococcosis 109
  diarrhoea 8
    chronic 108
  disease mechanisms 103–4
  early disease 104–5
  economic impact 102–3
  education 102
  epidemiology 100
  epilepsy 325
  fever 27, 29–30, 106–8
  histoplasmosis 296
  immunosuppression 103–4
  late disease 105
  latent phase 104
  leishmaniasis coinfection 73, 75
  malaria 62
  malignancy association 104
  meningitis
    cryptococcal 220
    epidemiology 212
  myelopathy 230–1
  natural history 104–5
  pathogen associations 103
  pathogenesis 103
  pneumonia 201
  risk 196, 198
  prophylaxis 109–10
  pruritis 105
  public health 40
  refugees 332, 334
  respiratory illness 12, 13, 16
  risk factors 101
  risk with blood transfusion 39

seroconversion 30, 104
  time to death 105
seroprevalence 100
sexually transmitted infections
  101
skin problems 105–6
staging 104
strains 103
surveillance 100
syndromic management 46–7
testing 28, 99–100
toxoplasmosis 109
transmission 100–2
  infected blood 101–2
  sexual 100–1
  vertical 101
tuberculosis 13, 16, 29, 30, 88,
  96–7
  association 103
  chemoprophylaxis 110
  clinical presentation 89
  coinfection 106–7
  early disease 104–5
  viral load assays 100
  voluntary testing and counselling
    (VTC) facilities 99, 102
  wasting 108
  see also AIDS
hookworm 35, 174, 176–7
  clinical features 176–7
  dog 34, 284
  epidemiology 176
  investigations 177
  iron deficiency anaemia 304
  larva migrans 176, 284
  management 177
  prevention 177
  tropical pulmonary eosinophilia
    differential diagnosis 209
house dust mites 322
human herpes virus 8 (HHV8) 109
human resources/training,
    humanitarian emergencies
    332
human T cell lymphotropic virus I
    (HTLV-1) 231
humanitarian emergencies 329–33
  communicable disease control
    332
  coordination 332
  emergency phase 329–32
  epidemics control 332
  food 330–1
  health care 331–2
  human resources/training 332
  initial assessment 330
  measles immunization 330
  nutrition 330–1

post-emergency phase 332–3
  public health 332
  sanitation 330
  shelter 331
  site planning 331
  water supply 330
hydatid cyst disease 156, 193–4
hydatid disease, alveolar 194–5
hydralazine 320
hydrocoele, filariasis 116, 119
hydroxycobalamin injections 306
hygiene
  amoebiasis prevention 158
  ascariasis 174
  giardiasis 169
  helminth infections 174
  shigellosis 161–2
hygiene hypothesis of asthma 321
Hymenolepis diminuta
    (hymenolepiasis) 173
Hymenolepis nana (hymenolepiasis)
    173
hyperosmolar non-ketotic coma
    315
hypertension 313, 318–20
  care 320
  causes 318–19
  complications 319
  education 320
  epidemiology 318
hypertensive crisis 319, 320
  management 319–20
  prevalence 318
  risk factors 318–19
hyperventilation 22
hypochlorhydria 5
hypoglycaemia
  cholera 163
  diabetes mellitus 315
  malaria 63
    P. falciparum 59, 61

imipenem 276
immunosuppression
  Chagas' disease reactivation 128
  cough 16
  malaria 62
impetigo, bullous 35
infants
  exposure to sandfly bites 82
  hepatitis B 182
    vaccination 185
  low birth weight in malaria 61,
    68
infections
  asylum seekers 334
  diabetes mellitus 315
  log–normal distribution 132

multidrug-resistant 96, 162,
    334
  refugees 334
  see also abscess; fungal infections
infectious mononucleosis
  brucellosis differential diagnosis
    241
  fever 27
infective endocarditis see bacterial
    endocarditis
insect bites 32, 35
  see also scorpion stings
insulin therapy 316
Integrated Management of
    Childhood Illness (IMCI;
    WHO) 14
interferon α (IFN-α) 183, 186–7
internally displaced persons 328
intestinal fluke 192
intestinal obstruction 3
  ascariasis 175
intracranial pressure, raised
  cerebral malaria 60
  cryptococcal meningitis 220
  lung fluke 208
  viral encephalitis 225
iodoquinol 157
iron
  chelation 301–2
  deficiency 304–5
  supplements 305
isoniazid 110
  tuberculosis 91
    preventive therapy 94–5
Isospora belli 108, 169, 170
itraconazole 296, 298
ivermectin
  filariasis 118–19
  larva migrans 35, 286
  onchocerciasis 115
  strongyloidiasis 290
  trichuriasis 178

Japanese encephalitis 221, 222–3,
    224, 226
  clinical features 221, 224
  prevention 224
  public health 224
Jarisch–Herxheimer reaction 268,
    274
jaundice 181
  malaria 57
  oriental liver fluke 191–2
jellyfish, venomous 310
jiggers 282–3

K39 test 76
kala-azar see leishmaniasis

Kaposi's sarcoma 34, 98, 104,
    108–9
  nodules 34
  pulmonary 106, 109
Katayama fever 35, 133, 135
  oriental liver fluke (similar
    syndrome) 191
Kérandel's sign 122
Kernig's sign 21
ketoacidosis see diabetic
    ketoacidosis
Koplick's spots 27
kwashiorkor, pneumonia risk 198

La Crosse encephalitis 223, 225
lactase deficiency 4
lagophthalmos 148
lamivudine 183
large bowel, diarrhoea 5, 6
larva currens 34, 35, 284
  clinical features 289
  rash 290
larva migrans 34, 35
  cutaneous 284–5
  hookworm 176, 284
  ocular 178–9
  prevention 284–5
  visceral 178
laryngotracheobronchitis 12
Lassa fever 35, 255, 259
  management 258
latrines, refugee camps 10
Latrodectus (widow spiders) 310
left ventricular hypertrophy 319
Leishmania 73, 80
  lifecycle 73–4
leishmaniasis, cutaneous 80–2
  diffuse 80–1, 82
  investigations 81–2
  management 82
  nodules 34
  prevention 82
  sporotrichosis differential
    diagnosis 298
leishmaniasis, lupoid 81
leishmaniasis, mucocutaneous 80,
    81
  treatment 82
leishmaniasis, post-kala-azar dermal
    34, 74, 78
leishmaniasis, visceral 73–9
  clinical features 74
  epidemiology 73
  fever 27
  hepatomegaly 74
  HIV coinfection 73, 75
  host elimination 78
  investigations 75–6

lymphadenopathy 74
  management 76–7
  massive tropical splenomegaly 49
  parasitological evidence 76
  prevention 78–9
  splenic aspirate 76
  splenomegaly 74
  vector elimination/avoidance
    78–9
  viscerotropic 75
leishmaniasis recidivans 80, 81
leishmanin test 81–2
Lemierre's syndrome 29
lepromin test 144
leprosy 19, 27, 141–50
  anaesthesia 143, 144
  BCG vaccine 149
  borderline 144
  borderline lepromatous 144–5
  borderline states 143
  borderline tuberculoid 144
  cell-mediated immunity 141, 143
  chemotherapy 146–7
  children 149
  classification 143–5
  clinical features 143
  control 149–50
  delayed hypersensitivity reaction
    147–8
  diagnosis 145–6
  differential diagnosis 146
  disability prevention 148–9
  elimination 149–50
  epidemiology 141
  eye 145
  foot care 149
  granuloma formation 142
  hand care 149
  immune response 142–3
  incidence 141
  lepromatous 81, 143, 145
  management 146–8
  multibacillary disease 143, 145,
    147
    children 149
  multidrug therapy 147
  neuritis 147, 148
  nodules 34
  paucibacillary 143, 145, 147
    children 149
  pigmentation change 34
  polar 144
  pregnancy 148–9
  prevention 149–50
  reconstructive surgery 148
  slit skin smears 146
  tuberculoid 142–3, 144
  type I reactions 147–8

women 148–9
  see also erythema nodosum
    leprosum (ENL) reactions
Leptospira interrogans (leptospirosis)
    272
leptospirosis 272–4
  aseptic meningitis 273
  bacteraemic 273
  clinical syndromes 273
  diagnosis 273–4
  epidemiology 272
  icteric 273
  pathology 272–3
  prophylaxis 274
  pulmonary syndrome 273
  transmission 272
  treatment 274
leukaemia 38
  chronic myeloid 49
  fever 27
Lhermitte's sign 231
lichen planus 34
liver
  metastases 156
  schistosomal fibrosis 3, 135
liver abscess
  amoebic 27, 106, 153, 155–6
    aspiration 158
    complications 156
    differential diagnosis 156
    drainage 158
    pneumonia differential
      diagnosis 199
    treatment 157, 158
  ascariasis 175
liver cancer, cholangiocarcinoma
    192
  see also hepatocellular carcinoma
liver failure, dengue 265
liver flukes 190–2
  clinical features 190–1
  investigations 191
  management 191
  oriental 191–2
  prevention 191
Loa loa infection 115
Löffler's syndrome 175
louse
  body 267, 270, 286–7
  head 287
  infestation in refugees/asylum
    seekers 334
  pubic 286–7
louse-borne relapsing fever 267–9,
    287
  clinical features 268
  diagnosis 268
  epidemiology 267

fatality rate 268
pathology 267–8
prevention 268–9
treatment 268
louse-borne typhus 270, 287
Loxosceles (recluse spider) 310
lumbar puncture 24–5, 28
pyogenic meningitis 214
lung fluke 206–8
distribution 207
lungs/lung disease
histoplasmosis 295
hookworm larval migration 176
hydatid cyst disease 194
leptospirosis pulmonary
syndrome 273
lung fluke 206
pathology 156
pneumonia risk 198
schistosomiasis 134
transthoracic aspirate 201
lupus vulgaris 82
lymphadenitis, tuberculosis 88, 90
lymphadenopathy
African tick typhus 271
brucellosis 241
fever 27
filariasis 116
leishmaniasis 74
scrub typhus 271
lymphoma
fever 27
massive tropical splenomegaly
49
primary CNS 104

macrophages 103
Madura foot 297–8
Madurella mycetomatis (mycetoma)
297, 298
malabsorption 4
diarrhoea 5, 6
malaria 55–71
anaemia 57
asylum seekers 334
cerebral 59–60
protective effect of ascariasis
176
chemoprophylaxis 68–9
children 39
clinical features 57
complicated 58, 59
congenital 61
control 71
diagnosis 62–3
diarrhoea 5
drug resistance 67
epidemiology 69–71

fever 27, 28, 30, 57
classical stages 57–8
G6PD protection 302
global eradication 71
HIV infection 30, 62
hyper-reactive splenomegaly 50
immune disorders 62
immunity 61–2, 70
immunosuppression 62
individual precautions 71
intermittent presumptive therapy
68
jaundice 57
leptospirosis differential diagnosis
273
life cycle 55–7
massive tropical splenomegaly
49
measuring in community 70–1
morbidity 70
mortality 70, 85
placental 68
pneumonia risk 198
pregnancy 61, 62
intermittent presumptive
therapy 68
protective factors 62
quartan 57
refugees 334
relapse 68
severe 58, 59
splenomegaly 57
stable 70
subtertian 57
tertian 57
treatment 63–8
chemotherapy 64–7
combination therapy 67
supportive 63–4
unstable 70–1
untreated 58
malathion 287
malignancy, cachexia 34
malnutrition
humanitarian emergencies
330–1
tuberculosis 88
viral encephalitis 225
Malta fever see brucellosis
Mantoux test 85
marasmus, pneumonia risk 198
Marburg fever 255, 259–60
marrar 190
Mazzotti reaction 115
measles 27, 28
immunization 330
mortality 85
pneumonia risk 198

mebendazole
alveolar hydatid disease 195
ascariasis 175
hookworm 177
trichuriasis 178
visceral larva migrans 178
Médecins sans Frontières (MSF)
329
mefloquine
malaria prophylaxis 69
malaria therapy 65–6
megaoesophagus 3
melarsoprol 124–5
melioidosis 275–6
meningism 17, 20
coma 23
pneumonia 199
meningitis 19
aseptic in leptospirosis 273
cryptococcal 17, 28, 219–20
HIV infection 220
pyogenic meningitis differential
diagnosis 213
raised intracranial pressure
220
epilepsy 325
meningococcal 215–16
control of epidemics 217–18
pneumonia 199
pyogenic 211–18
clinical features 213
complications 216–17
diagnosis 214–15
differential diagnosis 213
empirical therapy 216
epidemiology 211–12
management 215–17
vaccination 212
tuberculosis 229, 230
typhoid 247
vaccination 213
meningitis belt 211, 212
meningococcus/meningococcal
disease 25, 211–12
control of epidemics 217–18
management 215–16
meningoencephalitis 226
brucellosis 241
louse-borne typhus 270
menorrhagia 304
mesenteric adenitis 3
metabolic acidosis 12
cholera 164
P.falciparum malaria 59, 60
metformin 316, 317
methyldopa 320
metriphonate 138
metronidazole 157, 158, 168

microbicides, vaginal 102
microsporidia 169
mid-upper arm circumference 331
migrants
    louse-borne typhus 270
    urbanization 313
milk, brucellosis 240
miltefosine 77
minocycline 147
mite typhus 270
mites, scabies 286
Montenegro test 81–2
mosquitoes
    anopheline 55, 56
    arboviral encephalitis 221, 224,
        225
    arboviruses 252
    control 119
    dengue fever 262, 265
    filariasis vector 116, 118
    Japanese encephalitis 221, 224
    yellow fever 265
motor axonal neuropathy, acute
    (AMAN) 227, 228
motor neurones, spastic paralysis
    229
mouth
    fever 27
    Kaposi's sarcoma 109
    see also candidiasis
mumps 241
Murray Valley encephalitis 223,
    225
mycetoma 297–8
mycobacteria
    atypical infection 298
    faecal 108
    other than tuberculosis (MOTT)
        84
Mycobacterium avium, disseminated
    103
Mycobacterium leprae (leprosy)
    141
    cell-mediated immunity 141
    delayed hypersensitivity reaction
        147–8
    detection 145–6
    microbiology 141–2
Mycobacterium marinum (fish tank
    granuloma) 298
Mycobacterium tuberculosis
    (tuberculosis) 84
Mycobacterium tuberculosis MTB
    complex 87
    culture 90
    delayed-type hypersensitivity
        reaction 85, 87
    microbiology 84

purified protein derivatives (PPD)
    85, 90
Mycobacterium ulcerans (Buruli
    ulcer) 279
mycosis, endemic 15
myelitis 19
myelodysplastic syndrome 38
myelofibrosis 49
myiasis 282–3
    furuncular 34
    nasal 283

nails, fungal infections 106
nasopharyngitis 190–1
natural disasters 329
Necator americanus (hookworm)
    174, 176–7
neck stiffness 21
needlestick injuries
    HIV prophylaxis 110
    HIV transmission 102
Neisseria meningitidis 211
    see also meningococcus
nephropathy, diabetic 315
nephrosis, malarial 62
nephrotic syndrome 198
neuritis, leprosy 147, 148
neurocysticercosis 325
neurological disorders 17–25, 27
    causes 18
    chronic 17
    examination 20–2
    pathology 19
    space occupying lesions 19
    syndromes 17
neurological injury 12
neurological signs, focal 17
    coma 23
neuropathy 19
    diabetic 315
neuropsychiatric disorder,
        brucellosis 241
neuroschistosomiasis 135, 137,
    138
neurotoxicity, snake bites 308,
    309
nevirapine 102
niclosamide 173
nifurtimox 125, 128
nitazoxanide 158, 170
    ascariasis 175
    cestode infections 173
    giardiasis 168
    liver fluke 191
    trichuriasis 178
nitrosamines, dietary 192
Nocardia brasiliensis (mycetoma)
    297

nocardiosis 298
nodules, subcutaneous in
    onchocerciasis 114, 115
non-communicable diseases
    312–26
    demographic transition 313
    mortality 312, 313
    spectrum 312
non-governmental organizations
    (NGOs) 329
non-Hodgkin's lymphoma 104
non-steroidal anti-inflammatory
    drugs (NSAIDs) 304
    asthma trigger 321
nose, mucosal leishmaniasis 81
nutrition, humanitarian emergencies
    330–1

obesity 313, 316
    asthma incidence 321, 322
    hypertension 318
occupation, respiratory illness 12
oculocephalic reflex 22
oculovestibular reflex 22
oesophageal carcinoma 3
oesophageal varices 3
oesophagitis 3
ofloxacin 147
older people
    pneumonia risk 198
    refugees 328–9
Onchocerca volvulus 112
onchocerciasis 32, 112, 113,
    114–16
    clinical features 112, 114
    control 115–16
    diagnosis 114–15
    distribution 112, 113
    epidemiology 112
    finding microfilariae 114–15
    nodules 34
    pigmentation change 34
    subcutaneous nodules 114, 115
    treatment 115
O'nyong nyong 252
Opisthorchis sinensis (liver fluke)
    191
Opisthorchis viverrini (liver fluke) 191
oral hypoglycaemic agents 316
oral rehydration solution 7
    cholera 165–6
    refugee camps 10
orchitis, brucellosis 241
Orientia tsutsugamushi (scrub
    typhus) 270
osteomyelitis 247
oxamniquine 138
oxygen supplementation 203

pain response 22–3
pancreatitis
    ascariasis 175
    oriental liver fluke 192
paracetamol 26
paragonomiasis 15
paralysis 17
    acute flaccid 226–8
        anterior horn cell damage
            226–7, 228
        immune-mediated causes
            227–8
        rabies 228, 233
        spastic 229–31
            causes 229–31
paratyphoid 245, 250–1
    mode of infection 245
paravertebral plexus,
        schistosomiasis 135
parkinsonian syndrome 224
paromomycin 157, 170
parotitis 161
particle agglutination tests, HIV
        testing 99
pathogens
    opportunistic 103
    stool 108
Pediculus capitis (head louse) 287
Pediculus humanus corporis (body
        louse) 267, 270, 286, 287
pellagra 34
pelvic inflammatory disease 27
pemphigus 35
penicillin
    meningococcal disease 215–16
    pneumococcal disease 216
    pneumonia 203, 204
    resistance 204
    sickle cell disease 301
    tropical ulcer 277
pentamidine
    trypanosomiasis 124
    visceral leishmaniasis 77
peptic strictures 3
peptic ulceration 3
peripheral nervous system, leprosy
        143
peritonitis
    amoebiasis 154
    fever 27
    granulomatous in ascariasis 175
permethrin 286, 287
pets 179
phaeochromocytoma 318
pharyngitis 27, 29
phenobarbital 325, 326
phenytoin 325, 326
Phoneutria (banana spider) 310

Phthirus pubis (pubic louse) 286
phytates 304
pigmentation change in skin 34
piles 161
pinworm 174
piperazine 175
pityriasis versicolor 34
Plasmodium falciparum 55, 56, 57
    complicated malaria 58, 59
    hyperpyrexia 58
    immunity 61–2
    infection peculiarities 58–61
    protective factors 62
    untreated attack 58
Plasmodium malariae 55
    untreated attack 58
Plasmodium ovale 55
    relapse 68
    untreated attack 58
Plasmodium vivax 55, 61
    protective factors 62
    relapse 68
    untreated attack 58
pleural aspiration 201
pleural effusion 15–16
pleural fluid 14
pleurisy, tuberculosis 88, 90
pneumococci/pneumococcal
        disease 216
    epidemiology 212
    see also pneumonia,
        pneumococcal, Streptococcus
        pneumoniae
Pneumocystis carinii pneumonia
        (PCP) 103, 106
    HIV infection in children 109
    prevention 204
pneumonia 29, 196–205
    bacterial 15
    children 196, 198, 200, 203
    clinical features 198–9
    complications 204
    diarrhoea 5
    differential diagnosis 199
    epidemiology 196, 198
    fatality rate 198
    fever 27
    HIV infection 106, 196, 198,
        201
    investigations 199–201
    louse-borne typhus 270
    management 201, 202, 203–4
    microbiology 196, 197–8
    pneumococcal 13, 16, 30, 104
        prevention 110, 204
    prevention 204
    risk factors 198
    typhoid lobar 247

vaccination 204
viral encephalitis 225
pneumonitis
    ascaris 175
    hookworm 176
poisoning
    food poisoning 5
    pneumonia differential diagnosis
        199
poliomyelitis 226
polymyositis 247
polyneuropathy, acute inflammatory
        demyelinating (AIDP) 227,
        228
porphyria 35
portal hypertension 27
Pott's disease 229
poverty and tuberculosis 88
praziquantel 138, 325
    cestode infections 173
    cysticercosis 173
    lung fluke 208
    oriental liver fluke 192
prazosin 310
pregnancy
    brucellosis 243
    HIV testing 99
    HIV transmission 101, 102
    hypoglycaemia in P. falciparum
        malaria 60
    iron deficiency anaemia 305
    leprosy 148–9
    malaria 61, 62
        intermittent presumptive
            therapy 68
        pigmentation 34
        tetanus vaccination 238–9
primaquine 68
proctitis, granular 161
proguanil 68–9
protein-losing enteropathy 161
pruritis, HIV infection 105
pseudopapilloma 133
psychosis, typhoid fever 27
public health
    ascariasis 175–6
    brucellosis 243–4
    Buruli ulcer 281
    catastrophic emergencies 329
    giardiasis 169
    hookworm 177
    humanitarian emergencies 332
    Japanese encephalitis 224
    schistosomiasis 139
    sexually transmitted infections
        40, 47
    shigellosis 161–2
    toxocariasis 179

pulmonary eosinophilia
  pneumonia differential diagnosis
    199
  see also tropical pulmonary
    eosinophilia
pupils, light reaction 22
Putsi fly 282
Puumula virus 260
pyrantel pamoate
  ascariasis 175
  hookworm 177
pyrazinamide 91
pyrexia of unknown origin 30–1,
  241
pyrimethamine 170

quinfamide 157
quinine 65

rabies 19, 232–6
  clinical features 232–3
  diagnosis 233, 234
  dumb 232
  furious 232
  immunization 234
    pre-exposure 236
  paralytic 228, 233
  postexposure treatment 234
  postexposure vaccination
    234–6
  postmortem examination 233
  precautions with patients 233
  prevention 236
  treatment 233–6
rabies immune globulin 234
rash
  HIV infection
    children 109
    seroconversion 104
  larva currens 290
  meningococcal 25, 27
  onchocerciasis 112
  schistosomiasis 132
  typhus
    African tick 271
    louse-borne 270
    scrub 271
rectal prolapse 6, 161, 177
rectal strictures, amoebiasis 154
red cells
  sickle haemoglobin 300
  β-thalassaemia 301
refugees 328–34
  children 333–4
  cholera 166
  diarrhoea 5, 8–9
  health care programme 333–4
  HIV infection 332

humanitarian emergencies
    329–33
  infectious diseases 334
  louse-borne typhus 270
  sexually transmitted infections
    332
  shigellosis 162
  torture 334
  trauma 334
  tuberculosis 332
rehydration
  in cholera 164–6
  see also oral rehydration solution
Reiter's syndrome, post-dysenteric
    161
relapsing fevers 267–9, 287
renal disease
  hypertension 318, 319
  typhoid 247
reserpine 320
respiratory infections, acute 85
  HIV infection 106
respiratory tract disorders 11–16
  assessment 11–13
  examination 12–13
  investigation 13–14
  presentations 14–16
  see also lungs/lung disease;
    pneumonia
respiratory tract infection, lower
    14–15
resuscitation, pneumonia 203
retinopathy
  cerebral malaria 59–60
  diabetic 315
Reye-like syndrome 265
Reye's syndrome 26
rheumatic heart disease 27
ribavirin 187
  viral haemorrhagic fevers 258,
    259, 260
Rickettsia 270–1
  fever 29
  tick-borne 27
Rickettsia prowazekii 270
rifampicin
  brucellosis 242, 243
  leprosy 146, 147
  pregnancy 149
  tuberculosis 91
Rift Valley fever 255, 260
  Councilman bodies 265
  management 258
Ringer's lactate solution 165
river blindness see onchocerciasis
Rock fever see brucellosis
Rocky Mountain spotted fever 253,
    271

Ross River virus 252
roundworms, intestinal 35
rubella 27

St Louis encephalitis 223, 224
salbutamol 322, 323
Salmonella
  diarrhoea 5
  non-typhi 103, 106, 107
  septicaemia 30
Salmonella paratyphi (paratyphoid)
    245
Salmonella typhi (typhoid) 245
salt intake, increased 318–19
sandflies 73, 74, 78–9, 82
sanitation
  amoebiasis prevention 158
  giardiasis 169
  helminth infections 174
  humanitarian emergencies 330
  oriental liver fluke 192
  shigellosis 161–2
Sarcoptes scabiei (scabies) 286
scabies 32, 286
  onchocerciasis differential
    diagnosis 114
  refugees/asylum seekers 334
scarification 182
Schistosoma, life cycle 129, 131
Schistosoma haematobium 129, 130
  clinical features 133–4
  diagnosis 136, 137
  eggs 133
  life cycle 129, 131
Schistosoma intercalatum 129
Schistosoma japonicum 129, 130
  clinical features 135
  diagnosis 136, 137
  eggs 133
  life cycle 129, 131
Schistosoma mansoni 6, 129, 130
  clinical features 134–5
  diagnosis 136, 137
  eggs 133
  life cycle 129, 131
schistosomiasis 128–39
  acute 132–3, 137, 138
  Ag–Ab complex formation 133
  asymptomatic infection 138
  bladder calcification 134
  cercarial penetration 132
  clinicopathological features
    132–5
  contact reduction 139
  cor pulmonale 134
  diagnosis 136–7, 139
  distribution 129, 130
  epidemiology 131–2

exposure 131–2
fever 26, 27
genital 134
granuloma 134
hepatic fibrosis 3, 135
immunity 131–2
immunodiagnostic tests 136–7
investigations 136–7
Katayama syndrome 35, 133, 135
log–normal distribution 132
management 137–9
mass chemotherapy 139
massive tropical splenomegaly 49
parasitology 129, 131
prevention 139
progression to disease 132
public health 139
pulmonary complications 15
reservoir hosts 131
snail clearance 139
tissue reaction to retained eggs
   133
transmission 131
treatment monitoring 138–9
urticaria 35
water contamination reduction
   139
scorpion stings 309–10
scrub typhus 270–1
   leptospirosis differential diagnosis
      273
seizures
   P.falciparum malaria 59
   viral encephalitis 225
   see also epilepsy
selective feeding programmes (SFPs)
   331
Seoul virus 260
septicaemia
   fever 28–9
   melioidosis 276
   meningococcal 24, 35, 215
   paratyphoid 251
   Salmonella 30
   splenectomy 51
sexually transmitted infections
   40–8
   asylum seekers 334
   hepatitis B 182
   HIV risk 101
   HTLV-1 231
   local adaptations for management
      42
   partner notification/treatment
      47
   public health 40, 47
   refugees 332, 334
   syndromes 40, 41

syndromic management 40, 43,
   45–8
   treatment 102
   treatment-seeking behaviour 47
   WHO flow charts 42–3, 44, 45,
      46
shelter, humanitarian emergencies
   331
shiga toxin 160–1
Shigella 160
   multidrug resistance 162
shigellosis 5, 160–2
   clinical features 160–1
   epidemic 162
   investigation 161
   management 161
   prevention 161–2
   public health 161–2
shingles 105
shock, hypovolaemic 163, 164, 258
sickle cell crisis 301
sickle cell disease 300–1
   clinical features 300–1
   diagnosis 301
   management 301
   pneumonia prevention 204
   pneumonia risk 198
sickle cell trait 62
Simulium (blackflies) 112
Sin Nombre virus 260
skin 284–5
   bullae 35
   creeping eruptions 34, 176
   diseases 32–5
   fever 27
   infections in HIV 105–6
   itching 32, 286
   leprosy 143
   nodules 34
   onchocerciasis 112, 114
      diagnosis 114–15
   papules 34
   pigmentation change 34
   ulcers 32, 33, 80
      amoebiasis 155
   see also larva currens; larva
      migrans; rash
sleeping sickness see
   trypanosomiasis
slim disease 108
small bowel
   biopsy 168
   malabsorption 5, 6
   secretory diarrhoea 5
smoking 12
   asthma trigger 321
snake bite 35, 307–9
   diagnosis 307–8

epidemiology 309
first aid 307
management 308–9
prevention 309
sodium stibogluconate 76–7
sowda 114
spacer devices 323
spider bites 310
spinal cord
   compression 229
   subacute combined degeneration
      231
spinal epidural abscess 230
spleen aspirate 76
splenectomy 50–1
splenomegaly 49–51
   differential diagnosis 74–5
   fever 27
   filariasis 116
   hyper-reactive malarial 62
   leishmaniasis 74
   malaria 49, 50, 57
   massive tropical 49–50, 62
   relapsing fevers 268
Sporothrix schenckii (sporotrichosis)
   298
sputum examination 13–14
sputum staining 201
squamous cell carcinoma 278
staphylococcal infection, skin sepsis
   27
Staphylococcus aureus
   pyogenic meningitis 211
   spinal epidural abscess 230
Staphylococcus epidermidis 211
status asthmaticus 322
   treatment 323
status epilepticus 325, 326
steroids
   inhaled 322, 323
   meningococcal disease 215
   pyogenic meningitis 216–17
   tuberculosis 88
      adjunctive 94
   typhoid 250
Stevens–Johnson syndrome 107
stomach
   gastric carcinoma 3
   gastric ulcers 3
   Giardia lamblia colonization
      167
stools, ricewater 164
streptococcal infection, skin sepsis
   27
Streptococcus pneumoniae 15, 196
   chloramphenicol activity 203
   HIV association 103
   penicillin resistance 204

Streptococcus pneumoniae (*cont.*)
   pyogenic meningitis 211
   symptoms 199
Streptomyces somaliensis (mycetoma)
   297
streptomycin
   Madura foot 298
   tuberculosis 91, 92
strictures 161
   amoebiasis 154
stridor 11, 15
string test 168, 290
stroke 319, 325
Strongyloides fülleborni (swollen belly
   syndrome) 288
Strongyloides stercoralis
   (strongyloidiasis) 4, 174,
   288–91
   acute 289
   autoinfection 288–9
   chronic 289
   clinical features 289
   control 291
   diagnosis 290
   epidemiology 291
   hookworm differential diagnosis
      177
   hyperinfection 288, 289–90
      prevention 291
      steroids 324
   larva currens 34, 35, 284, 289
      rash 290
   treatment 290–1
   tropical pulmonary eosinophilia
      differential diagnosis 209
sulfadoxine–pyrimethamine 66
sulphonylureas 315, 316, 317
sulphur ointment 286
sunburn 35
supratentorial focal damage 20
suramin 124
swimmer's itch 132
swollen belly syndrome 288
Symmers pipestem fibrosis 135
syphilis 42, 46
   fever 27
   gumma 156
   HIV infection risk 101

T-helper cells 103
tachypnoea 12
Taenia saginata (tapeworm) 171,
   172
Taenia solium (tapeworm) 171–2
taeniasis 171–2
tannates 304
tapeworms 171–3
tattooing 182

tendon transfers 148
Ternidens diminutus 177
tetanospasmin 237
tetanus 237–9
   breathing difficulties 12
   cephalic 238
   clinical manifestations 237–8
   diagnosis 238
   epidemiology 238–9
   generalized 237
   localized 238
   neonatal 237–8
   prevention 238–9
   treatment 238
   tropical ulcer 278
   vaccination 238–9
tetanus immunoglobulin 238
tetanus toxoid 238–9
tetracycline 157
   African tick typhus 271
   brucellosis 242
   louse-borne typhus 270
   pneumonia 203
   resistance 271
   scrub typhus 271
α-thalassaemia 37
β-thalassaemia 301–2
   malaria protection 62
thioacetazone 91, 92
   toxicity in HIV 107
thiazides 319
threadworm 174
thrifty genotype theory 315
thyroid disease
   breathing difficulties 12
   vitamin B_{12} deficiency 306
tiabendazole
   chromoblastomycosis 298
   strongyloidiasis 290–1
   visceral larva migrans 178
tick-borne encephalitis *223*, 224–5
tick-borne relapsing rever 267
   clinical features 268
   diagnosis 268
   epidemiology 267
   fatality rate 268
   pathology 267–8
   prevention 269
   treatment 268
tick paralysis 228
tick typhus 271
tinidazole 157
tine test 85
tinea 106
torture 334
toxaemia in typhoid 247
toxic megacolon 161
Toxocara canis (toxocariasis) 178–9

Toxocara catis (toxocariasis)
   178–9
toxoplasmosis
   fever 27, 30
   HIV association 103–4, 109
trauma
   epilepsy 325
   refugees 334
travellers
   hepatitis A 181
   hepatitis E 187
   rabies immunization 236
trematodes 190
trench fever 287
triatomine bugs 128
Trichomonas vaginalis
   (trichomoniasis) 42–3
   HIV infection risk 101
Trichostrongylus 177
Trichuris trichuria (whipworm) 6,
   174, 177–8
triclabendazole 191, 208
trimethoprim–sulfamethoxazole *see*
   co-trimoxazole
tropical pulmonary eosinophilia 15,
   116, 209–10
   asthma differential diagnosis
      321
tropical spastic paraparesis 231
tropical sprue 4, 306
tropical ulcer 277–8
Trypanosoma brucei gambiense 120,
   121, 122, 123
   control 126
   epidemiology 125
   surveillance 126
   treatment 124–5
Trypanosoma brucei rhodesiense 120,
   121, 122–3
   control 126
   epidemiology 125–6
   surveillance 126
   treatment 124–5
Trypanosoma cruzi 127
   see also Chagas' disease
trypanosomiasis, African 120–6
   control 126
   cure monitoring 125
   diagnosis 123–4
   epidemiology 125–6
   fever 27
   immune response 122
   local effects 120–1
   pathogenesis 122
   relapse 125
   skin itching 32
   surveillance 126
   systemic effects 121–2

treatment 124–5
vector 125
control 126
trypanosomiasis, South American
see Chagas' disease
tsetse fly 120, 125
Tsutsugamushi fever (scrub typhus)
270
tubercle bacilli 90
tuberculoma 229
tuberculosis 13, 84–97
amoebic liver abscess differential
diagnosis 156
amoeboma differential diagnosis
155
annual risk of infection (ARI)
95
asylum seekers 334
BCG vaccination 94
brucellosis differential diagnosis
241, 242
case-finding 96
chemotherapy 90–4
clinical features 88–90
clinical manifestations 87
corticosteroid therapy 88
adjunctive 94
cough 11, 15
cutaneous 82
diagnosis 96, 97
disseminated 27, 107, 108
DOTS-Plus pilot schemes 96
DOTS strategy 95–6
epidemiology 84–5, 86, 87–8
HIV effect 88
extrapulmonary disease 88
clinical features 89
fever 27
HIV association 103
HIV infection 13, 16, 29, 30, 88,
96–7
association 103
clinical presentation 89
coinfection 106–7
early disease 104–5
TB chemoprophylaxis 110
hospital admission 93–4
immunity 85
incidence 85, 86
infection 85
intestinal 4
investigations 90
isolation 93–4
malnutrition 88
management 90–4
meningitis 229, 230
mortality 85
multidrug-resistant 96

pneumonia differential diagnosis
199
pneumonic illness 196
postprimary lesions 87
poverty 88
prevention 94–6
progression to disease 87
pulmonary disease 88
differential diagnosis 89–90
signs/symptoms 89
pyogenic meningitis differential
diagnosis 213, 214
refugees 332, 334
risk factors 88
spastic paralysis 229–30
sporotrichosis differential
diagnosis 298
systemic symptoms 88–9
transmission 85
treatment
monitoring 92–3
outcomes 93, 94
regimens 92, 93, 96
vitamin B$_{12}$ deficiency 306
Tumbu fly 282
Tunga penetrans (chiggers) 282
typhoid fever 27, 30, 245–50
asylum seekers 334
carrier state 250
clinical signs 246
complications 246–7
diagnosis 247–8
leptospirosis differential diagnosis
273
mode of infection 245
psychosis 27
refugees 334
treatment 248, 249, 250
vaccine 250
typhoid nodules 246
typhus
African tick 271
louse-borne 270, 287
scrub 270–1
leptospirosis differential
diagnosis 273

ulcerative colitis, post-dysenteric
155
ulcers
Buruli 279–81
chronic 278
genital 101
guinea worm 294
recurrent 278
skin 32, 33, 80
amoebiasis 155
tropical 277–8

UN High Commission for Refugees
(UNHCR) 328–9, 333
Universal Declaration of Human
Rights 329
urbanization 318–19
ureters, schistosomiasis 134
urethral discharge 42, 43
urinary bladder calcification 134,
137
uropathy, obstructive 133–4, 137
urticaria 35
guinea worm 292

vaccines/vaccination
cholera 166
dengue 265
hepatitis B 185
HIV 102
leprosy 149
measles 330
pneumococcal 204–5
pneumonia 204
pyogenic meningitis 212, 213
rabies postexposure 234–6
tetanus 238–9
typhoid fever 250
yellow fever 265
vaginal discharge 42, 44
vaginal flora 101
valproate 325
varicella zoster virus 103
Venezuelan equine encephalitis
223, 225
Vibrio cholerae (cholera) 163
Eltor 166
serotype 0139 166
viral haemorrhagic fever 27, 35,
254–60
diagnosis 256, 257, 258
differential diagnosis 258
epidemiology 254
identification 254, 256
management 254, 258–9
nosocomial spread 258–9
pathogenesis 254
South American 259
transmission 256
vitamin A supplements 28, 161, 330
vitamin B$_{12}$ deficiency 37, 231, 305,
306
vitamin deficiencies 19
vitiligo 34, 306

wasting in HIV infection 108
children 109
water supply
amoebiasis prevention 158
chlorination 139

water supply (*cont.*)
    filtration 139, 294
    giardiasis 169
    guinea worm 294
    humanitarian emergencies 330
    refugee camps 10
weight-for-height (W/H) index
        330
weight loss, HIV infection 30

Weil–Felix test 270
Weil's disease 273
West Nile virus *223*, 224, 226
westernization 313
wheeze 11, 15
    asthma 321
whipworm 174
whooping cough 11
Widal test 247

women, leprosy 148–9
Wuchereria bancrofti (filariasis) 116,
        209

yaws 34
yellow fever 252, 255, 262, *263*,
        265

Zenker's degeneration 247